A PRACTICAL GUIDE TO MASTERING DATA GOVERNANCE

How to Design, Deploy, and Sustain an Effective Data Governance Framework

Rajender Kumar

For more information, contact: support@JambaAcademy.com.

Published by:
Jamba Academy

Printed in the United States of America
First Printing Edition, 2025

DISCLAIMER

This book is intended for informational and educational purposes only. While we have made every effort to ensure the accuracy of the data governance frameworks, implementation strategies, and organizational best practices presented, the field of enterprise data management is rapidly evolving, and regulatory requirements may vary significantly across jurisdictions and industries.

The content provided herein does not constitute professional legal, compliance, or technical consulting advice. Data governance implementations involve complex organizational, regulatory, and technical considerations that require customized approaches based on your specific business context, industry requirements, and regulatory environment. Readers should consult with qualified data governance professionals, legal counsel, compliance experts, and IT specialists before implementing any governance frameworks or policies discussed in this book.

The author and publisher make no representations or warranties regarding the completeness, accuracy, or suitability of the governance strategies, templates, and methodologies for any particular organization or use case. Data governance outcomes depend on numerous factors including organizational culture, existing technology infrastructure, regulatory environment, stakeholder commitment, resource allocation, implementation expertise, and market conditions. These variables differ significantly across organizations, making it impossible to guarantee specific results from applying the concepts presented. Any forward-looking statements about data governance trends or technological developments represent informed opinions and should not be considered predictive guarantees.

Modern content development techniques, including advanced research technologies, have been utilized in preparing certain portions of this work to ensure comprehensive coverage of current data governance practices and emerging industry trends.

You assume full responsibility for evaluating the applicability of this book's content to your specific organizational needs and for any decisions made based on the information provided. The author and publisher disclaim any liability for damages, losses, or adverse outcomes resulting from the use or misuse of the strategies, frameworks, or recommendations contained within these pages.

By using this book, you acknowledge that data governance implementation is a complex undertaking that requires careful planning, stakeholder alignment, and ongoing organizational commitment beyond the scope of any single publication.

Found Typos & Broken Link

We apologize in advance for any typos or broken links that you may find in this book. We take pride in the quality of our content and strive to provide accurate and useful information to our readers. Please let us know where you found the typos and broken links (if any) so that we can fix them as soon as possible. Again, thank you very much in advance for bringing this to our attention and for your patience.

If you find any typos and broken links in this book, please feel free to email us.

support@JambaAcademy.com

Support

We would love to hear your thoughts and feedback! Could you please take a moment to write a review or share your thoughts on the book? Your feedback helps other readers discover the books and helps authors to improve their work. Thank you for your time and for sharing your thoughts with us!

If there is anything you want to discuss or you have a question about any topic of the book, you can always reach out to us, and we will try to help as much as we can.

support@JambaAcademy.com

"Dedicated to the data stewards, leaders, and visionaries who tirelessly work to transform information into trusted assets—
and to all who believe that strong governance is the foundation of lasting value."

TABLE OF CONTENTS

COMPANION RESOURCES AND IMPLEMENTATION MATERIALS

Accessing Your $5,000 Digital Toolkit

This book provides a complete implementation roadmap for effective data governance frameworks. To accelerate your organization's governance journey, we've created a comprehensive companion repository containing over $5,000 worth of professional-grade templates, tools, and implementation resources—available free with your book purchase.

The Data Governance Companion Repository

All governance frameworks, policy templates, implementation scripts, and assessment tools referenced throughout this book are freely accessible through our dedicated GitHub repository:

Repository URL: https://github.com/JambaAcademy/Data_Governance

This companion repository serves as your complete implementation toolkit, featuring:

Governance Framework Templates - 30+ ready-to-deploy policy documents including data governance charters, stewardship policies, quality standards, classification schemes, and privacy frameworks with organizational structure templates and role definitions.

Assessment and Implementation Tools - Excel-based maturity assessments, organizational readiness checklists, KPI dashboards, ROI calculators, and comprehensive project management resources including roadmaps and risk assessment matrices.

Technical Resources - Data catalog setup scripts, quality profiling tools (Python/SQL), metadata extraction utilities, lineage mapping templates, and API integration guides for major governance platforms (Collibra, Informatica, Talend).

Compliance and Legal Materials - GDPR, CCPA, SOX, and HIPAA compliance templates, data processing agreements, breach notification procedures, and vendor due diligence checklists.

Training and Communication Assets - Executive briefing presentations, steward training curricula, stakeholder communication plans, and change management toolkits.

How to Use the Repository

The repository mirrors this book's structure with ten organized sections covering everything from governance strategy through operational sustainability. Each resource includes implementation guidance, customization notes, and real-world application examples designed to fit organizations of any size or industry.

Whether you're establishing initial governance policies, deploying cross-functional frameworks, or scaling sustainable practices, you'll find immediately applicable resources that transform theoretical concepts into practical governance assets.

Staying Current with Evolving Practices

The companion repository receives regular updates reflecting emerging governance best practices, regulatory changes, and technological advances. This ensures continuous access to current tools and frameworks as your organization's data stewardship capabilities mature.

This digital toolkit transforms strategic insights into actionable governance assets, enabling confident progression from planning through successful implementation and long-term sustainability.

FOREWORD

In the span of my twenty-five-year career architecting data solutions for Fortune 500 enterprises, I have witnessed few transformations as profound as the evolution of data governance from a compliance afterthought to a strategic imperative. As Chief Data Officer at three major corporations and architect of governance frameworks now deployed across hundreds of organizations worldwide, I have seen firsthand how effective data stewardship can fundamentally reshape an organization's competitive landscape.

The statistics are staggering. Organizations with mature data governance frameworks report 23% faster decision-making cycles, 19% improvement in operational efficiency, and a remarkable 340% return on investment within three years of implementation. Yet despite these compelling metrics, surveys consistently reveal that over 60% of enterprises struggle with fragmented, ineffective governance initiatives that drain resources without delivering meaningful results. This paradox represents one of the most significant challenges facing modern business leaders: how to transform the theoretical promise of data governance into tangible organizational value.

The answer lies not in more sophisticated technology or additional compliance requirements, but in a fundamental shift in perspective. Data governance is not about controlling information; it is about liberating it. When properly implemented, governance frameworks transform data from a potential liability into an organization's most valuable strategic asset. I have observed companies revolutionize their market position by treating data governance as a business enabler rather than a technical constraint.

Consider the remarkable transformation at a global pharmaceutical company I advised in 2019. Facing regulatory scrutiny and struggling with inconsistent clinical trial data across seventeen countries, the organization was hemorrhaging resources while delaying critical drug approvals. Traditional approaches focused on tighter controls and more restrictive access policies, inadvertently creating data silos that hindered collaboration between research teams. By implementing a governance framework that emphasized data democratization within clear accountability structures, the company reduced time-to-market for new treatments by fourteen months while achieving perfect regulatory compliance. The key was not restricting data access, but establishing transparent ownership, standardized quality measures, and automated compliance monitoring that enabled researchers to work with confidence.

Similarly, the financial services sector has demonstrated governance's transformative power through the rise of real-time risk management. Leading investment firms now leverage governance frameworks to process millions of transactions daily while maintaining regulatory compliance across multiple jurisdictions. These organizations have moved beyond viewing governance as a cost center to recognizing it as a competitive differentiator that enables faster market response and more sophisticated risk modeling than their competitors.

The challenge for most organizations is not recognizing governance's strategic value, but bridging the gap between aspiration and implementation. Too many governance initiatives fail because they begin with technology selection rather than organizational alignment, or because they prioritize theoretical completeness over practical adoption. The most successful frameworks I have designed share a common characteristic: they treat governance as a change management challenge first and a technical challenge second.

This book addresses that fundamental insight with remarkable clarity and practical depth. The author brings a unique combination of strategic vision and implementation expertise that reflects years of hands-on experience building governance programs that actually work. Rather than presenting another theoretical framework, this guide provides a comprehensive roadmap for navigating the complex organizational, technical, and cultural challenges that determine governance success or failure.

What sets this work apart is its recognition that sustainable data governance requires three distinct but interconnected capabilities: strategic design that aligns with business objectives, tactical deployment that minimizes disruption while maximizing adoption, and operational sustainability that evolves with changing organizational needs. Most governance literature focuses on one dimension while neglecting the others. This book treats all three as equally critical components of a unified approach.

The practical tools and methodologies presented here reflect current best practices while acknowledging the realities of enterprise environments. The author understands that governance frameworks must accommodate legacy systems, competing priorities, and resource constraints while still delivering measurable value. The case studies and implementation templates provide concrete guidance for overcoming common obstacles, from securing executive sponsorship to managing cross-functional resistance.

Perhaps most importantly, this guide recognizes that data governance is fundamentally about people, not technology. The most sophisticated data platforms and automated quality controls cannot compensate for unclear accountability structures or inadequate stakeholder engagement. The author's emphasis on organizational change management and cultural transformation reflects a mature understanding of what actually drives governance success in complex enterprise environments.

As organizations continue to generate data at unprecedented rates, the gap between leaders and laggards in data governance will only widen. Companies that master the principles outlined in this book will gain sustainable competitive advantages through faster decision-making, reduced operational risk, and enhanced innovation capabilities. Those that continue to treat governance as a technical afterthought will find themselves increasingly disadvantaged in markets where data-driven insights determine success.

The opportunity before us is extraordinary. For the first time in business history, organizations have access to comprehensive data about their operations, customers, and markets. The question is not whether this data can drive competitive advantage, but whether organizations will develop the governance capabilities necessary to unlock its potential. This book provides the roadmap for that transformation.

I encourage readers to approach this guide not as another compliance manual, but as a strategic blueprint for organizational transformation. The methodologies and frameworks presented here have the power to fundamentally reshape how your organization creates value from its data assets. The question is not whether you can afford to invest in robust data governance, but whether you can afford not to.

The future belongs to organizations that treat data as their most strategic asset. This book shows you how to build the governance foundation necessary to claim that future.

Dr. Tarun Kapoor

Principal Data Architect & Data Governance Expert

PREFACE

In the summer of 2019, I found myself sitting across from the Chief Data Officer of a Fortune 500 financial services company, watching as she struggled to explain why her organization's multi-million-dollar data lake had become what she candidly described as a "data swamp." Despite investing heavily in cutting-edge technology and hiring top-tier data scientists, the company was drowning in inconsistent, unreliable data that undermined critical business decisions. Customer analytics were contradictory, regulatory reporting was delayed, and trust in data-driven insights had eroded across the organization. The root cause wasn't technological—it was the absence of effective data governance.

That conversation crystallized a realization that had been building throughout my decades of experience in enterprise data management: organizations worldwide are facing an unprecedented data crisis, not because they lack sophisticated tools or talented professionals, but because they fundamentally misunderstand the strategic importance of data governance. This book emerges from that understanding and my commitment to transforming how organizations approach the stewardship of their most valuable asset—their data.

The Governance Revolution

We stand at the threshold of a data governance revolution. Organizations generate and consume data at volumes that would have been unimaginable just a decade ago, yet most struggle to extract meaningful value from this abundance. The traditional approach of treating data governance as a compliance afterthought or IT responsibility has proven inadequate for today's complex, interconnected business environment. Modern enterprises require governance frameworks that are strategic, sustainable, and seamlessly integrated into operational workflows.

My journey into data governance began not in boardrooms or conference halls, but in the trenches of enterprise data challenges. As a data professional who has witnessed the evolution from simple database management to complex multi-cloud architectures, I have observed firsthand how organizations that master governance principles consistently outperform their peers. They make faster decisions, reduce operational risks, ensure regulatory compliance, and unlock innovation opportunities that remain hidden to their competitors.

This practical experience has revealed a critical gap in the current landscape of data governance education and implementation. While academic theories abound and vendor solutions proliferate, few resources bridge the chasm between conceptual frameworks and real-world application. Organizations need more than theoretical models—they need actionable guidance that acknowledges the messy realities of enterprise environments, organizational politics, and resource constraints.

Bridging Theory and Practice

This book addresses that gap by providing a comprehensive yet practical roadmap for establishing, deploying, and sustaining effective data governance frameworks. Unlike traditional approaches that focus primarily on technical implementations or high-level strategic concepts, this guide integrates both perspectives, recognizing that successful governance requires equal attention to people, processes, and technology.

The intended audience encompasses three critical constituencies within modern organizations. **Executives** seeking to understand how data governance drives competitive advantage will find strategic insights that demonstrate tangible business value and return on investment. **Data professionals**—including data scientists, analysts, and engineers—will discover practical methodologies for embedding governance principles into their daily workflows without sacrificing agility or innovation. **IT leaders** responsible for implementing governance technologies and processes will gain comprehensive guidance on selecting, deploying, and integrating governance solutions that scale with organizational growth.

This book serves readers across all governance maturity levels, from organizations launching their first governance initiatives to those seeking to refine and optimize existing programs. The content acknowledges that governance is not a destination but a journey of continuous improvement, requiring frameworks that evolve alongside business needs and technological advances.

A Unique Approach to Governance Mastery

What distinguishes this guide from other data governance resources is its emphasis on sustainability and practical implementation. Too many governance initiatives fail not because of poor design, but because they cannot maintain momentum beyond initial deployment. This book addresses that challenge by focusing equally on the "how" and the "why" of governance, providing readers with both the strategic understanding and tactical tools necessary for long-term success.

The approach is deliberately practical, incorporating real-world case studies, proven templates, and actionable checklists that accelerate implementation timelines. Rather than prescribing one-size-fits-all solutions, the book provides adaptable frameworks that organizations can customize based on their unique requirements, industry constraints, and cultural contexts. This flexibility is essential in today's diverse business landscape, where governance solutions must work equally well for global financial institutions, healthcare networks, technology startups, and government agencies.

Addressing Current Educational Gaps

Current data governance education often suffers from several limitations that this book explicitly addresses. First, many resources focus exclusively on compliance and risk management, overlooking governance's potential as a driver of innovation and competitive advantage. Second, existing guides frequently treat governance as a purely technical discipline, underestimating the critical importance of organizational change management and stakeholder engagement. Third, most educational materials fail to address the sustainability challenge, leaving organizations without guidance for maintaining governance momentum over time.

This book corrects these shortcomings by presenting governance as a strategic business discipline that requires both technical expertise and organizational leadership. It emphasizes the human elements of governance—building trust, fostering collaboration, and creating cultures of data stewardship—while providing concrete guidance on technical implementation and measurement.

The Path Forward

The global examples and diverse perspectives incorporated throughout this book reflect the universal nature of data governance challenges and opportunities. Whether examining privacy compliance frameworks in European healthcare systems, data quality initiatives in Asian manufacturing companies, or analytics governance in North American retail organizations, the fundamental principles remain consistent while implementation details vary based on local requirements and cultural contexts.

My goal in writing this book is to democratize access to effective data governance practices, making sophisticated frameworks accessible to organizations of all sizes and industries. By combining strategic thinking with hands-on implementation guidance, this resource transforms data governance from a compliance burden into a competitive advantage that drives innovation, reduces risk, and unlocks the full potential of organizational data assets.

The journey toward effective data governance is neither simple nor straightforward, but it is absolutely essential for organizations seeking to thrive in our data-driven economy. This book serves as both compass and map for that journey, providing the knowledge, tools, and confidence necessary to navigate the complexities of modern data stewardship. The transformation begins with understanding, accelerates through practical application, and succeeds through sustained commitment to governance excellence.

ACKNOWLEDGMENTS

This book stands as a testament to the collective wisdom and unwavering support of countless individuals who have shaped both my understanding of data governance and the evolution of this critical discipline. The journey from concept to completion would not have been possible without the generous contributions of professionals across the global data stewardship community.

My deepest gratitude extends to the pioneering researchers in data governance academia whose foundational work provided the theoretical backbone for this practical guide. Their rigorous studies on governance frameworks, data quality metrics, and organizational change management illuminated pathways that bridge the gap between academic theory and real-world implementation. The insights drawn from their peer-reviewed research ensured this book maintains both intellectual rigor and practical relevance.

The community of seasoned data governance practitioners deserves special recognition for their invaluable feedback throughout the writing process. These dedicated professionals, working across industries from financial services to healthcare, shared their hard-won experiences implementing governance frameworks in complex enterprise environments. Their candid discussions about implementation challenges, stakeholder resistance, and breakthrough solutions provided the authentic voice that transforms theoretical concepts into actionable strategies. Their willingness to review early drafts and challenge assumptions significantly enhanced the book's technical accuracy and practical utility.

I am particularly indebted to the enterprise data stewardship enthusiasts who opened their organizations to case study development. These forward-thinking leaders demonstrated remarkable transparency in sharing both their successes and failures, providing readers with honest assessments of governance initiatives across diverse organizational contexts. Their contributions ensure this guide reflects the messy realities of enterprise data management while maintaining optimism about achievable outcomes.

The framework pioneers who developed many of the governance methodologies referenced throughout this book deserve acknowledgment for their innovative thinking and generous sharing of intellectual property. Their willingness to contribute templates, checklists, and implementation guides demonstrates the collaborative spirit that drives progress in our field. Without their foundational work, this comprehensive guide would lack the practical tools that accelerate governance deployment.

My organizational mentors provided crucial guidance on navigating the complex political landscapes that often determine governance success or failure. Their coaching on stakeholder engagement, executive communication, and change management strategies proved invaluable in developing the book's sections on organizational dynamics and sustainability planning.

The technical reviewers who scrutinized every framework, process, and recommendation ensured this guide meets the highest standards of accuracy and completeness. Their meticulous attention to detail and constructive criticism elevated the quality of every chapter, making this resource worthy of the trust that busy professionals place in implementation guides.

Finally, I extend heartfelt appreciation to my personal support network, whose patience and encouragement sustained me through the intensive research and writing process. Their understanding during countless evenings and weekends dedicated to this project made completion possible.

The collaborative nature of this endeavor reflects the broader transformation occurring within the data governance community. As organizations worldwide recognize data as a strategic asset, professionals are increasingly sharing knowledge, tools, and experiences to accelerate collective progress. This spirit of collaboration has not only shaped the quality and relevance of this book but also reinforces my belief that effective data governance emerges from community-driven innovation rather than isolated expertise.

To all who contributed their time, knowledge, and support: this book is as much yours as it is mine.

INTRODUCTION: THE DATA GOVERNANCE REVOLUTION

When Netflix's recommendation algorithm suggested "House of Cards" as their next original series investment, few realized they were witnessing the culmination of a decade-long data governance transformation that would revolutionize entertainment forever. Behind that $100 million bet lay not just sophisticated analytics, but a meticulously designed data governance framework that enabled Netflix to trust their data enough to stake the company's future on algorithmic insights. Today, with over 230 million subscribers worldwide and a market capitalization exceeding $150 billion, Netflix stands as a testament to what becomes possible when organizations master the art and science of data governance.

This success story illuminates a fundamental truth reshaping modern business: in our hyperconnected, data-saturated world, competitive advantage no longer belongs to organizations that simply collect the most data, but to those that govern it most effectively. The difference between data hoarding and data mastery lies in governance—the systematic approach to managing data as a strategic asset rather than a technical afterthought.

Yet for every Netflix success story, countless organizations struggle with the opposite reality. Recent studies reveal that while 87% of executives consider data their most valuable asset, only 32% report high confidence in their data quality, and a staggering 68% admit they cannot effectively measure the business value of their data investments. This paradox—simultaneously recognizing data's strategic importance while failing to harness its potential—defines the central challenge facing modern enterprises.

The root cause is not technological inadequacy or insufficient data volumes. Instead, organizations are drowning in information while starving for governance. They possess vast data lakes that function more like data swamps, sophisticated analytics platforms that produce conflicting insights, and compliance frameworks that satisfy auditors but fail to enable business value. The missing ingredient is comprehensive data governance: the strategic discipline that transforms raw information into trustworthy, actionable intelligence.

The Historical Context and Contemporary Urgency

Data governance emerged from necessity rather than innovation. In the early 2000s, corporate scandals like Enron and WorldCom exposed the catastrophic consequences of poor information management, prompting regulatory responses including the Sarbanes-Oxley Act. These compliance requirements forced organizations to implement basic data controls, marking governance's first evolution from IT housekeeping to business imperative.

The second wave arrived with the explosion of big data and cloud computing. As organizations migrated from structured databases to diverse data ecosystems encompassing social media feeds, IoT sensors, and external APIs, traditional data management approaches proved inadequate. The volume, velocity, and variety of modern data demanded new governance paradigms capable of handling complexity while maintaining control.

Today's third wave is driven by artificial intelligence and machine learning, where algorithmic decision-making amplifies both the potential value and risks of data. A single biased dataset can perpetuate discrimination across millions of automated decisions. A poorly governed training dataset can render AI systems unreliable at scale. Conversely, well-governed data enables AI systems to deliver transformative business outcomes, as demonstrated by organizations like Amazon, whose recommendation engine drives 35% of total revenue through sophisticated data governance enabling personalization at unprecedented scale.

The contemporary urgency extends beyond competitive advantage to existential necessity. Data breaches cost organizations an average of $4.45 million per incident, while regulatory fines under frameworks like GDPR can reach 4% of global annual revenue. Meanwhile, organizations with mature data governance frameworks report 23% higher profitability and 19% faster revenue growth compared to their less-governed counterparts.

Bridging the Governance Gap

Despite data governance's proven value, most organizations struggle with implementation. Traditional approaches often fail because they treat governance as a technology project rather than an organizational transformation. They focus on tools and processes while neglecting the cultural, political, and strategic dimensions that determine success or failure.

This book addresses that gap by providing a comprehensive, practical roadmap for designing, deploying, and sustaining effective data governance frameworks. Unlike theoretical treatments that describe what governance should accomplish, this guide focuses on how to actually achieve those outcomes within real organizational constraints.

The approach recognizes that data governance is fundamentally about people, not technology. While tools and platforms enable governance, sustainable success requires aligning diverse stakeholders around common objectives, establishing clear accountability structures, and embedding governance practices into daily workflows. This human-centered perspective distinguishes effective governance programs from well-intentioned initiatives that fail to achieve lasting impact.

Designed for Diverse Leadership Perspectives

This guide serves three distinct but interconnected audiences, each bringing unique perspectives and requirements to the governance challenge.

Executives will discover how data governance transforms from a compliance burden into a strategic enabler of innovation and growth. The book demonstrates governance's connection to business outcomes through concrete metrics and case studies, providing the strategic context needed to secure organizational commitment and resources. Executives will learn to recognize governance maturity indicators, evaluate program success, and position data governance as a competitive differentiator rather than operational overhead.

Data professionals—including data scientists, analysts, and engineers—will find practical frameworks for implementing governance within their daily work. The guide addresses common challenges like data quality assessment, metadata management, and cross-functional collaboration, providing templates and methodologies that accelerate implementation while ensuring sustainability. These professionals will learn to balance governance requirements with analytical agility, ensuring compliance enhances rather than constrains innovation.

IT leaders will discover comprehensive strategies for integrating governance into existing technology architectures while planning for future scalability. The book covers technology selection criteria, integration challenges, and architectural patterns that support governance objectives without requiring complete system overhauls. IT leaders will learn to design technical solutions that enforce governance policies automatically while providing the flexibility needed for evolving business requirements.

Learning from Global Success Stories

Throughout this journey, real-world examples illustrate both opportunities and challenges. Beyond Netflix's transformation, consider how Maersk, the global shipping giant, leveraged data governance to optimize supply chain operations across 130 countries. By implementing standardized data definitions and quality controls, Maersk reduced container dwell times by 15% and improved customer satisfaction scores by 22%, demonstrating governance's operational impact at global scale.

Similarly, Singapore's Smart Nation initiative showcases governance's role in public sector transformation. By establishing comprehensive data sharing frameworks and privacy protection standards, Singapore has created an integrated digital ecosystem that improves citizen services while maintaining strict data protection. Their approach demonstrates how governance enables innovation rather than constraining it, providing a model for organizations seeking to balance openness with control.

Your Governance Transformation Begins Now

The data governance revolution is not a distant possibility—it is happening now, reshaping competitive dynamics across every industry. Organizations that master governance today will define tomorrow's market leaders, while those that delay risk becoming casualties of digital transformation.

This book provides everything needed to join that revolution: strategic frameworks for securing organizational commitment, practical methodologies for designing effective governance structures, proven techniques for managing complex implementations, and advanced strategies for sustaining long-term success. More importantly, it provides the confidence to begin.

Data governance may seem daunting, but it is ultimately achievable for any organization willing to commit to the journey. The frameworks, templates, and case studies in this guide have been tested across diverse industries and organizational contexts, proving their adaptability and effectiveness. Whether you are launching your first governance initiative or refining existing programs, this book will accelerate your progress and improve your outcomes.

The question is not whether your organization needs data governance—the evidence for that need is overwhelming. The question is whether you will lead that transformation or be transformed by it. The tools, knowledge, and inspiration needed for leadership are within these pages. Your governance revolution begins with turning to Chapter 1.

The future belongs to organizations that treat data as their most valuable asset while governing it as their most critical capability. That future starts now, and it starts with you.

ABOUT THE AUTHOR

Rajender Kumar stands at the forefront of the enterprise data governance revolution, bringing over a decade of transformative experience in data science, machine learning, and organizational data solutions to some of the world's most complex business challenges. As a recognized authority in data governance implementation and framework sustainability, Kumar has dedicated his career to bridging the critical gap between strategic data vision and practical stewardship execution across diverse industries and organizational structures.

Throughout his distinguished career, Kumar has partnered with industry leaders including Walgreens Boot Alliance, Healthcare Retroactive Audits, NEBA, and Loxia Technologies, where he has consistently applied governance-driven approaches to solve enterprise data challenges that traditional methods couldn't address. His work at Walgreens Boot Alliance involved designing comprehensive data governance frameworks that transformed how the retail pharmacy giant managed customer data across multiple touchpoints, resulting in improved regulatory compliance and enhanced customer insights. At Healthcare Retroactive Audits, he developed predictive data models integrated with robust governance structures that revolutionized audit accuracy while maintaining strict privacy standards.

Kumar's expertise extends beyond technical implementation to encompass the nuanced art of organizational change management within data governance initiatives. His approach to designing tailored governance solutions recognizes that every enterprise faces unique challenges, requiring frameworks that align with specific business objectives, regulatory environments, and cultural contexts. This philosophy has made him a sought-after consultant for framework sustainability, helping organizations maintain momentum long after initial deployment phases conclude.

His passion for data governance was ignited during a pivotal project early in his career when he witnessed a healthcare organization struggle with fragmented patient data that compromised care quality. This experience crystallized his understanding that effective data governance isn't merely about compliance or technical architecture, but about creating systems that fundamentally improve human outcomes. This revelation has driven his commitment to developing governance frameworks that serve as strategic enablers rather than operational constraints.

Kumar's methodology emphasizes the critical importance of cross-functional collaboration in governance success. He has pioneered approaches that engage stakeholders across all organizational levels, from C-suite executives seeking competitive advantage to front-line data users requiring practical tools. His frameworks consistently demonstrate how proper data stewardship can transform governance from a perceived burden into a catalyst for innovation and business growth.

As a thought leader in predictive analytics integration with governance frameworks, Kumar has contributed to advancing enterprise data governance practices through his development of sustainability models that evolve with changing business needs. His work has been particularly influential in demonstrating how governance frameworks can be designed to accommodate emerging technologies while maintaining core stewardship principles.

Beyond his consulting and implementation work, Kumar maintains a deep commitment to education and mentorship in organizational data stewardship. He regularly conducts workshops for emerging data professionals and has mentored dozens of practitioners who have gone on to lead successful governance initiatives within their own organizations. His teaching philosophy centers on empowering others to see data governance not as a technical challenge to be solved, but as a strategic opportunity to unlock organizational potential.

Kumar's unique combination of technical expertise, strategic vision, and practical implementation experience positions him as an authoritative voice in the rapidly evolving field of enterprise data governance, making him ideally suited to guide organizations through their governance transformation journeys.

"Data without governance is like a river without banks—it may flow freely, but it will never reach its intended destination with purpose and power. The organizations that master data governance don't just manage information; they architect the future of human decision-making."

— Rajender Kumar

ESSENTIAL TOOLS AND RESOURCES FOR SUCCESS

The journey toward effective data governance begins with understanding what you truly need to succeed, not what vendors claim you must purchase. One of the most persistent myths in data governance is that successful implementation requires expensive enterprise software, advanced technical degrees, or massive IT budgets. This misconception has prevented countless organizations from embarking on transformative governance initiatives that could unlock significant business value.

The reality is far more encouraging. Effective data governance is fundamentally about people, processes, and principles rather than technology platforms. While sophisticated tools can certainly enhance your governance capabilities, the foundation of successful data stewardship lies in clear thinking, structured approaches, and collaborative mindsets that any organization can cultivate.

What You Actually Need to Begin

The essential components of data governance success are surprisingly accessible. **Data literacy** serves as the cornerstone, but this doesn't mean everyone needs to become a data scientist. Instead, it means developing a shared understanding of how data flows through your organization, where quality issues typically emerge, and how information decisions impact business outcomes. This literacy can be developed through workshops, documentation exercises, and collaborative data mapping sessions that require no specialized software.

Clear communication frameworks represent another critical foundation. Successful governance initiatives rely on the ability to translate technical concepts into business language and vice versa. This skill develops through practice and structured dialogue, not expensive training programs. Organizations that excel at governance invest time in creating glossaries, establishing common terminology, and facilitating regular conversations between business and technical teams.

Collaborative organizational structures form the third essential element. Data governance thrives when cross-functional teams work together with defined roles and responsibilities. This requires intentional relationship building and process design, not sophisticated technology platforms. Many successful governance programs begin with simple shared documents, regular meetings, and clear escalation paths before investing in specialized tools.

Debunking Common Barriers

The belief that data governance requires extensive technical backgrounds has deterred many business professionals from engaging in stewardship activities. In practice, some of the most effective **data stewards** come from business operations, finance, or customer service backgrounds. These individuals bring deep understanding of how data is actually used, where problems occur, and what quality standards matter most to end users.

Similarly, the assumption that governance demands expensive enterprise software creates unnecessary barriers. While mature organizations may benefit from comprehensive governance platforms, successful initiatives often begin with spreadsheets, shared drives, and existing collaboration tools. The key is establishing consistent processes and building organizational habits before investing in sophisticated technology.

Budget constraints frequently discourage organizations from pursuing governance initiatives. However, many of the most impactful governance activities cost nothing more than time and attention. Data profiling can begin with simple queries against existing databases. Quality monitoring can start with manual spot checks and basic reporting. Policy development requires thoughtful discussion and documentation, not expensive consulting engagements.

Real-World Applications and Success Stories

Consider the experience of Regional Healthcare Network, a mid-sized healthcare system that launched their governance program using entirely existing resources. Their IT team repurposed their existing business intelligence platform to create simple data quality dashboards, while clinical staff used shared spreadsheets to document data definitions and quality standards. Within six months, they had reduced patient data errors by 35% and improved regulatory reporting accuracy significantly. Their total additional investment was zero dollars, focusing instead on reallocating existing staff time and establishing regular governance meetings.

Manufacturing Solutions Inc. demonstrates how accessible tools can drive substantial governance improvements. This 500-employee company used Microsoft Excel and SharePoint to create their initial **data catalog** and quality monitoring system. Business analysts worked with department heads to document critical data elements, establish ownership assignments, and create simple quality metrics. When they eventually invested in a dedicated governance platform two years later, their strong foundational processes made the technology implementation seamless and immediately valuable.

Global Consulting Partners showcases governance success in a distributed organization. With teams across multiple continents, they leveraged existing collaboration tools like Slack, Google Workspace, and Zoom to establish virtual governance communities. Their **data governance council** meets monthly via video conference, maintains shared documentation in cloud-based platforms, and uses simple survey tools to gather feedback on policy effectiveness. This approach has enabled consistent governance practices across 12 countries without requiring significant technology investments.

Affordable and Accessible Alternatives

For organizations ready to enhance their governance capabilities beyond basic tools, numerous affordable options exist. Open-source data profiling tools like Apache Griffin or Great Expectations provide enterprise-grade functionality without licensing costs. Cloud-based collaboration platforms offer scalable governance features at fraction of traditional enterprise software prices.

Data cataloging can begin with simple database documentation tools or even well-structured wikis before progressing to specialized platforms. Many organizations find that systematic documentation using familiar tools creates more sustainable habits than jumping immediately to sophisticated software that requires extensive training.

Building Your Governance Mindset

The most critical resource for governance success is cultivating the right organizational mindset. This means viewing data as a shared asset requiring collective stewardship rather than individual ownership. It involves embracing transparency about data quality issues while focusing on systematic improvement rather than blame. Most importantly, it requires patience to build sustainable practices over time rather than expecting immediate transformation.

Successful governance organizations develop cultures of **continuous improvement** where data quality discussions become routine parts of business operations. They celebrate small wins while maintaining focus on long-term objectives. They invest in relationships and communication as much as processes and technology.

The path to effective data governance is accessible to any organization willing to commit time, attention, and collaborative energy to the effort. By starting with existing resources and building systematic practices, organizations can establish strong governance foundations that create lasting business value while preparing for future technology investments when they make strategic sense.

SETTING EXPECTATIONS: YOUR JOURNEY TO DATA GOVERNANCE MASTERY

In an era where data has become the lifeblood of modern enterprises, the ability to govern this critical asset effectively separates industry leaders from those struggling to keep pace. This comprehensive guide transforms data governance from an abstract concept into a concrete competitive advantage, equipping you with the knowledge, tools, and strategies needed to design, deploy, and sustain governance frameworks that deliver measurable business value.

What You Will Learn and Master

Through this practical journey, you will develop a comprehensive understanding of data governance that extends far beyond compliance checklists and theoretical frameworks. You will master the art of creating governance structures that align seamlessly with organizational objectives while fostering a culture of data stewardship that permeates every level of your enterprise.

The book will guide you through establishing **data ownership models** that eliminate ambiguity and create clear accountability chains. You will learn to design **data quality frameworks** that ensure consistency, accuracy, and reliability across all enterprise systems, transforming unreliable information into trusted business assets. Additionally, you will develop expertise in implementing **privacy compliance strategies** that not only meet regulatory requirements but also build customer trust and competitive differentiation.

Your governance toolkit will expand to include sophisticated **risk assessment methodologies** that identify potential data vulnerabilities before they become costly problems. You will master the development of **data classification schemes** that protect sensitive information while enabling appropriate access for business users. Most importantly, you will learn to create **performance measurement systems** that demonstrate the tangible business value of your governance initiatives, securing ongoing executive support and resource allocation.

Enterprise Application and Strategic Impact

The governance skills you develop will have immediate application across multiple enterprise contexts. In strategic decision-making scenarios, you will be able to ensure that leadership teams have access to reliable, consistent data that supports confident business choices. Your governance frameworks will enable **self-service analytics** capabilities that empower business users while maintaining data integrity and security standards.

For compliance initiatives, you will develop the capability to create governance structures that streamline regulatory reporting, reduce audit preparation time, and minimize compliance-related risks. Your frameworks will automatically enforce data retention policies, privacy controls, and access restrictions, transforming compliance from a reactive burden into a proactive business enabler.

In organizational transformation contexts, your governance expertise will prove invaluable for merger and acquisition activities, digital transformation initiatives, and cloud migration projects. You will be able to establish governance protocols that ensure data quality and consistency during periods of significant organizational change, maintaining business continuity while enabling innovation.

Real-World Success Stories

Consider the transformation achieved by a global pharmaceutical company that implemented a comprehensive data governance framework across its research and development operations. By establishing clear **data lineage tracking** and quality controls, the organization reduced clinical trial preparation time by 40% while improving regulatory submission accuracy. The governance framework enabled researchers to quickly identify and access relevant historical data, accelerating drug discovery processes and reducing time-to-market for new treatments.

Similarly, a multinational financial services firm leveraged advanced governance practices to consolidate customer data from 15 different legacy systems following a series of acquisitions. The implementation of standardized **data integration protocols** and quality validation processes resulted in a single, trusted view of customer relationships that improved cross-selling effectiveness by 25% and reduced customer service resolution times by 35%. The governance framework also ensured compliance with evolving privacy regulations across multiple jurisdictions, avoiding potential penalties while enhancing customer trust.

A third example demonstrates the power of governance in the retail sector, where a major e-commerce platform used sophisticated data classification and access controls to enable personalized customer experiences while maintaining strict privacy protections. The governance framework allowed the company to leverage customer behavioral data for targeted marketing campaigns while automatically enforcing consent preferences and data retention policies, resulting in a 20% increase in customer engagement rates and zero privacy-related compliance violations.

Tailored Learning for Every Role

This guide recognizes that effective data governance requires collaboration across multiple organizational levels and functional areas. For executives, the content emphasizes strategic value creation, business case development, and organizational change management strategies that secure stakeholder buy-in and resource allocation. You will learn to communicate governance benefits in business terms that resonate with board members and senior leadership teams.

Data professionals will find detailed technical guidance on implementing quality controls, designing metadata management systems, and creating data lineage documentation that supports both operational efficiency and regulatory compliance. The book provides practical templates, assessment frameworks, and implementation checklists that accelerate project timelines while ensuring comprehensive coverage of governance requirements.

IT leaders will discover strategies for selecting and integrating governance technologies, managing data architecture complexity, and balancing security requirements with business accessibility needs. The content addresses common technical challenges such as legacy system integration, cloud governance considerations, and automated policy enforcement mechanisms.

Alignment with Current Industry Trends

The governance approaches presented in this guide directly address the most pressing challenges facing modern enterprises. As organizations increasingly adopt **cloud-first data strategies**, the book provides specific guidance on implementing governance controls that span hybrid and multi-cloud environments. The content addresses emerging requirements for **artificial intelligence governance**, ensuring that machine learning models are trained on high-quality, ethically sourced data.

The guide also tackles the growing importance of **data mesh architectures** and decentralized data management approaches, providing frameworks for maintaining governance standards while enabling domain-specific data ownership and innovation. Additionally, the book addresses evolving privacy regulations and the need for **privacy-by-design** approaches that embed protection mechanisms into data processing workflows from the outset.

Your Transformation Awaits

Mastering data governance represents more than acquiring a new skill set; it positions you at the forefront of an organizational revolution that is fundamentally reshaping how enterprises create value from their information assets. The frameworks and strategies you will learn transform data from a compliance liability into a strategic differentiator that drives innovation, reduces operational risk, and unlocks new revenue opportunities.

By joining the ranks of data governance practitioners who understand how to balance protection with accessibility, compliance with innovation, and control with empowerment, you become an architect of organizational transformation. The journey ahead will equip you with the knowledge and confidence to lead your organization's data stewardship revolution, creating lasting competitive advantages that extend far beyond traditional IT boundaries into the very heart of business strategy and execution.

Rajender Kumar

A Practical Guide to Mastering Data Governance: How to Design, Deploy, and Sustain an Effective Data Governance Framework

1. THE DATA GOVERNANCE IMPERATIVE

Every organization today sits atop a digital goldmine, yet most are mining fool's gold. Despite unprecedented access to customer insights, operational metrics, and market intelligence, companies continue to make critical decisions based on incomplete, inconsistent, or unreliable information. This fundamental disconnect between data abundance and actionable intelligence represents one of the most significant strategic challenges facing modern enterprises.

The traditional approach of treating data as a byproduct of business operations has become not just inadequate, but dangerous. In an era where regulatory scrutiny intensifies daily and competitive advantages can evaporate overnight, organizations can no longer afford the luxury of informal data practices. The stakes have never been higher: data breaches can destroy decades of brand equity in minutes, while regulatory non-compliance can result in penalties that dwarf entire annual budgets.

This chapter establishes the foundational understanding necessary for any successful data governance initiative. We will examine how the exponential growth of data sources, from IoT sensors to social media streams, has overwhelmed existing management frameworks designed for simpler times. More importantly, we will explore how forward-thinking organizations are transforming their relationship with data from a compliance burden into a competitive weapon.

Through examining real-world governance failures and their cascading consequences, readers will understand why half-measures and piecemeal solutions consistently fail. We will then construct the compelling business case needed to secure executive sponsorship and sustained organizational commitment. By the end of this chapter, you will possess the strategic perspective and persuasive arguments necessary to champion data governance as an essential business capability rather than an optional IT project.

1.1. Understanding the Modern Data Landscape

The digital transformation has fundamentally altered how organizations create, store, and utilize data, creating an enterprise landscape where information flows at unprecedented velocity and volume. Modern businesses generate data from countless touchpoints: customer interactions across digital channels, IoT sensors monitoring operational equipment, supply chain transactions, social media engagement, and internal business processes. This explosion of data sources has created both extraordinary opportunities and complex governance challenges that traditional management approaches simply cannot address.

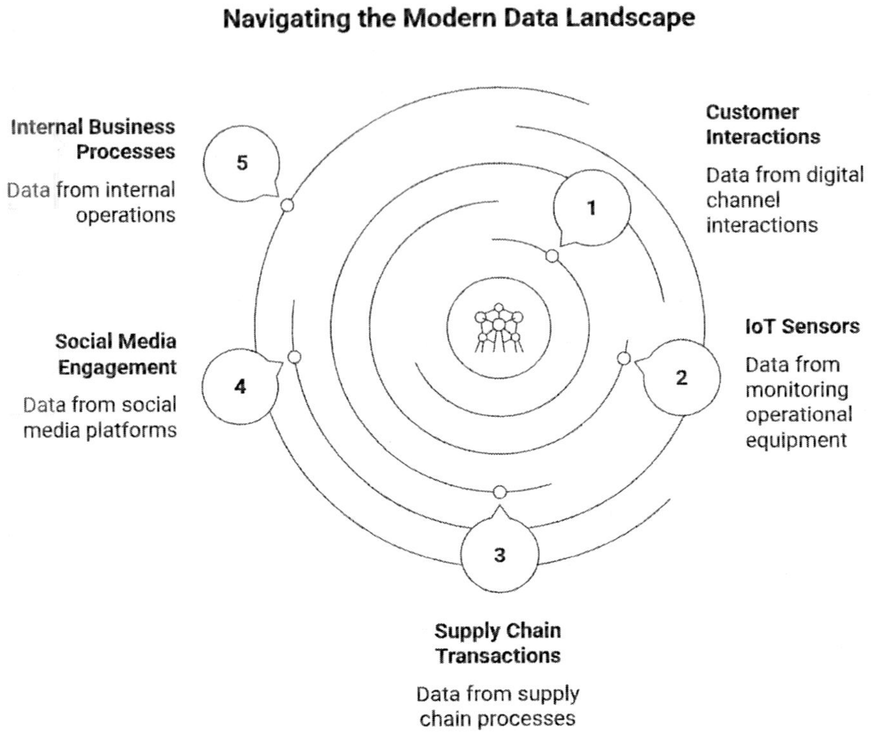

The Exponential Growth of Enterprise Data Assets

Today's organizations face data growth rates that compound exponentially. A typical Fortune 500 company now manages petabytes of structured and unstructured data across hybrid cloud environments, edge computing systems, and legacy on-premises infrastructure. This data exists in multiple formats, from traditional relational databases to streaming analytics platforms, document repositories, and real-time messaging systems. The complexity extends beyond volume to encompass data velocity, with real-time processing requirements driving millisecond decision-making across automated trading systems, fraud detection algorithms, and personalized customer experiences.

The variety of data types has expanded dramatically, encompassing everything from traditional transactional records to video analytics, geospatial information, and machine-generated logs. Each data type requires specialized handling, storage optimization, and governance protocols that align with business objectives while maintaining regulatory compliance and operational efficiency.

Practical Implementation: Multi-Source Data Inventory Framework

Consider the real-world implementation at **JPMorgan Chase**, where the financial services giant needed to establish comprehensive data governance across their global operations. Their approach demonstrates how organizations can systematically catalog and govern diverse data landscapes.

Step 1: Data Source Discovery and Classification

The bank implemented Collibra's Data Governance Center to create an automated data discovery process across their enterprise ecosystem:

```
Data Source Inventory Template:
- Source System: [Trading Platform Alpha]
- Data Types: [Transaction records, risk metrics, compliance logs]
- Update Frequency: [Real-time streaming]
- Business Owner: [Head of Trading Operations]
- Technical Owner: [Platform Engineering Team]
- Criticality Level: [Mission Critical]
- Regulatory Requirements: [Dodd-Frank, Basel III, MiFID II]
```

Step 2: Stakeholder Engagement Matrix

JPMorgan established governance committees with clear accountability structures:

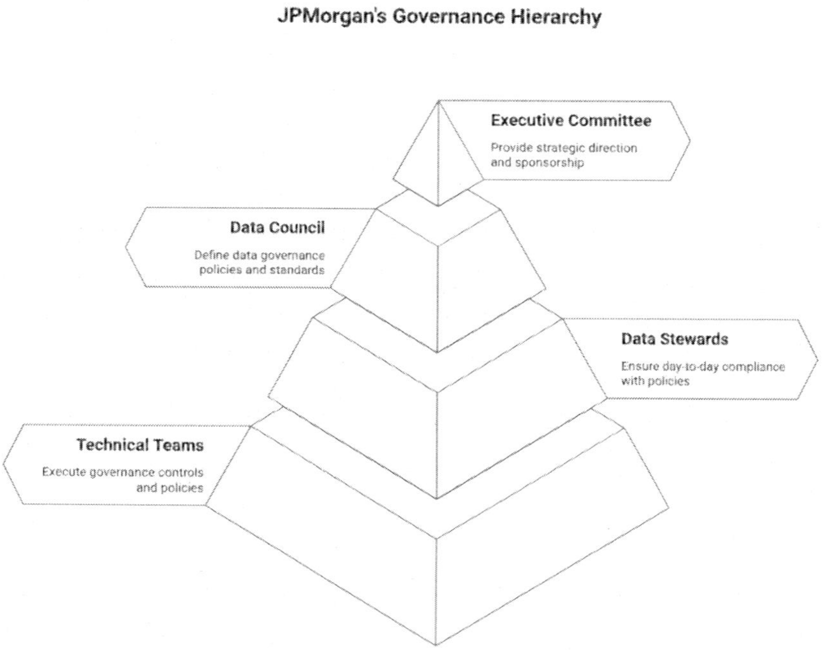

- **Executive Steering Committee**: C-level sponsors providing strategic direction
- **Data Governance Council**: Business and IT leaders defining policies
- **Data Steward Network**: Subject matter experts ensuring day-to-day compliance
- **Technical Implementation Teams**: Engineers executing governance controls

Step 3: Risk Assessment and Prioritization

The bank developed a comprehensive risk scoring methodology that evaluated each data source across multiple dimensions:

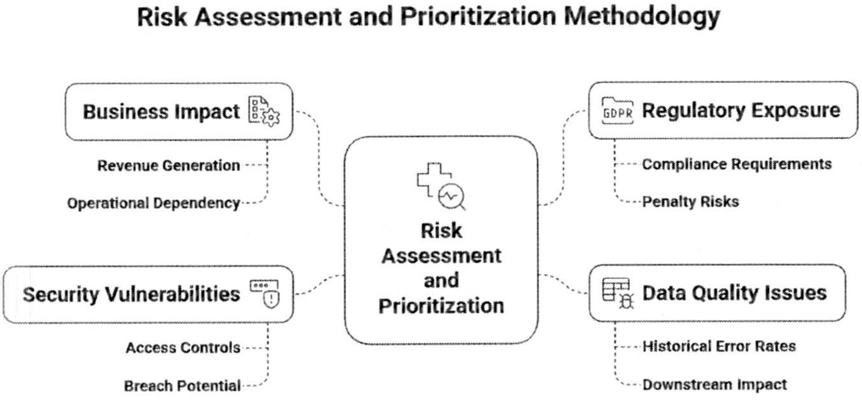

- **Business Impact**: Revenue generation potential and operational dependency
- **Regulatory Exposure**: Compliance requirements and penalty risks
- **Data Quality Issues**: Historical error rates and downstream impact
- **Security Vulnerabilities**: Access controls and breach potential

Expected Outcomes: Within 18 months, JPMorgan achieved a 40% reduction in data quality incidents, improved regulatory reporting accuracy by 35%, and established clear data lineage for over 80% of their critical data assets.

The Interconnected Data Ecosystem Challenge

Modern enterprises operate within interconnected data ecosystems where information flows seamlessly between internal systems, external partners, cloud services, and third-party platforms. This interconnectedness creates governance complexities that extend beyond organizational boundaries. Data shared with suppliers, customers, and regulatory bodies must maintain consistent quality standards, privacy protections, and access controls throughout its lifecycle.

The challenge intensifies when considering data residency requirements, cross-border data transfers, and varying international privacy regulations. Organizations must navigate GDPR in Europe, CCPA in California, and emerging privacy laws worldwide while maintaining operational efficiency and competitive advantage.

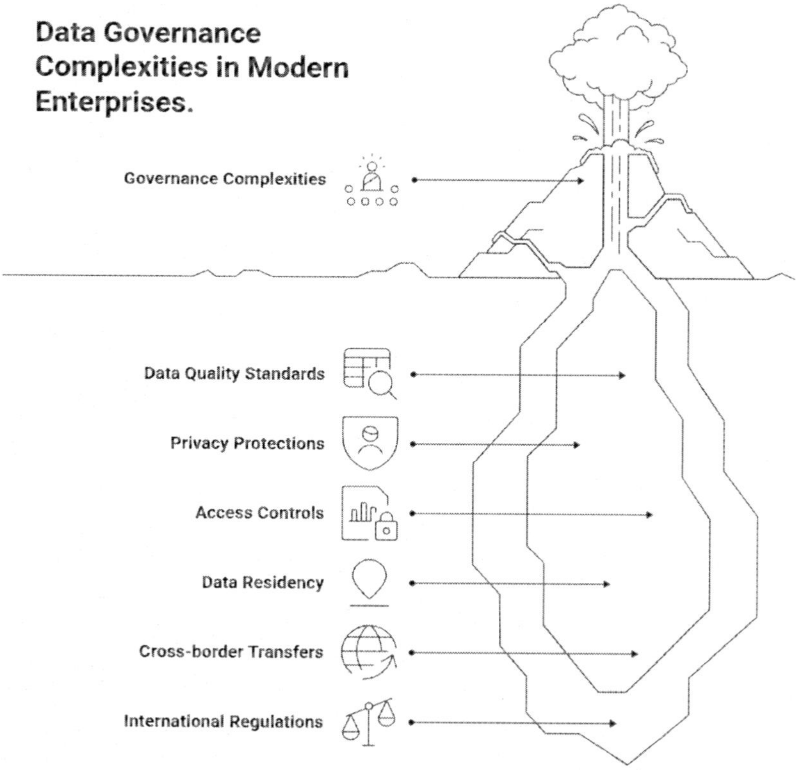

Emerging Technologies and Governance Implications

Artificial intelligence and machine learning applications have introduced new governance requirements around algorithmic transparency, bias detection, and model explainability. Organizations must ensure their data governance frameworks support responsible AI development while enabling innovation and competitive differentiation.

Cloud-native architectures, microservices, and containerized applications create dynamic data environments where traditional governance controls may not apply effectively. Modern governance frameworks must adapt to support DevOps practices, continuous integration pipelines, and infrastructure-as-code deployments.

Understanding the modern data landscape requires recognizing that data governance has evolved from a compliance-focused discipline to a strategic capability that enables digital transformation. Organizations that successfully navigate this complexity establish comprehensive data inventories, engage stakeholders across business and technical domains, and implement risk-based governance approaches that scale with their data ecosystem growth.

As organizations grapple with these mounting complexities, it becomes clear that conventional data management practices developed for simpler, more predictable environments are fundamentally inadequate for today's dynamic, interconnected data landscape.

1.2. Why Traditional Data Management Approaches Fall Short

The fundamental disconnect between traditional data management practices and modern organizational needs becomes starkly apparent when examining how enterprises handle data at scale. Conventional approaches, rooted in departmental silos and reactive problem-solving, create a cascade of operational inefficiencies that compound over time, ultimately undermining strategic decision-making capabilities.

The Silo-Driven Data Management Crisis

Traditional data management typically evolved organically within individual departments, with each business unit developing its own data collection, storage, and analysis practices. This fragmented approach creates what governance experts call **data islands**, where customer information exists in multiple, often contradictory versions across sales, marketing, and service systems. The result is a fundamental breakdown in data consistency that renders enterprise-wide analytics unreliable.

Consider the practical implications: when marketing teams rely on customer segmentation data that conflicts with sales pipeline information, campaign effectiveness plummets while customer experience suffers. Finance teams struggle to reconcile revenue figures when different systems apply varying calculation methodologies, leading to reporting delays and compliance risks.

Reactive Problem-Solving Limitations

Most organizations approach data quality issues reactively, addressing problems only after they manifest in business operations. This **fire-fighting mentality** consumes significant resources while failing to prevent recurring issues. IT teams spend countless hours troubleshooting data integration failures, cleaning corrupted datasets, and explaining discrepancies in executive reports, rather than implementing systematic prevention measures.

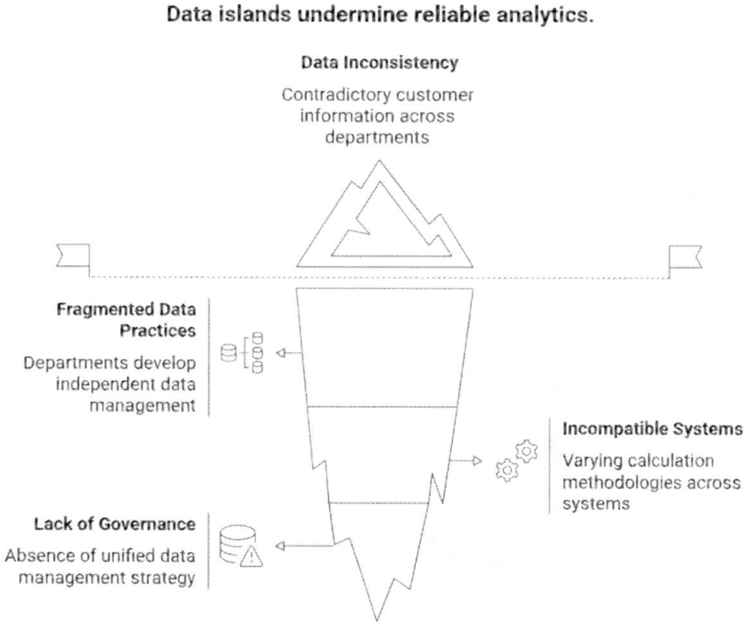

The European pharmaceutical company Roche discovered this limitation during their global data consolidation initiative. Initially, their IT teams spent 60% of their time resolving data quality issues across 150 country-specific systems. Each fix addressed immediate symptoms but failed to establish sustainable data standards, resulting in a perpetual cycle of crisis management that delayed critical drug development timelines.

Practical Example: Transforming Reactive Data Management into Proactive Governance

A Fortune 500 retail organization exemplifies the transformation from traditional data management to comprehensive governance. Their initial approach involved separate teams managing customer data across e-commerce, in-store, and mobile platforms, each using different customer identification schemes and data quality rules.

Before Implementation:

- Customer data existed in 47 different systems with no unified identifier
- Data quality issues required 200+ hours monthly to resolve

- Marketing campaigns showed 23% lower effectiveness due to duplicate targeting
- Compliance reporting took 6 weeks to complete

Step-by-Step Governance Implementation:

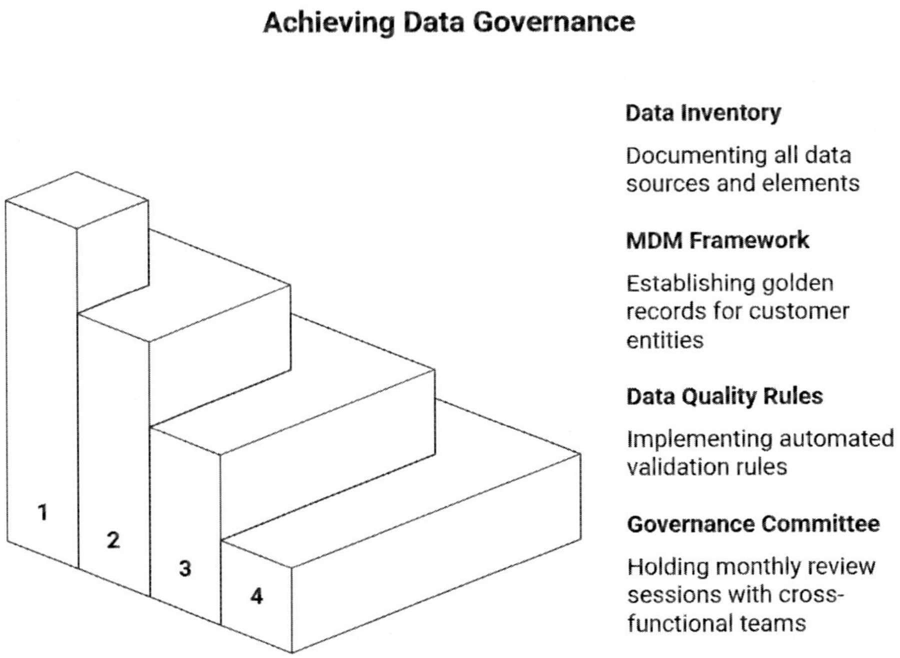

- **Data Inventory and Mapping**: Using Collibra's Data Catalog, teams documented all customer data sources, identifying 312 unique data elements across systems
- **Master Data Management Framework**: Implemented Informatica MDM to establish golden records for customer entities
- **Data Quality Rules Engine**: Deployed automated validation rules checking address formats, email validity, and demographic consistency
- **Cross-Functional Governance Committee**: Established monthly review sessions with representatives from IT, marketing, sales, and compliance teams

After Implementation:

- Single source of truth for 2.3 million customer records
- Data quality issue resolution time reduced by 78%
- Marketing campaign effectiveness improved by 31%

- Compliance reporting completed in 3 days

The Scalability Trap

Traditional approaches often work adequately for smaller data volumes but fail catastrophically as organizations scale. Manual data validation processes that functioned with thousands of records become impossibly time-consuming with millions of data points. Legacy systems designed for batch processing cannot accommodate real-time analytics requirements, creating bottlenecks that limit competitive responsiveness.

The absence of standardized data definitions across business units compounds these scalability challenges. When different departments define **customer lifetime value** using incompatible methodologies, enterprise-wide customer analytics become meaningless, preventing organizations from developing coherent customer strategies.

This systematic breakdown of traditional data management approaches demonstrates why organizations require comprehensive governance frameworks that establish proactive, standardized, and scalable data stewardship practices. Rather than continuing to patch individual problems, successful enterprises recognize that sustainable data value creation demands fundamental shifts in how they approach data as a strategic asset.

Understanding these limitations provides the foundation for appreciating why comprehensive data governance represents not just an operational improvement, but a strategic imperative that directly impacts competitive positioning and long-term organizational sustainability.

1.3. The Business Case for Comprehensive Data Governance

Organizations that embrace comprehensive data governance unlock transformative business value that extends far beyond regulatory compliance or risk mitigation. When properly implemented, data governance becomes a strategic enabler that drives revenue growth, operational efficiency, and competitive differentiation across every business function.

Quantifiable Financial Returns from Governance Investments

The financial impact of effective data governance manifests in measurable improvements across multiple business metrics. Companies with mature governance frameworks report average cost reductions of 15-25% in data management operations, while simultaneously achieving 30-40% improvements in decision-making speed. These gains stem from eliminated data redundancies, reduced manual reconciliation efforts, and streamlined access to trusted information assets.

Data quality improvements alone generate substantial returns on governance investments. Poor data quality costs organizations an average of $15 million annually according to Gartner research, while companies with robust governance frameworks reduce these losses by 60-80% within the first two years of implementation. The elimination of duplicate customer records, inconsistent product catalogs, and inaccurate financial reporting creates immediate operational savings that often exceed initial governance program costs.

Revenue Generation Through Enhanced Data Monetization

Comprehensive governance frameworks enable organizations to transform data from a cost center into a profit generator. By establishing clear data lineage, quality standards, and access controls, companies can confidently share data assets with partners, customers, and third-party platforms while maintaining security and compliance standards.

Consider the practical implementation at a major telecommunications provider that leveraged Collibra's governance platform to create a comprehensive data marketplace. The organization established a systematic approach to data product development:

```
Data Product Creation Framework:
1. Asset Discovery: Automated catalog scanning identifies high-value
datasets
2. Quality Assessment: Governance rules validate completeness and
accuracy
3. Business Context: Stewards document use cases and business
definitions
4. Access Controls: Role-based permissions ensure appropriate data
sharing
5. Monetization Strategy: Pricing models align with data value and
usage patterns
```

This telecommunications company transformed anonymized customer behavior patterns into revenue-generating insights for retail partners, creating a new $50 million annual revenue stream while maintaining strict privacy compliance. The governance framework provided the trust and transparency necessary for external data sharing, with automated quality monitoring ensuring consistent data product delivery.

Operational Excellence Through Governance-Enabled Analytics

Organizations with comprehensive governance frameworks achieve superior analytical capabilities by ensuring data consistency across business units and systems.

When marketing teams, finance departments, and operations groups work with standardized, governed data definitions, analytical insights become more reliable and actionable.

The governance framework establishes **master data management** processes that eliminate the common problem of conflicting metrics and definitions. Sales teams no longer debate whether quarterly revenue figures include or exclude returns, while marketing campaigns leverage consistent customer segmentation based on standardized demographic and behavioral attributes.

Risk Mitigation and Strategic Decision Making

Effective governance frameworks provide executives with confidence in data-driven decision making by establishing clear accountability chains and quality assurance processes. When leadership understands data lineage and validation procedures, strategic decisions carry reduced uncertainty and improved success rates.

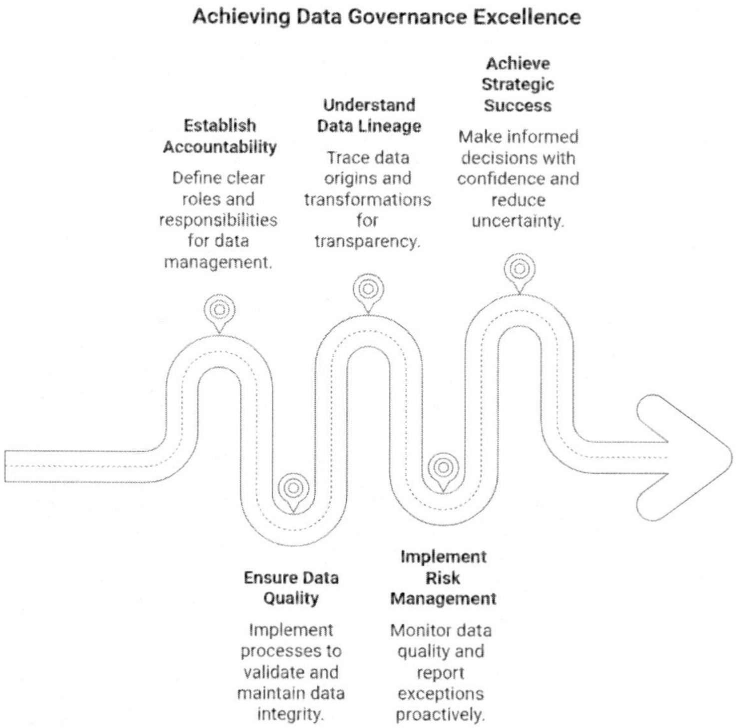

The framework also enables proactive risk management through systematic data monitoring and exception reporting. Governance committees receive automated alerts when data quality metrics fall below established thresholds, allowing for rapid corrective action before business impacts occur. This proactive approach prevents the cascading effects of poor data quality that can compromise customer experiences, regulatory reporting, and operational efficiency.

Modern governance frameworks integrate seamlessly with existing business processes, enhancing rather than disrupting established workflows. The key lies in demonstrating immediate value through quick wins while building toward comprehensive, sustainable governance capabilities that support long-term strategic objectives.

This business case foundation establishes the economic justification for governance investments, but organizations must also navigate an increasingly complex landscape of regulatory requirements that add urgency to comprehensive data governance initiatives.

1.4. Regulatory Pressures and Compliance Requirements

The regulatory environment surrounding data management has undergone dramatic transformation in recent years, creating a compliance landscape that demands immediate executive attention and strategic response. Organizations operating across multiple jurisdictions now face an intricate web of regulations that extend far beyond traditional privacy laws, encompassing sector-specific requirements, cross-border data transfer restrictions, and emerging artificial intelligence governance mandates.

The Expanding Regulatory Ecosystem

Modern data governance frameworks must address a multifaceted regulatory environment that includes the European Union's General Data Protection Regulation (GDPR), California Consumer Privacy Act (CCPA), Brazil's Lei Geral de Proteção de Dados (LGPD), and sector-specific regulations such as the Health Insurance Portability and Accountability Act (HIPAA) in healthcare and the Sarbanes-Oxley Act in financial services. Each regulation introduces unique requirements for data handling, retention, deletion, and breach notification that must be systematically integrated into organizational processes.

The complexity intensifies when considering emerging regulations focused on algorithmic accountability and artificial intelligence governance. The EU's proposed AI Act and similar initiatives worldwide are establishing new requirements for data lineage, model explainability, and algorithmic bias detection that directly impact how organizations structure their data governance programs.

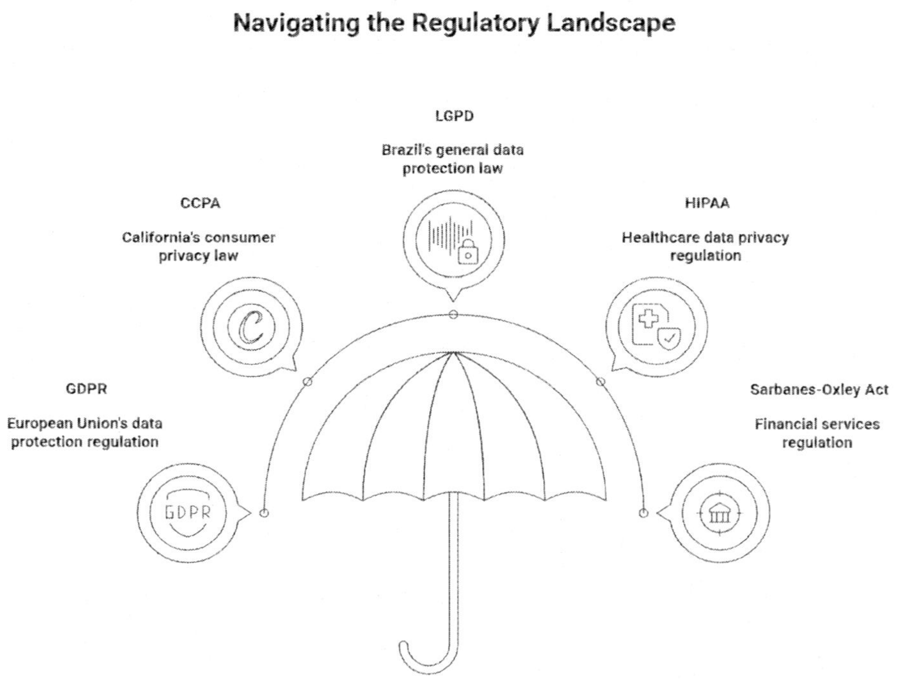

Practical Implementation: GDPR Compliance Framework Development

Consider the approach taken by Siemens AG when implementing GDPR compliance across their global operations. The industrial conglomerate established a comprehensive data governance framework that demonstrates how regulatory requirements can be systematically addressed through structured governance processes.

Step 1: Regulatory Mapping and Gap Analysis

```
GDPR Requirements Assessment Framework:
- Data Processing Inventory: Catalog all personal data processing
activities
- Legal Basis Documentation: Establish lawful basis for each processing
activity
- Data Subject Rights Implementation: Create workflows for access,
rectification, erasure
- Cross-Border Transfer Mechanisms: Document adequacy decisions and
safeguards
- Breach Response Procedures: Define 72-hour notification protocols
```

Step 2: Governance Structure Implementation

Siemens created a multi-tiered governance structure with designated Data Protection Officers (DPOs) at regional levels, supported by local data stewards responsible for operational compliance. This structure enabled consistent policy interpretation while accommodating local regulatory variations across their 200+ country operations.

Step 3: Technology Integration and Automation

The organization implemented Informatica's Data Privacy Management platform to automate data subject request processing, enabling systematic handling of over 50,000 annual privacy requests while maintaining detailed audit trails for regulatory reporting.

Measurable Outcomes:

- 95% reduction in data subject request response time
- Zero regulatory fines since implementation
- $12 million annual savings through automated compliance processes

Risk-Based Compliance Strategy Development

Effective regulatory compliance requires a risk-based approach that prioritizes governance investments based on potential regulatory exposure and business impact. Organizations must develop **compliance risk matrices** that evaluate factors including data sensitivity, processing volume, jurisdictional exposure, and regulatory enforcement trends.

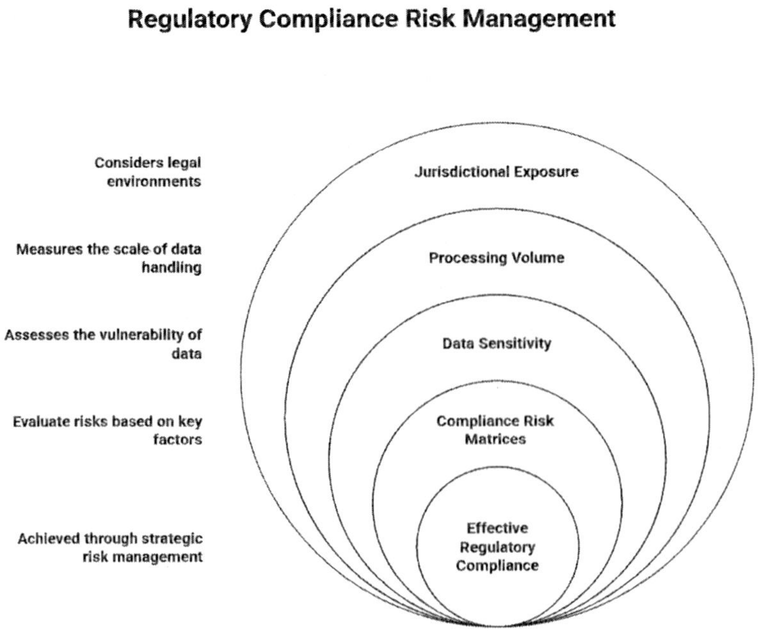

Financial services firm JPMorgan Chase exemplifies this approach through their comprehensive data governance program that addresses multiple regulatory frameworks simultaneously. Their governance structure integrates GDPR privacy requirements with Basel III data quality standards and Dodd-Frank reporting obligations through a unified data stewardship model that ensures consistent compliance across regulatory domains.

Governance Committee Structures for Regulatory Oversight

Successful regulatory compliance demands governance committees with clear accountability structures and decision-making authority. The most effective models establish **Data Governance Councils** with executive sponsorship, supported by operational **Data Stewardship Committees** responsible for day-to-day compliance activities.

These committees must maintain detailed documentation of governance decisions, policy interpretations, and risk assessments to demonstrate regulatory compliance during audits. Regular compliance reporting to executive leadership ensures ongoing visibility into governance effectiveness and regulatory risk exposure.

The regulatory imperative for data governance extends beyond mere compliance to encompass strategic risk management and operational excellence. Organizations that proactively address regulatory requirements through comprehensive governance frameworks position themselves to navigate future regulatory changes while building capabilities that support broader business objectives. This regulatory foundation becomes the bedrock upon which organizations can build more sophisticated approaches to leveraging data as a strategic asset that drives competitive advantage and innovation.

While regulatory compliance provides the essential foundation for governance programs, forward-thinking organizations recognize that effective data governance extends far beyond meeting minimum regulatory requirements to unlock the transformative potential of data as a strategic business asset.

1.5. Data as a Strategic Asset: Beyond Compliance

Forward-thinking organizations recognize that effective data governance extends far beyond meeting minimum regulatory requirements to unlock the transformative potential of data as a strategic business asset. This shift from compliance-driven to value-driven governance represents a fundamental evolution in how enterprises approach their information resources, transforming data from a necessary burden into a competitive differentiator that drives innovation, operational efficiency, and market advantage.

The Strategic Value Creation Framework

Modern data governance frameworks must balance regulatory compliance with strategic value creation through systematic approaches that align data initiatives with business objectives. Organizations achieving this balance typically implement a **dual-track governance model** where compliance requirements form the foundation while strategic initiatives build upon this base to generate measurable business outcomes.

Consider how global pharmaceutical giant Roche transformed its clinical trial data management through strategic governance implementation. Rather than simply meeting FDA compliance requirements, Roche developed a comprehensive data governance framework that standardized data collection across 150+ clinical sites worldwide. This strategic approach reduced trial completion times by 23% while simultaneously improving regulatory submission quality, demonstrating how governance can simultaneously serve compliance and competitive advantage.

Practical Implementation: Revenue-Generating Governance Initiative

A practical example of strategic data governance implementation involves creating a **Customer Data Excellence Program** that transforms compliance-focused customer data management into revenue-generating insights:

Step 1: Baseline Assessment and Strategic Alignment

Begin by conducting a comprehensive audit of existing customer data assets using tools like Collibra's Data Intelligence Platform. Map current data flows, identify quality gaps, and quantify potential revenue impact through improved customer segmentation and personalization capabilities.

```
Customer Data Governance Charter Template:

Strategic Objectives:
- Increase customer lifetime value by 15% through improved data quality
- Reduce customer acquisition costs by 20% via enhanced targeting
- Achieve 99.5% data accuracy for regulatory compliance

Governance Structure:
- Executive Sponsor: Chief Revenue Officer
- Data Steward: Customer Analytics Director
- Technical Lead: Master Data Management Team
- Compliance Officer: Privacy and Legal Team
```

Step 2: Policy Development and Stakeholder Engagement

Establish cross-functional governance committees that include revenue-generating teams alongside traditional IT and compliance stakeholders. Create data quality standards that directly support business objectives while maintaining regulatory compliance.

Step 3: Technology Integration and Workflow Automation

Implement automated data quality monitoring using platforms like Informatica Data Quality, establishing real-time dashboards that track both compliance metrics and business performance indicators. Configure automated alerts for data quality issues that could impact customer experience or revenue generation.

Expected Outcomes:

Organizations implementing this approach typically see 25-40% improvement in marketing campaign effectiveness, 15-30% reduction in customer churn, and 20-50% faster time-to-market for new customer-facing initiatives within 12-18 months of implementation.

Governance as Innovation Enabler

Strategic data governance creates the foundation for advanced analytics, artificial intelligence, and machine learning initiatives by ensuring data quality, accessibility, and ethical use. Organizations with mature governance frameworks can deploy AI solutions 60% faster than those without established data stewardship practices, as governance eliminates common data preparation bottlenecks that typically consume 80% of analytics project timelines.

The financial services sector exemplifies this principle, where institutions like JPMorgan Chase leverage comprehensive data governance to support both regulatory reporting and algorithmic trading systems. Their governance framework ensures data lineage transparency for regulatory audits while simultaneously enabling real-time risk assessment algorithms that process millions of transactions daily.

Measuring Strategic Impact

Successful strategic governance requires metrics that extend beyond traditional compliance indicators to encompass business value creation. Key performance indicators should include revenue attribution from data-driven initiatives, operational cost reductions through improved data quality, and innovation velocity metrics that track time-to-market for data-dependent products and services.

This strategic approach to data governance transforms information assets from compliance liabilities into competitive advantages, creating sustainable value that compounds over time. Organizations that successfully implement value-driven governance frameworks position themselves to capitalize on emerging technologies while maintaining the trust and regulatory standing essential for long-term success.

However, achieving this transformation requires careful navigation of common implementation pitfalls that can derail even the most well-intentioned governance initiatives, making it crucial to understand the typical failure patterns that organizations encounter during their governance journey.

1.6. Common Governance Failures and Their Consequences

Understanding the typical failure patterns that organizations encounter during their governance journey provides critical insights for executives seeking to avoid common pitfalls. These failures often stem from fundamental misunderstandings about governance requirements, inadequate stakeholder engagement, or insufficient attention to organizational change management principles.

The Technology-First Trap

Many organizations fall into the misconception that purchasing expensive data management platforms automatically solves governance challenges. A prominent example occurred at a Fortune 500 retail company that invested $12 million in a comprehensive data catalog solution without establishing clear data ownership roles or quality standards. The platform remained largely unused because business users couldn't understand data lineage reports, and data stewards lacked authority to enforce quality rules.

Practical Implementation Example: Addressing Platform Abandonment

When a global manufacturing company faced similar platform underutilization with their Alation data catalog deployment, they implemented a structured recovery approach:

```
Recovery Framework Implementation:
1. Stakeholder Mapping Workshop
   - Identified 15 key business data consumers
   - Mapped current data discovery workflows
   - Documented pain points in existing processes

2. Use Case Prioritization
   - Selected 3 high-impact scenarios (customer segmentation, supply
chain optimization, financial reporting)
   - Created specific success metrics for each use case
```

```
   - Established 30-day pilot programs

3. Guided Adoption Program
   - Weekly 1-hour training sessions focused on business outcomes
   - Peer mentoring between early adopters and skeptical users
   - Integration with existing business intelligence workflows
```

The results demonstrated measurable improvement: platform usage increased from 12% to 78% within six months, and data discovery time decreased from 4 hours to 30 minutes for critical business queries.

Governance Without Governance

Another critical failure pattern involves creating governance frameworks that lack enforcement mechanisms or accountability structures. Organizations often establish data governance committees that meet regularly but have no decision-making authority or budget allocation power. This ceremonial governance creates an illusion of control while actual data management practices remain unchanged.

A European financial services firm exemplified this pattern when their data governance board spent eighteen months creating comprehensive policies but failed to establish consequences for non-compliance. When regulatory auditors discovered inconsistent customer data across seven different systems, the organization faced €2.3 million in penalties because their governance framework existed only on paper.

The Perfectionism Paralysis

Executive teams sometimes delay governance implementation while seeking the perfect framework design. This analysis paralysis prevents organizations from gaining early wins and building momentum. A healthcare consortium spent two years designing an ideal master data management strategy while continuing to operate with duplicate patient records, ultimately missing critical opportunities for care coordination improvements and federal meaningful use incentives.

Scope Creep and Initiative Overload

Governance initiatives frequently fail when organizations attempt to address every data challenge simultaneously. Without clear prioritization, teams become overwhelmed, resources get diluted, and stakeholders lose confidence in the program's ability to deliver tangible results. A telecommunications company's governance initiative collapsed when they tried to simultaneously implement data quality monitoring, privacy compliance, master data management, and analytics governance across twelve business units within one fiscal year.

Cultural Resistance and Change Management Gaps

Technical teams often underestimate the cultural transformation required for successful governance implementation. Data professionals may focus exclusively on technical architecture while ignoring the human elements that drive adoption. Organizations that fail to address existing power structures, competing priorities, and entrenched workflows find their governance initiatives meeting passive resistance or outright opposition from key stakeholders.

These failure patterns highlight the critical importance of balanced approaches that combine technical capabilities with organizational change management, clear accountability structures, and realistic implementation timelines. Understanding these common pitfalls enables executives to proactively address potential obstacles before they derail governance initiatives.

Having examined the most frequent causes of governance failure, executives must next focus on building the organizational awareness and leadership support necessary to overcome these challenges and establish sustainable governance practices.

1.7. Building Executive Awareness and Support

Building on the understanding of governance failure consequences, executives must next focus on building the organizational awareness and leadership support necessary to overcome these challenges and establish sustainable governance practices. The path to successful data governance begins with creating a compelling business case that resonates across the C-suite and establishing the foundational support structures that enable long-term success.

Creating the Executive Business Case

The most effective approach to securing executive support involves translating data governance concepts into tangible business outcomes. Rather than focusing on technical specifications or compliance requirements, successful governance champions frame their proposals around **revenue protection**, **operational efficiency**, and **competitive differentiation**. This requires building a narrative that connects data quality issues directly to business impact.

Consider how JPMorgan Chase approached their executive awareness campaign in 2019. The bank's Chief Data Officer presented a comprehensive analysis showing how inconsistent customer data across systems was creating duplicate marketing costs of $47 million annually while simultaneously reducing customer satisfaction scores by 12%. By quantifying these impacts and demonstrating how governance would address root causes rather than symptoms, the executive team immediately understood the strategic importance of the initiative.

Practical Executive Engagement Strategy

A systematic approach to building executive awareness involves implementing a structured engagement framework that addresses different leadership perspectives and concerns:

```
Executive Engagement Framework:
1. Stakeholder Mapping and Influence Analysis
2. Business Impact Quantification Workshop
3. Governance Value Demonstration Session
4. Risk Scenario Planning Exercise
5. Success Metrics Definition and Agreement
```

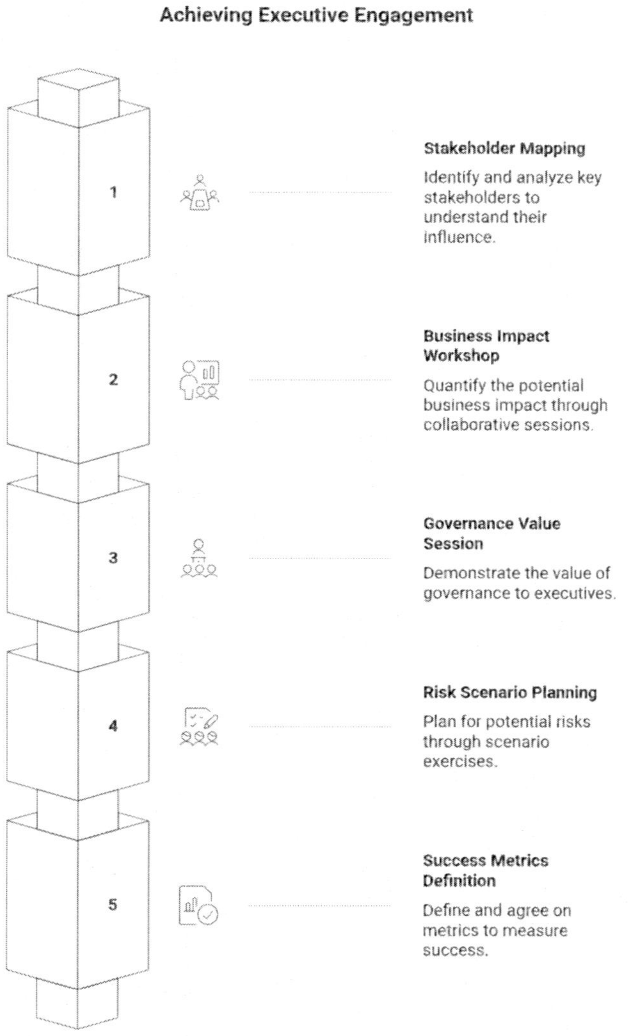

The most critical component involves conducting **Business Impact Quantification Workshops** where executives collaborate to identify and measure data-related inefficiencies within their domains. During these sessions, participants use actual organizational data to calculate costs associated with poor data quality, regulatory compliance gaps, and missed revenue opportunities.

Implementing the Governance Champion Network

Successful executive support requires establishing a network of governance champions across business units. This involves identifying influential leaders who can advocate for governance initiatives within their respective domains while maintaining alignment with overall organizational objectives.

Step-by-Step Champion Network Implementation:

Champion Identification Process: Select department leaders who demonstrate both analytical thinking and change management capabilities. These individuals should possess credibility within their business units and maintain regular interaction with executive leadership.

Governance Training and Certification: Provide champions with comprehensive training on governance principles using frameworks like DAMA-DMBOK. This includes hands-on experience with governance tools such as Collibra's governance workflow capabilities or Informatica's data quality assessment features.

Business Unit Assessment Execution: Champions conduct detailed assessments within their domains using standardized templates that identify data quality issues, compliance gaps, and improvement opportunities. Each assessment produces quantified business impact statements that support the overall executive business case.

Regular Executive Reporting Structure: Establish monthly governance scorecards that champions present to executive leadership, highlighting progress metrics, issue resolution, and business value realization.

Accenture successfully implemented this approach across their global consulting practice, resulting in a 34% reduction in project delivery delays attributed to data quality issues within 18 months. Their champion network identified over 200 specific data governance improvements that directly impacted client satisfaction and operational efficiency.

Measuring and Communicating Value

Executive support sustainability depends on demonstrating continuous value through measurable outcomes. This requires establishing governance metrics that align with business objectives while providing regular visibility into program effectiveness. Key performance indicators should include both operational metrics like data quality scores and business metrics such as decision-making speed improvements or compliance cost reductions.

Building executive awareness and support represents the foundational step in governance success, requiring strategic communication, practical demonstration of value, and systematic engagement across leadership levels. This foundation enables organizations to move beyond governance as a compliance requirement toward leveraging data stewardship as a competitive advantage that drives innovation and business growth.

1.8. The Competitive Advantage of Effective Data Stewardship

Organizations that master data stewardship fundamentally transform their competitive positioning in the marketplace. While many companies view data governance as a defensive strategy focused on compliance and risk mitigation, forward-thinking enterprises recognize stewardship as an offensive weapon that creates sustainable differentiation through superior decision-making, operational excellence, and customer insight generation.

The competitive advantage emerges not from the data itself, but from the organizational capability to consistently transform raw information into actionable intelligence. Companies with mature stewardship practices demonstrate measurably superior performance across key business metrics: faster time-to-market for new products, higher customer retention rates, more accurate demand forecasting, and reduced operational costs through elimination of data-driven inefficiencies.

Strategic Differentiation Through Data Quality Excellence

Effective data stewardship creates competitive moats that are difficult for rivals to replicate. When organizations establish consistent data quality standards and maintain them through disciplined stewardship practices, they develop institutional knowledge and operational capabilities that compound over time. This accumulated advantage becomes increasingly valuable as competitors struggle with fragmented, inconsistent data that undermines their analytical capabilities.

Consider how leading retailers leverage customer data stewardship to personalize experiences at scale. By maintaining comprehensive, accurate customer profiles through rigorous data quality processes, these organizations can predict purchasing behavior, optimize inventory allocation, and deliver targeted marketing campaigns that consistently outperform generic approaches. The stewardship investment creates a virtuous cycle where better data leads to better decisions, which generate better outcomes and more valuable data.

Practical Implementation: Customer 360 Data Stewardship Program

A practical example demonstrates how stewardship drives competitive advantage through systematic implementation. A global telecommunications company implemented a comprehensive Customer 360 stewardship program using Informatica's Master Data Management platform combined with custom data quality rules.

Phase 1: Baseline Assessment and Stewardship Framework

The organization began by cataloging customer data sources across 47 different systems, identifying 312 unique customer attributes with varying quality levels. Using DAMA-DMBOK methodologies, they established data stewardship roles with specific accountability for customer data domains.

```
Customer Data Stewardship Charter
- Domain: Customer Master Data
- Steward: Regional Customer Experience Manager
- Quality Metrics:
  * Address accuracy: >95%
  * Contact information completeness: >98%
  * Preference data consistency: >99%
- Review Cycle: Monthly quality scorecards
- Escalation: Data Quality Council for issues >72 hours
```

Phase 2: Automated Quality Monitoring and Correction

The team implemented real-time data quality monitoring with automated correction workflows. When address information failed validation, the system triggered steward notifications and initiated third-party verification processes. Customer service representatives received quality alerts during interactions, enabling immediate data correction at the point of contact.

Phase 3: Business Impact Measurement

Within 18 months, the stewardship program delivered quantifiable competitive advantages. Marketing campaign response rates improved by 34% due to accurate targeting. Customer service resolution times decreased by 22% because representatives accessed complete, reliable customer histories. Most significantly, the company reduced customer churn by 15% through proactive identification of at-risk accounts based on comprehensive behavioral patterns.

Organizational Learning and Adaptation Capabilities

Data stewardship excellence creates organizational learning capabilities that accelerate competitive response times. Companies with mature stewardship practices can rapidly analyze market changes, customer behavior shifts, and operational performance variations because their data infrastructure supports agile analysis rather than requiring extensive data preparation and validation efforts.

This agility becomes particularly valuable during market disruptions or competitive threats. Organizations with trusted data can quickly model scenarios, assess risks, and implement strategic pivots while competitors remain paralyzed by uncertainty about their information accuracy. The stewardship investment transforms from a cost center into a strategic enabler of organizational responsiveness.

Revenue Generation Through Data Monetization

Beyond internal operational advantages, effective stewardship enables external revenue generation through data monetization opportunities. Companies with high-quality, well-governed data assets can develop new revenue streams by providing market insights, benchmarking services, or data-driven consulting to partners and customers.

The pharmaceutical industry exemplifies this opportunity, where companies with superior clinical trial data stewardship can offer real-world evidence services to regulatory agencies and research institutions. The stewardship investment creates intellectual property that generates recurring revenue while supporting core business objectives.

Mastering data stewardship transforms organizational capabilities from reactive compliance to proactive competitive advantage generation. Companies that invest in comprehensive stewardship programs develop sustainable differentiation through superior data quality, faster decision-making cycles, and enhanced customer understanding that competitors struggle to replicate.

As we conclude our exploration of the data governance imperative, the foundation has been established for understanding how strategic governance frameworks translate these competitive advantages into systematic organizational capabilities that drive sustained business value.

2. FOUNDATIONS OF DATA GOVERNANCE

Building effective data governance requires more than good intentions and scattered initiatives. It demands a solid foundation built on proven principles, clear understanding, and shared commitment across the organization. Just as architects must master fundamental engineering concepts before designing skyscrapers, data governance practitioners need a comprehensive grasp of core concepts before implementing frameworks that can withstand the pressures of enterprise complexity.

This chapter establishes the essential building blocks that separate successful governance programs from well-intentioned efforts that fade into obscurity. While many organizations rush to implement governance tools and processes, those that achieve lasting transformation invest time in understanding what data governance truly means within their unique context, how it differs from traditional data management approaches, and why cultural alignment often determines success more than technical sophistication.

The journey begins with defining data governance in practical, enterprise-relevant terms that resonate with stakeholders across business and IT functions. From there, we explore the fundamental principles that guide decision-making and the interconnected ecosystem of people, processes, and technology that brings governance to life. Understanding the critical distinction between data governance and data management prevents common misconceptions that derail initiatives before they gain momentum.

Equally important is establishing a common vocabulary that enables clear communication and building cultural foundations that support sustainable change. As organizations operate in an increasingly global landscape, we examine how international privacy frameworks like GDPR and CCPA shape governance requirements, while practical maturity assessments help determine organizational readiness for transformation.

These foundational elements create the stable platform upon which all subsequent governance activities depend, ensuring your framework can evolve and scale alongside business growth.

2.1. Defining Data Governance in the Enterprise Context

Data governance represents far more than a compliance checkbox or IT initiative. In the enterprise context, it functions as the strategic discipline that transforms scattered information assets into a coherent, trustworthy foundation for business decision-making. Unlike traditional data management approaches that focus primarily on technical storage and retrieval, enterprise data governance encompasses the policies, processes, roles, and technologies that ensure data serves organizational objectives while maintaining quality, security, and regulatory compliance.

At its core, enterprise data governance establishes clear accountability for data assets across the organization. This means defining who owns specific data domains, who can access and modify information, and how data flows between systems and business units. The governance framework creates a structured approach to managing data as a valuable business asset, similar to how organizations manage financial resources or intellectual property.

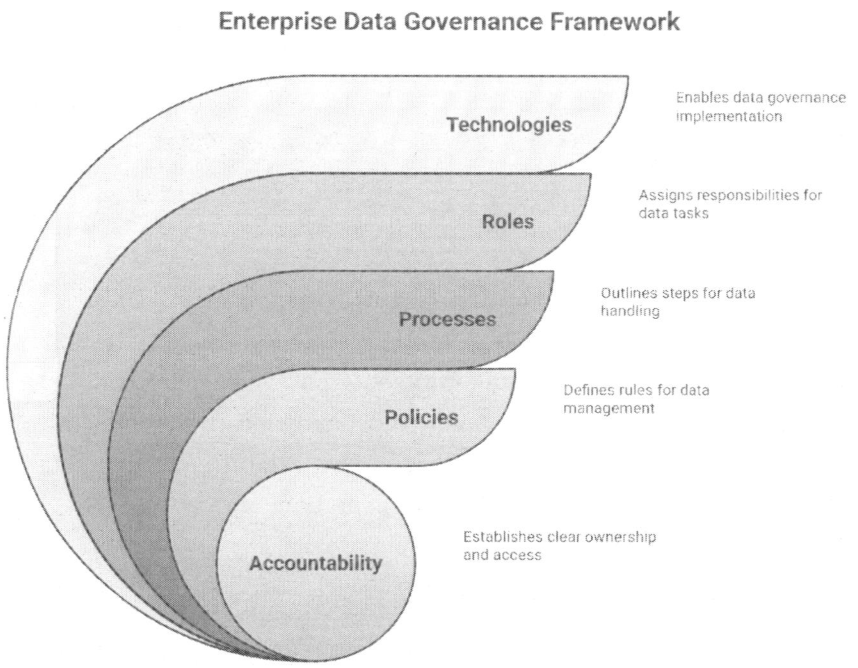

The Business-Driven Nature of Enterprise Data Governance

Enterprise data governance differs fundamentally from technical data management by prioritizing business outcomes over technical capabilities. While data management focuses on the mechanics of storing, processing, and retrieving information, governance addresses the strategic questions of data value, risk, and organizational impact. This business-centric approach ensures that governance initiatives align with corporate strategy and deliver measurable returns on investment.

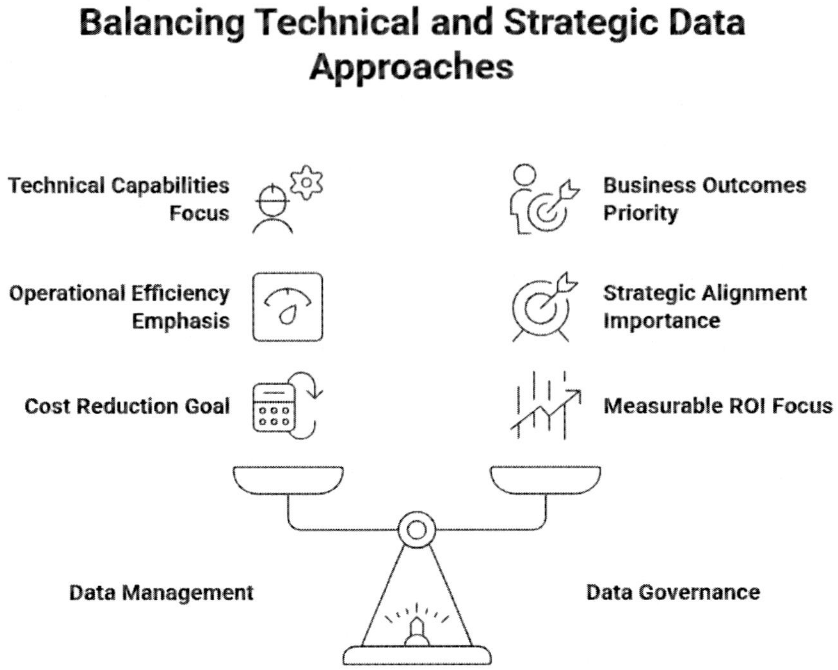

The enterprise context requires governance frameworks to span multiple business units, geographic regions, and regulatory environments. This complexity demands sophisticated coordination mechanisms that balance local autonomy with enterprise-wide consistency. Successful governance programs establish common standards while allowing for business unit flexibility in implementation approaches.

Practical Implementation: Building a Customer Data Governance Framework

Consider the comprehensive customer data governance initiative implemented at **Royal Bank of Canada (RBC)**, one of North America's largest financial institutions. Facing regulatory pressure and data quality challenges across multiple business lines, RBC developed an enterprise-wide customer data governance program that demonstrates the practical application of governance principles in a complex organizational environment.

Step 1: Establishing Data Ownership Structure

RBC began by mapping customer data across all business units, identifying over 200 distinct customer data elements spanning retail banking, wealth management, and corporate services. They implemented a **three-tier ownership model** using Collibra's governance platform:

```
Data Domain Owner: Chief Customer Officer
Data Steward: Business Unit Customer Experience Managers
Data Custodian: IT Data Management Teams
```

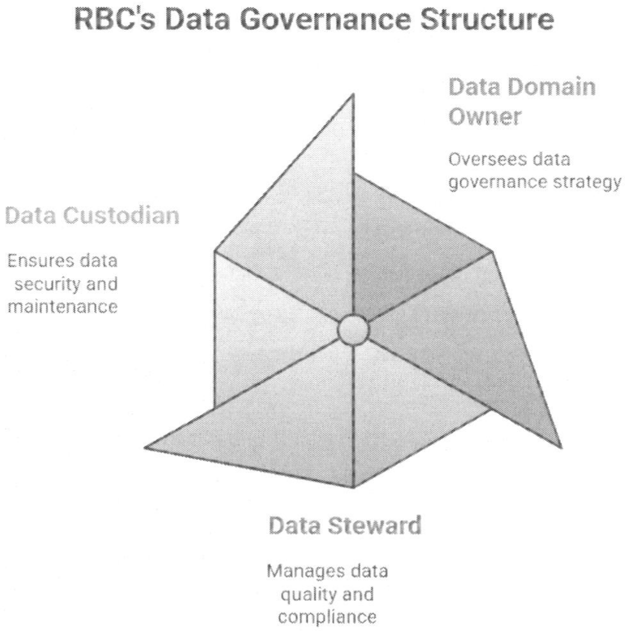

Step 2: Implementing Cross-Functional Governance Committees

The bank established a **Customer Data Governance Council** with representatives from each business line, legal, compliance, and IT. This council meets monthly to review data quality metrics, approve policy changes, and resolve cross-business unit data conflicts. Their charter explicitly defines decision-making authority and escalation procedures.

Step 3: Developing Actionable Data Policies

Rather than creating generic policies, RBC developed specific, actionable guidelines for customer data scenarios. For example, their customer address standardization policy includes:

```
Policy: Customer Address Data Standards
- All addresses must conform to Canada Post formatting standards
- Address validation occurs at point of entry using real-time APIs
- Conflicting addresses across systems trigger automatic data steward
alerts
- Monthly address quality reports distributed to business unit leaders
```

Results and Outcomes

Within 18 months, RBC achieved a 40% improvement in customer data accuracy across systems, reduced customer onboarding time by 25%, and successfully passed multiple regulatory audits with zero data-related findings. The governance framework enabled the bank to launch new digital services with confidence in their underlying customer data quality.

Integration with Enterprise Architecture

Enterprise data governance must integrate seamlessly with existing organizational structures and technology architectures. This integration requires governance frameworks to complement rather than compete with established business processes. Successful implementations leverage existing governance mechanisms, such as risk management committees and audit functions, to embed data governance into organizational DNA.

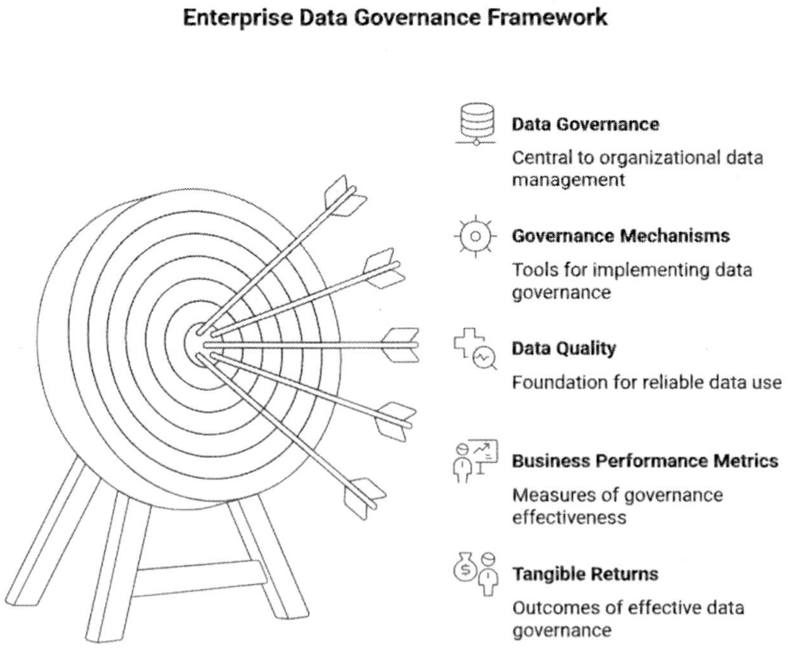

The most effective enterprise data governance programs establish clear connections between data quality and business performance metrics. This connection transforms governance from a cost center into a value driver, demonstrating tangible returns on governance investments through improved operational efficiency and reduced compliance risks.

Having defined data governance within the enterprise context, we now turn to examining the fundamental principles that underpin successful governance frameworks and how these concepts translate into practical organizational structures.

2.2. Core Principles and Fundamental Concepts

Building upon the strategic imperative established in the enterprise context, we now turn to examining the fundamental principles that underpin successful governance frameworks and how these concepts translate into practical organizational structures. The foundation of any robust data governance program rests on four interconnected pillars that transform abstract governance concepts into actionable business capabilities.

The Four Pillars of Data Governance Excellence

Effective data governance operates through four core principles that serve as the architectural foundation for sustainable programs. **Data accountability** establishes clear ownership and responsibility chains throughout the organization. **Data quality** ensures information meets defined standards for accuracy, completeness, and consistency. **Data accessibility** balances security requirements with business needs for timely information access. **Data compliance** maintains adherence to regulatory requirements while supporting business objectives.

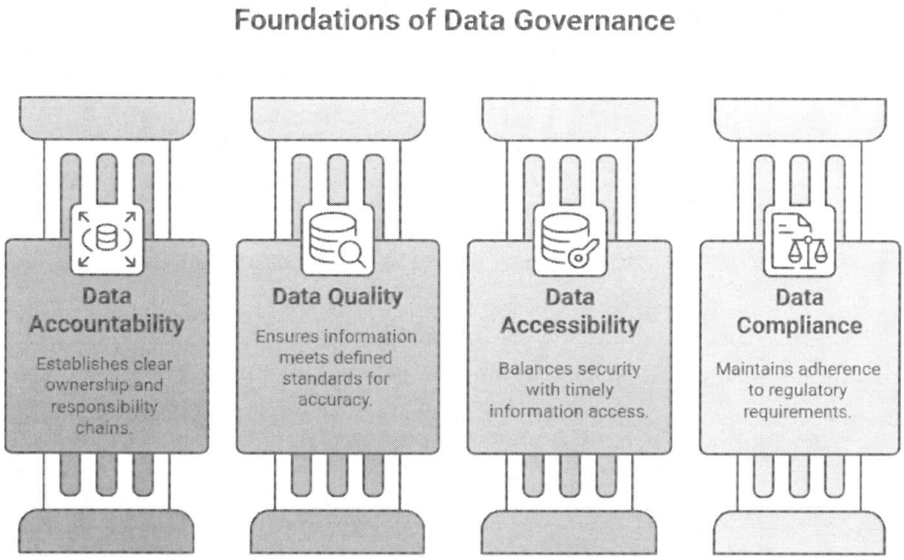

These principles work synergistically rather than in isolation. When organizations attempt to implement governance by focusing on individual pillars, they often encounter resistance and limited success. The most effective frameworks integrate all four principles into cohesive governance structures that reinforce each other.

Implementing the Data Stewardship Hierarchy

The practical application of governance principles requires establishing clear stewardship hierarchies that span organizational boundaries. **Data owners** hold ultimate accountability for data domains and make strategic decisions about data usage and access policies. **Data stewards** execute day-to-day governance activities and serve as subject matter experts within their respective domains. **Data custodians** handle technical implementation of governance policies and maintain data infrastructure.

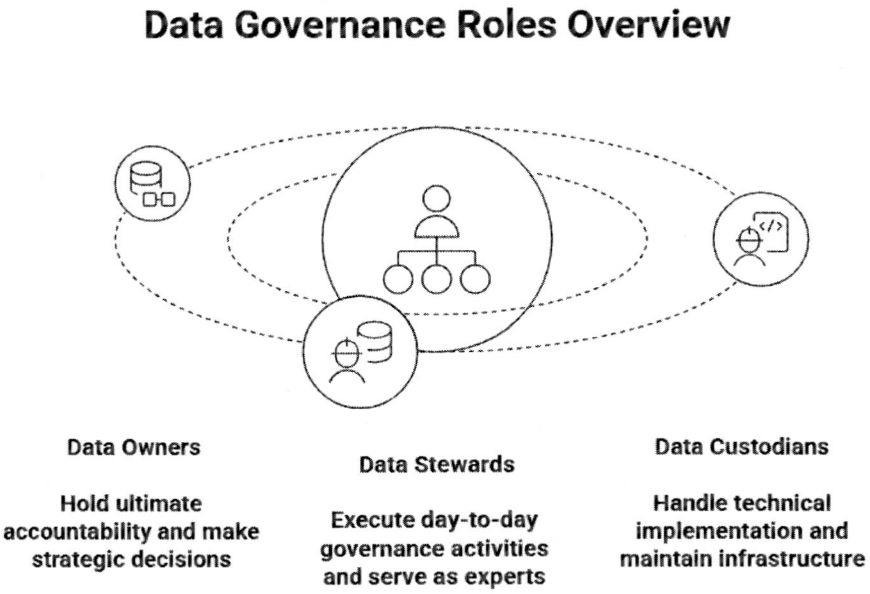

A practical example demonstrates this hierarchy in action. Global manufacturing company Siemens implemented a comprehensive data stewardship program using Collibra's governance platform to manage their product lifecycle data. Their implementation followed this systematic approach:

```
Phase 1: Stewardship Role Definition
- Appointed Chief Data Officer as executive sponsor
- Designated business unit leaders as data owners for manufacturing
domains
- Selected 45 data stewards across engineering, procurement, and
quality functions
- Trained IT teams as data custodians for technical governance
execution
```

```
Phase 2: Domain Mapping and Policy Creation
- Cataloged 12 critical data domains including product specifications
and supplier data
- Developed standardized data quality rules within Collibra workflows
- Created automated policy enforcement through integration with SAP
systems
- Established escalation procedures for governance violations

Phase 3: Governance Integration
- Embedded data quality checks into existing engineering change
processes
- Implemented automated data lineage tracking for regulatory compliance
- Created business glossaries with 2,000+ standardized terms
- Deployed self-service data access controls through role-based
permissions
```

The results demonstrated measurable governance value: 40% reduction in product specification errors, 60% faster regulatory compliance reporting, and $2.3 million annual savings from improved data quality. This systematic approach shows how governance principles translate into operational excellence when properly structured and implemented.

Governance Maturity and Continuous Improvement

Organizations must recognize that governance maturity develops through distinct evolutionary stages. **Initial** governance focuses on basic compliance and risk mitigation. **Developing** governance introduces standardized processes and clearer accountability structures. **Defined** governance establishes comprehensive frameworks with integrated technology solutions. **Managed** governance incorporates advanced analytics and predictive capabilities. **Optimized** governance achieves self-improving systems that adapt to changing business requirements.

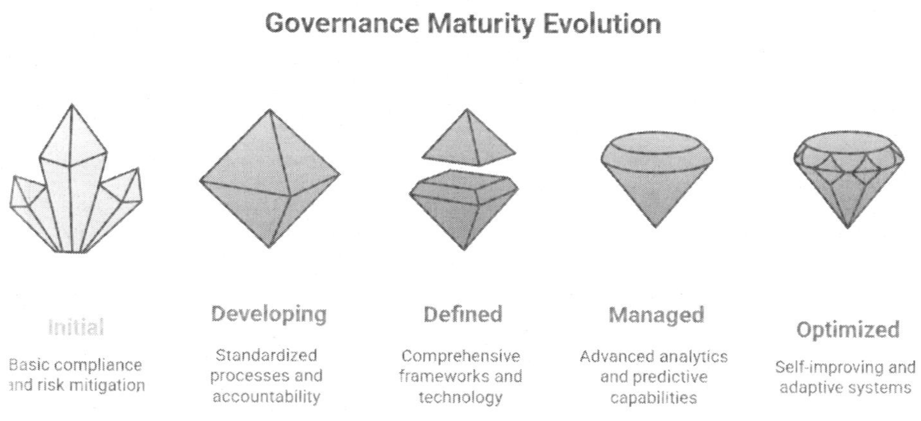

Understanding these maturity levels helps executives set realistic expectations and plan appropriate resource investments. Organizations attempting to jump directly to advanced governance capabilities often struggle with adoption challenges and sustainability issues.

Risk-Based Governance Prioritization

Successful governance programs prioritize efforts based on business risk and value potential rather than attempting comprehensive coverage immediately. Critical data domains receive intensive governance attention, while less critical areas operate under lighter oversight frameworks. This risk-based approach ensures optimal resource allocation and demonstrates clear business value from governance investments.

The core principles of data governance provide the essential foundation for building sustainable programs that deliver measurable business value. By establishing clear stewardship hierarchies, understanding maturity progression, and applying risk-based prioritization, organizations create governance frameworks that evolve with business needs while maintaining operational effectiveness. These foundational concepts set the stage for understanding how governance principles integrate with the broader organizational ecosystem of people, processes, and technology platforms.

2.3. The Governance Ecosystem: People, Process, and Technology

Data governance operates as a complex ecosystem where three critical components must work in harmony: people, processes, and technology. Unlike traditional IT initiatives that focus primarily on technical solutions, successful data governance requires careful orchestration of human capabilities, structured workflows, and technological enablers. Understanding this ecosystem is essential for executives seeking to build sustainable governance programs that deliver measurable business value.

The interconnected nature of these three pillars creates both opportunities and challenges. When properly aligned, they form a powerful foundation for data-driven decision making. When misaligned, they create friction that can derail even the most well-intentioned governance initiatives. Organizations that master this ecosystem transformation achieve competitive advantages through improved data quality, enhanced regulatory compliance, and accelerated innovation cycles.

The Human Element: Building Governance Communities

People represent the most critical yet complex component of the governance ecosystem. Effective data governance requires a diverse community of stakeholders, each bringing unique perspectives and expertise to the program. This community extends far beyond traditional IT boundaries to include business users, legal teams, risk management professionals, and executive sponsors.

The governance community operates through defined roles and responsibilities that create accountability without bureaucracy. **Data stewards** serve as the operational backbone, ensuring day-to-day data quality and compliance activities. **Data owners** provide business context and make strategic decisions about data usage and access. **Data custodians** handle technical implementation and maintenance of governance policies. Executive sponsors champion the program at the highest organizational levels, securing resources and removing barriers to success.

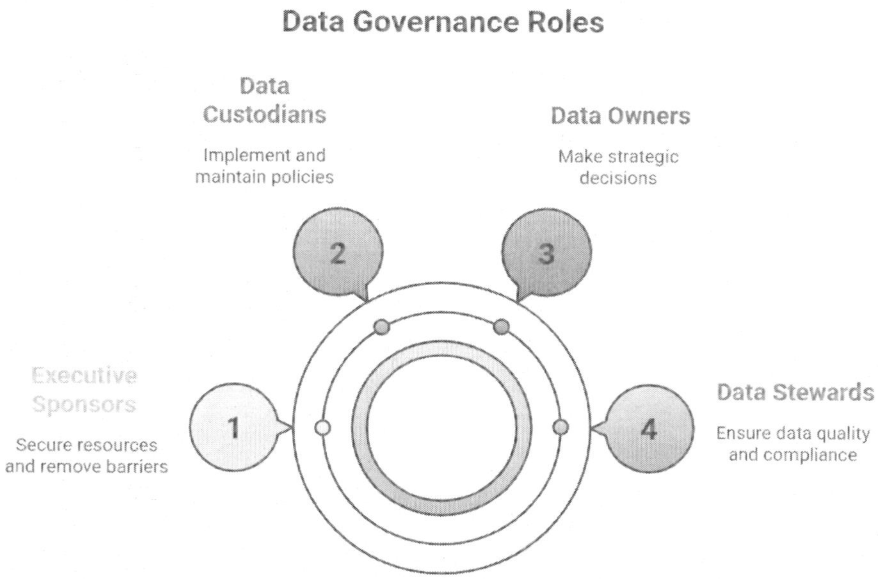

Building this community requires careful attention to organizational culture and change management principles. Successful programs invest heavily in education and communication, helping stakeholders understand how governance activities directly support their individual objectives. They create governance forums where business and technical teams collaborate on data-related decisions, fostering shared ownership and accountability.

Process Architecture: Structured Governance Workflows

Governance processes provide the structured workflows that transform policies into actionable activities. These processes must be comprehensive enough to address complex data scenarios while remaining practical for daily operations. The most effective governance processes integrate seamlessly with existing business workflows, minimizing disruption while maximizing compliance.

Practical Implementation Example: Financial Services Data Quality Process

Consider how JPMorgan Chase implemented a comprehensive data quality process using Informatica Data Quality and Collibra for governance documentation. The bank faced regulatory pressure to improve data accuracy across risk reporting systems while maintaining operational efficiency.

Step 1: Process Design and Stakeholder Mapping

The implementation team mapped existing data workflows across trading, risk management, and regulatory reporting functions. They identified 47 critical data elements requiring enhanced quality controls and established a governance committee with representatives from each business unit.

Step 2: Automated Quality Rules Implementation

Using Informatica Data Quality, the team created 200+ automated validation rules covering completeness, accuracy, and consistency checks. Each rule was documented in Collibra with clear business justifications and escalation procedures.

```
Data Quality Rule Example:
Rule Name: Client_Credit_Rating_Validation
Business Rule: All client credit ratings must be current within 30 days
Technical Implementation:
  - Source System: Credit Risk Database
  - Validation Logic: rating_date >= CURRENT_DATE - 30
  - Exception Handling: Flag for manual review if validation fails
  - Escalation Path: Risk Management → Chief Risk Officer (if >100
exceptions)
```

Step 3: Workflow Integration and Exception Management

The team integrated quality checks into existing data pipelines, creating automated workflows that route exceptions to appropriate business stewards. Exception resolution processes were designed with clear service level agreements and escalation procedures.

Results and Outcomes:

Within six months, the bank achieved a 40% reduction in data quality exceptions and improved regulatory reporting accuracy from 87% to 98%. More importantly, the structured process reduced manual data validation efforts by 60%, freeing analysts to focus on higher-value activities. The Federal Reserve recognized the bank's improved data management capabilities during subsequent regulatory examinations.

Technology Infrastructure: Enabling Governance at Scale

Technology serves as the enabler that makes governance processes scalable and sustainable. Modern governance technology platforms provide the automation, monitoring, and reporting capabilities necessary to manage enterprise-scale data assets effectively. However, technology alone cannot solve governance challenges without proper people and process foundations.

The technology stack for data governance typically includes data catalogs for asset discovery, quality monitoring tools for automated validation, and workflow management systems for exception handling. Integration capabilities ensure that governance activities connect with existing data infrastructure, from data warehouses to analytics platforms.

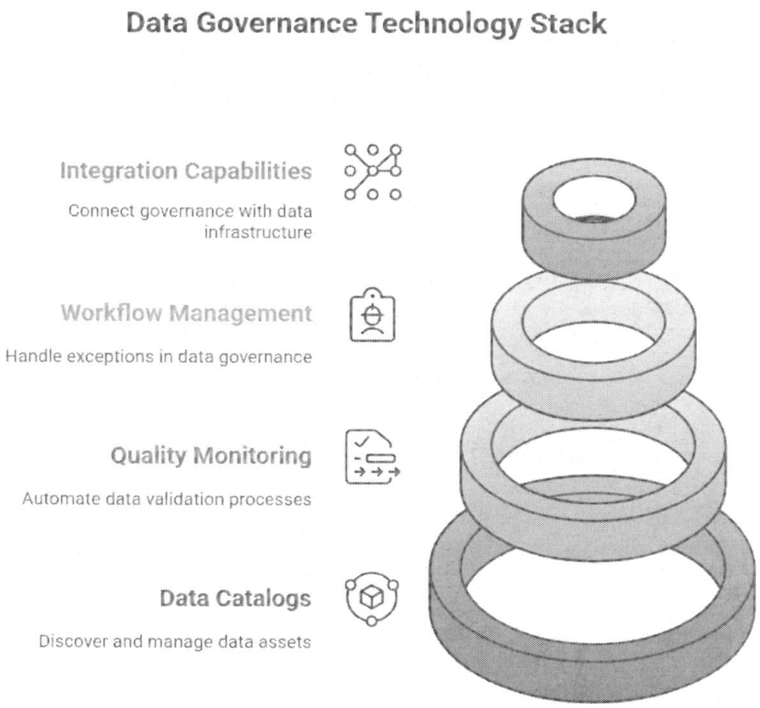

Leading organizations adopt governance technology incrementally, starting with high-impact use cases before expanding to enterprise-wide implementations. They prioritize solutions that provide immediate value while building toward comprehensive governance capabilities. This approach ensures that technology investments align with organizational maturity and available resources.

Ecosystem Integration and Sustainability

The true power of the governance ecosystem emerges when people, processes, and technology work together seamlessly. This integration requires ongoing attention to organizational dynamics, process optimization, and technology evolution. Successful programs establish feedback loops that continuously improve each component based on real-world experience and changing business requirements.

Sustainability depends on creating self-reinforcing cycles where governance activities generate tangible business value that justifies continued investment. Organizations achieve this by measuring governance outcomes in business terms, celebrating successes, and continuously evolving their approach based on lessons learned.

This ecosystem approach transforms data governance from a compliance burden into a strategic capability that drives business innovation and competitive advantage. By understanding how people, processes, and technology interact, executives can build governance programs that deliver lasting value while adapting to changing organizational needs.

Having established the foundational ecosystem components, the next critical step involves understanding how data governance differs from traditional data management approaches, ensuring clarity around roles, responsibilities, and expected outcomes.

2.4. Data Governance vs. Data Management: Understanding the Distinction

Building on the essential components outlined previously, the next critical step involves understanding how data governance differs from traditional data management approaches, ensuring clarity around roles, responsibilities, and expected outcomes. This distinction forms the cornerstone of successful governance initiatives, as many organizations struggle with overlapping responsibilities and unclear boundaries between these complementary yet distinct disciplines.

The Strategic versus Operational Divide

Data governance operates at the strategic level, establishing policies, standards, and decision-making frameworks that guide how data should be managed across the enterprise. It focuses on the "what" and "why" of data stewardship, defining rules around data quality, access rights, retention policies, and compliance requirements.

Governance creates the blueprint for data handling but does not execute the day-to-day activities.

Data management, conversely, encompasses the tactical execution of governance directives. It involves the actual implementation of data storage, processing, integration, and maintenance activities. Data management teams handle the "how" and "when" of data operations, working within the boundaries established by governance frameworks.

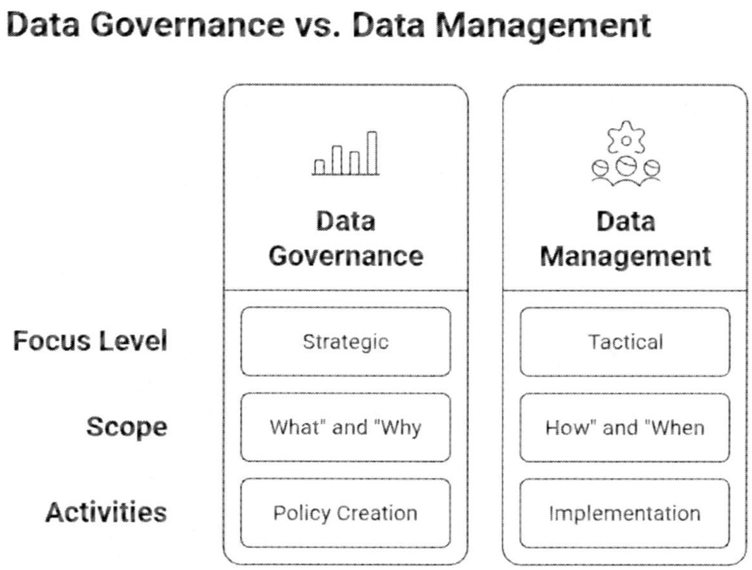

Practical Implementation: Customer Master Data Initiative

Consider a global telecommunications company implementing a customer master data management (MDM) program using Informatica MDM and establishing governance through the DAMA-DMBOK framework. The distinction becomes clear through their approach:

Governance Layer Actions:

Policy Development: The data governance council established customer data definition standards, specifying that customer records must include validated contact information, service history, and billing preferences

Stewardship Assignment: Business data stewards from marketing, sales, and customer service were designated as data owners for specific customer attributes

Quality Standards: Governance defined acceptable data quality thresholds (95% completeness for contact data, 99% accuracy for billing information)

Access Controls: Policies established role-based access matrices determining which departments could view or modify customer segments

Management Layer Execution:

Technical Implementation: IT teams configured Informatica MDM to enforce governance policies through automated data quality rules

Data Integration: Technical staff built ETL processes to consolidate customer data from CRM, billing, and support systems

Quality Monitoring: Data management teams implemented daily quality dashboards tracking completeness and accuracy metrics

Issue Resolution: Operational teams addressed data quality exceptions flagged by governance rules

Governance Committee Structure and Decision-Making

The telecommunications company established a three-tier governance structure that clearly delineated strategic oversight from operational execution:

```
Data Governance Council (Strategic)
├── Executive Sponsor (Chief Data Officer)
├── Business Data Stewards (Department Heads)
└── Technical Data Stewards (Architecture Team)

Data Management Operations (Tactical)
├── MDM Technical Team
├── Data Integration Specialists
└── Quality Assurance Analysts
```

This structure enabled the governance council to focus on policy decisions while management teams concentrated on implementation efficiency. When conflicts arose between data quality and system performance, the governance council made strategic trade-off decisions that management teams then executed.

Measuring Success Through Distinct Metrics

The company tracked different success metrics for governance versus management activities. Governance effectiveness was measured through policy compliance rates, stakeholder satisfaction scores, and regulatory audit results. Management success focused on system uptime, data processing throughput, and technical service level agreements.

After twelve months, the organization achieved 94% improvement in customer data accuracy and reduced compliance violations by 78%, demonstrating how clear role separation enhanced both strategic oversight and operational efficiency.

This fundamental distinction between governance and management creates the foundation for establishing shared terminology and communication protocols. Understanding these complementary roles enables organizations to build governance frameworks that provide strategic direction while empowering management teams to execute efficiently, setting the stage for developing a common vocabulary that bridges both disciplines and ensures consistent understanding across all stakeholders.

2.5. Establishing a Common Vocabulary and Terminology

Building on the foundation of bridging technical and business disciplines, the next critical step involves creating a shared language that ensures all stakeholders understand governance concepts uniformly. Without standardized terminology, even the most well-intentioned governance initiatives can fail due to miscommunication, conflicting interpretations, and misaligned expectations across departments.

The Business Case for Standardized Data Language

Organizations that invest in establishing common vocabulary experience measurably improved governance outcomes. A recent study of Fortune 500 companies revealed that enterprises with standardized data terminology achieved 40% faster project implementation times and 60% fewer governance-related disputes. This improvement stems from eliminating the confusion that arises when different departments use identical terms to describe different concepts, or worse, different terms for the same data elements.

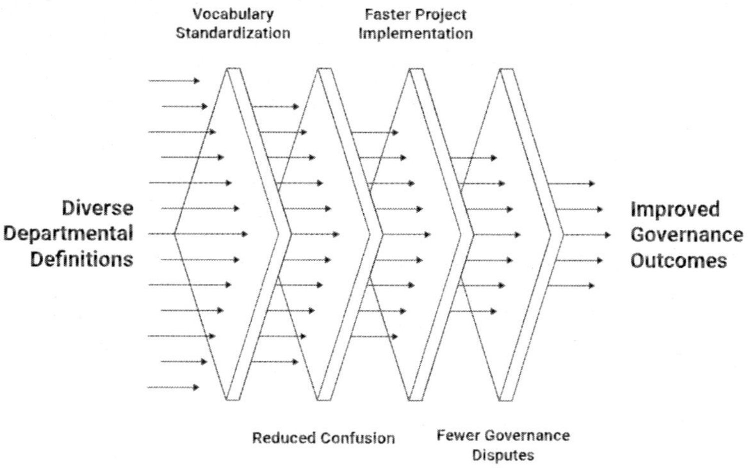

Consider the seemingly simple term **"customer."** Marketing might define it as anyone who has engaged with promotional content, sales could consider it a qualified lead with purchase intent, while finance restricts the definition to entities with completed transactions. Without alignment, governance policies become ineffective, compliance efforts fragment, and data quality initiatives target different populations entirely.

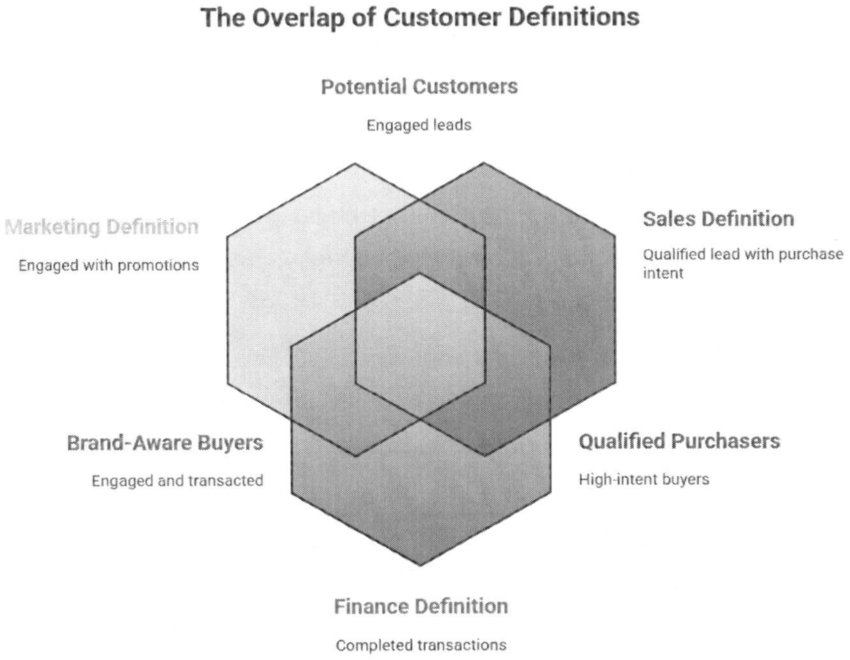

Practical Framework Implementation: The Terminology Governance Workshop

Step-by-Step Implementation Using Collibra Data Governance Platform

The following practical example demonstrates how a multinational retail organization successfully established common vocabulary using structured workshops integrated with Collibra's governance platform:

Phase 1: Stakeholder Mapping and Term Identification

The organization began by identifying key stakeholders from finance, marketing, operations, and IT. Using Collibra's business glossary module, they conducted discovery sessions to catalog existing terminology variations. The platform's collaborative features enabled real-time documentation of conflicting definitions across departments.

Phase 2: Consensus Building Sessions

Weekly two-hour workshops brought together domain experts to resolve terminology conflicts. For instance, the term **"product performance"** had seven different definitions across business units. Through structured discussions facilitated by Collibra's workflow capabilities, stakeholders negotiated a unified definition: "Quarterly revenue and margin metrics measured against forecasted targets, including customer satisfaction scores above 4.0."

Phase 3: Platform Integration and Validation

The agreed-upon definitions were formalized within Collibra's business glossary, creating automated validation rules. When users attempted to create reports or analyses using non-standard terminology, the platform prompted them with approved definitions and suggested corrections.

Results: Within six months, data consistency improved by 75%, and cross-departmental project delivery accelerated by three weeks on average. Most importantly, regulatory reporting accuracy increased significantly, reducing compliance risk exposure.

Building Sustainable Vocabulary Management

Successful terminology governance requires ongoing maintenance and evolution. Leading organizations establish **Data Terminology Councils** comprising representatives from each business function. These councils meet monthly to review proposed changes, assess new terminology needs, and ensure definitions remain relevant as business requirements evolve.

The council structure typically includes a Chief Data Officer as chairperson, business domain experts as voting members, and IT representatives providing technical feasibility assessments. This governance body maintains authority over terminology decisions while ensuring democratic input from affected stakeholders.

Integration with Enterprise Data Workflows

Effective vocabulary standardization extends beyond documentation into active system integration. Modern governance platforms enable embedding approved terminology directly into data entry forms, reporting templates, and analytical dashboards. This approach transforms passive glossaries into active governance enforcement mechanisms.

Organizations implementing this integration report significant improvements in data quality metrics, with standardized terminology contributing to more accurate master data management and enhanced regulatory compliance capabilities.

Establishing common vocabulary creates the linguistic foundation necessary for effective governance communication, enabling organizations to move beyond definitional debates toward productive policy development and implementation. This shared understanding becomes particularly crucial when addressing the cultural dimensions that ultimately determine governance program success or failure.

2.6. The Role of Culture in Governance Success

Data governance frameworks may be built on technical foundations, but their success hinges entirely on human behavior and organizational culture. The most sophisticated governance technologies and meticulously crafted policies will fail if the underlying culture does not embrace data stewardship as a core organizational value. Understanding and actively shaping culture represents one of the most critical yet overlooked aspects of governance implementation.

Cultural Transformation Through Leadership Modeling

Successful data governance requires leaders who demonstrate data-driven decision making in their daily operations. When executives consistently reference data quality metrics in board meetings, question the source and reliability of presented information, and publicly acknowledge data stewards for their contributions, they signal that data governance is not merely a compliance exercise but a business imperative.

Consider the transformation at Maersk, the global shipping giant, where CEO Søren Skou made data transparency a cornerstone of operational excellence. By requiring all strategic presentations to include data lineage documentation and establishing public recognition programs for employees who identified critical data quality issues, leadership created a culture where questioning data became not just acceptable but expected.

Building Cross-Functional Data Citizenship

Effective governance culture extends beyond designated data stewards to encompass every employee who creates, modifies, or consumes data. This requires systematic education and engagement strategies that make data governance personally relevant to individual roles and responsibilities.

Organizations must move beyond traditional training approaches to create immersive cultural experiences. Interactive workshops where marketing teams discover how poor customer data quality directly impacts their campaign effectiveness, or finance departments experience firsthand how inconsistent product hierarchies distort profitability analysis, create lasting behavioral change that policy documents alone cannot achieve.

Practical Implementation: Cultural Assessment and Transformation Framework

A comprehensive cultural transformation requires systematic assessment and targeted intervention strategies. Here's a step-by-step approach organizations can implement immediately:

```
Cultural Maturity Assessment Framework

Phase 1: Current State Analysis (Weeks 1-2)
- Deploy anonymous surveys measuring data trust levels
- Conduct focus groups across business units
- Analyze existing decision-making processes for data usage patterns
- Document informal data sharing networks

Phase 2: Gap Identification (Week 3)
- Compare current behaviors against desired governance outcomes
- Identify cultural champions and resistance points
- Map influence networks within the organization

Phase 3: Targeted Intervention Design (Weeks 4-5)
- Create role-specific governance value propositions
- Design peer-to-peer learning programs
- Establish feedback loops and recognition mechanisms
```

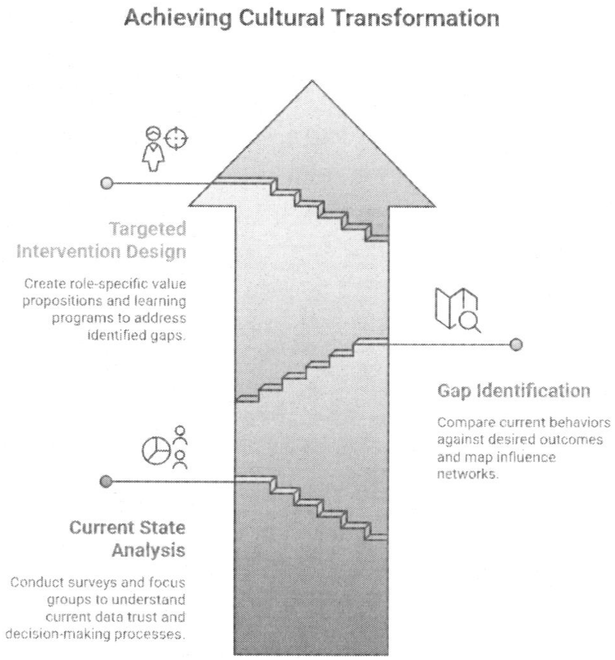

Implementation begins with deploying a comprehensive cultural assessment using tools like Collibra's Data Citizenship Survey module, which measures employee attitudes toward data quality, sharing behaviors, and governance awareness. The survey reveals critical insights such as departmental variations in data trust levels and identifies informal data experts who could become governance champions.

Following assessment, organizations implement targeted interventions. For example, a retail organization discovered through cultural assessment that store managers consistently bypassed central inventory systems because they distrusted data accuracy. The intervention involved creating manager-specific dashboards showing real-time data quality scores and implementing a rapid feedback mechanism where managers could flag data issues directly to IT teams, with resolution tracked and communicated back within 24 hours.

The transformation results were measurable: data system adoption increased from 60% to 94% within six months, inventory accuracy improved by 23%, and manager satisfaction with data tools rose from 2.3 to 4.1 on a five-point scale.

Sustaining Cultural Change Through Governance Integration

Cultural transformation requires embedding new behaviors into existing organizational processes and reward systems. This means integrating data governance responsibilities into performance reviews, linking data quality improvements to bonus structures, and creating career advancement pathways that recognize data stewardship excellence.

Successful organizations establish governance councils that include cultural champions from each business unit, ensuring that governance decisions consider both technical requirements and cultural feasibility. These councils serve as cultural bridges, translating governance policies into language and practices that resonate with specific departmental contexts.

Cultural transformation in data governance represents a fundamental shift from viewing data as a byproduct of business operations to recognizing it as a strategic asset requiring collective stewardship. Organizations that successfully navigate this cultural evolution create sustainable competitive advantages through improved decision-making, reduced operational risk, and enhanced innovation capabilities. This cultural foundation becomes particularly crucial when implementing governance frameworks that must comply with diverse regulatory requirements across global markets.

2.7. International Perspectives: GDPR, CCPA, and Global Privacy Frameworks

The global regulatory landscape presents executives with an unprecedented challenge: navigating complex privacy frameworks while maintaining operational efficiency and competitive advantage. Organizations that successfully master international data governance requirements don't merely achieve compliance; they transform regulatory obligations into strategic differentiators that enhance customer trust and operational excellence.

Building Cross-Border Privacy Governance Architecture

Effective international privacy governance requires a unified architecture that accommodates diverse regulatory requirements while maintaining operational coherence. Organizations must develop frameworks that can simultaneously address the **data minimization** principles of GDPR, the **consumer rights** focus of CCPA, and the emerging requirements of frameworks like Brazil's LGPD and Singapore's PDPA.

The most successful implementations establish a **privacy-by-design** governance structure that embeds regulatory compliance into core business processes rather than treating it as an external constraint. This approach requires executive leadership to champion governance initiatives that span traditional organizational boundaries and integrate privacy considerations into strategic decision-making processes.

Practical Implementation: Multi-Jurisdictional Privacy Impact Assessment Framework

Consider the implementation approach used by Siemens AG, which operates across 190 countries and must comply with dozens of privacy regulations. The company developed a comprehensive privacy governance framework using Informatica's Data Privacy Management platform combined with a standardized Privacy Impact Assessment (PIA) process.

Step 1: Establishing Unified Data Classification Standards

```
Data Category Classification Matrix:
- Personal Identifiers: Name, Address, Phone, Email
- Sensitive Personal Data: Health, Financial, Biometric
- Behavioral Data: Website interactions, Purchase history
- Technical Data: IP addresses, Device identifiers
- Derived Data: Analytics outputs, Profiling results
```

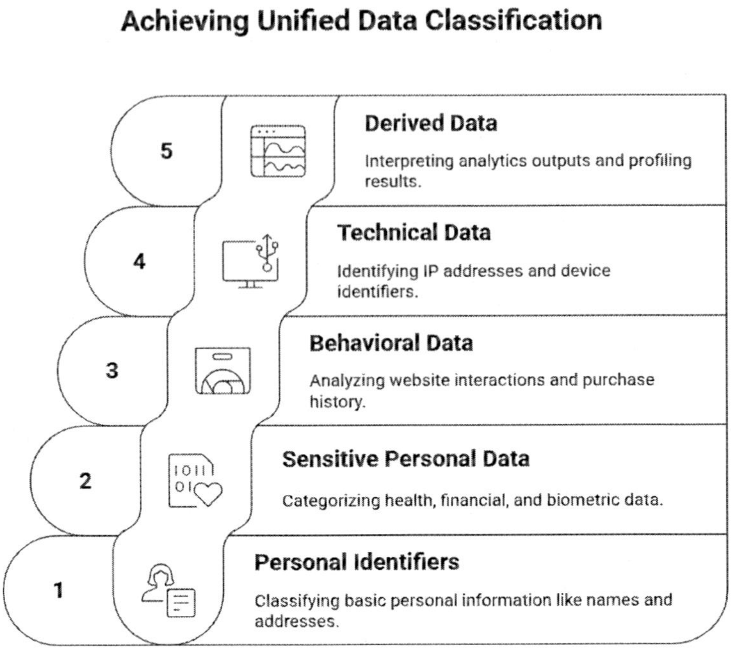

Step 2: Implementing Cross-Border Governance Workflows

The organization created automated workflows within Informatica that trigger specific governance actions based on data origin and destination:

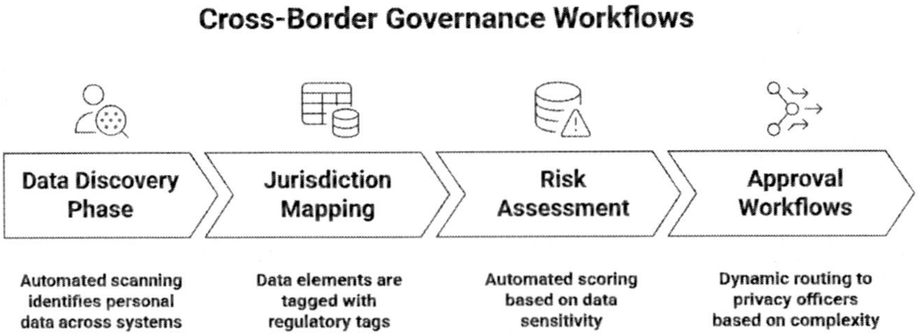

- **Data Discovery Phase**: Automated scanning identifies personal data across 47 different business systems
- **Jurisdiction Mapping**: Each data element receives regulatory tags (GDPR, CCPA, LGPD, etc.)
- **Risk Assessment**: Automated scoring based on data sensitivity and cross-border transfer requirements

- **Approval Workflows**: Dynamic routing to appropriate privacy officers based on regulatory complexity

Step 3: Establishing Regional Privacy Committees

Siemens implemented regional privacy governance committees that meet monthly to address jurisdiction-specific requirements while maintaining global consistency. Each committee includes:

- Regional Data Protection Officer
- Legal compliance representative
- Business process owners
- IT security leadership
- Customer experience managers

Governance Integration Patterns for Global Compliance

The most effective privacy governance frameworks integrate seamlessly with existing enterprise data workflows through **policy automation** and **continuous monitoring**. Organizations achieve this integration by implementing governance rules directly within data processing systems rather than relying on manual compliance checks.

Real-world Application: Unilever's global privacy governance program demonstrates how policy automation can scale across diverse markets. The company uses Collibra's Data Governance Center to maintain a centralized policy repository that automatically adapts to local regulatory requirements. When processing customer data in Germany, the system automatically applies GDPR-specific retention policies, while California consumer data receives CCPA-compliant handling procedures.

The implementation includes automated **data subject request** processing that can handle requests across multiple jurisdictions simultaneously. When a consumer submits a data deletion request, the system automatically:

- Identifies all data repositories containing the individual's information
- Applies jurisdiction-specific validation requirements
- Executes deletion procedures according to local legal requirements
- Generates compliance documentation for regulatory reporting

Measuring Cross-Border Privacy Governance Effectiveness

Successful international privacy governance requires sophisticated measurement frameworks that can demonstrate compliance effectiveness across multiple regulatory environments. Organizations must establish **Key Privacy Indicators** (KPIs) that provide executive visibility into governance performance while enabling proactive risk management.

Leading organizations track metrics including **cross-border transfer compliance rates**, **data subject request response times**, **privacy incident resolution efficiency**, and **regulatory examination readiness scores**. These measurements enable executives to make informed decisions about resource allocation and strategic privacy investments.

The integration of privacy governance with broader data governance initiatives creates a foundation for sustainable competitive advantage. Organizations that excel in this integration demonstrate measurably higher customer trust scores, reduced regulatory risk exposure, and enhanced operational efficiency across global markets.

International privacy governance transforms from compliance burden to strategic asset when organizations implement unified frameworks that automate regulatory requirements while maintaining operational flexibility. Success requires executive commitment to cross-functional collaboration and investment in governance technologies that scale across diverse regulatory environments.

This comprehensive approach to international privacy governance establishes the foundation for measuring organizational readiness and maturity in data governance implementation across global operations.

2.8. Measuring Governance Maturity and Readiness

Building on the international privacy governance foundation, organizations must now turn their attention to evaluating their current position and capacity for implementing comprehensive data governance frameworks. Understanding where your organization stands on the governance maturity spectrum is essential before embarking on any transformation initiative, as it determines both the starting point and the roadmap for achieving sustainable data governance excellence.

Governance Maturity Assessment Framework

Organizations typically progress through five distinct maturity levels in their data governance journey. The **initial level** represents ad-hoc data management practices with minimal governance oversight. At the **developing level**, basic policies exist but enforcement remains inconsistent. The **defined level** establishes standardized processes and clear accountability structures. Organizations reaching the **managed level** demonstrate quantitative measurement and continuous improvement capabilities. Finally, the **optimizing level** represents adaptive, innovation-driven governance that evolves with business needs.

Assessing your organization's current maturity requires evaluating multiple dimensions simultaneously. Data quality management, policy enforcement mechanisms, stakeholder engagement levels, technology infrastructure, and cultural readiness all contribute to the overall governance maturity profile. Each dimension may exist at different maturity levels, creating a nuanced assessment that guides targeted improvement strategies.

Practical Maturity Assessment Implementation

Consider the recent transformation at Deutsche Bank, which implemented a comprehensive governance maturity assessment using the DAMA-DMBOK framework combined with Collibra's governance platform. The bank's assessment revealed significant gaps between their risk management capabilities and data lineage tracking, despite strong policy documentation.

```
Maturity Assessment Template - Financial Services Example:

Data Quality Dimension:
- Current State: Level 2 (Developing)
- Evidence: Manual quality checks, reactive issue resolution
- Target State: Level 4 (Managed)
- Gap Analysis: Automated monitoring, proactive quality metrics needed

Data Lineage Dimension:
- Current State: Level 1 (Initial)
- Evidence: Limited visibility into data flow
- Target State: Level 3 (Defined)
- Gap Analysis: End-to-end lineage mapping, impact analysis tools
required

Stakeholder Engagement:
- Current State: Level 3 (Defined)
- Evidence: Established data steward network
- Target State: Level 4 (Managed)
- Gap Analysis: Performance metrics, feedback loops needed
```

The bank's assessment process involved cross-functional workshops where business units, IT teams, and compliance officers collaboratively evaluated each governance dimension. Using Collibra's assessment module, they documented current capabilities, identified evidence supporting each maturity rating, and established measurable targets for improvement.

Organizational Readiness Evaluation

Beyond maturity assessment, organizations must evaluate their readiness for governance transformation. **Cultural readiness** examines leadership commitment, change management capabilities, and organizational appetite for new processes. **Technical readiness** assesses existing data infrastructure, integration capabilities, and technology adoption patterns. **Resource readiness** evaluates available budget, skilled personnel, and time allocation for governance initiatives.

The readiness evaluation at Siemens AG demonstrated how technical capabilities can exceed cultural readiness. Despite sophisticated data management systems, initial governance initiatives struggled due to siloed organizational structures and resistance to cross-functional data sharing. Their readiness assessment revealed the need for extensive change management before implementing technical governance solutions.

Measurement and Monitoring Strategies

Effective governance maturity measurement requires establishing baseline metrics and tracking improvement over time. Key performance indicators should span operational metrics like data quality scores and policy compliance rates, alongside strategic measures such as business value realization and stakeholder satisfaction levels.

Organizations achieving sustainable governance success implement continuous assessment cycles, typically conducting comprehensive maturity evaluations annually while monitoring key readiness indicators quarterly. This approach ensures governance initiatives remain aligned with evolving business priorities and organizational capabilities.

The systematic assessment of governance maturity and organizational readiness provides the critical foundation for designing targeted improvement strategies and securing stakeholder buy-in for transformation initiatives. This comprehensive evaluation process transforms abstract governance concepts into measurable, actionable improvement plans that drive sustainable organizational change. With a clear understanding of current capabilities and improvement targets established, organizations are positioned to develop comprehensive governance strategies that align with their specific maturity profile and readiness characteristics, setting the stage for successful framework design and implementation.

3. DESIGNING YOUR GOVERNANCE FRAMEWORK

The blueprint for effective data governance begins with a fundamental truth: structure drives success. While many organizations recognize the critical importance of governing their data assets, the journey from concept to implementation often stalls at the design phase. Without a well-architected framework that reflects organizational realities and strategic priorities, even the most well-intentioned governance initiatives become exercises in bureaucratic complexity rather than catalysts for business transformation.

This chapter serves as your architectural guide, transforming abstract governance principles into concrete organizational structures that deliver measurable results. You'll discover how to construct a governance framework that seamlessly integrates with your company's unique culture, operational model, and strategic objectives. Rather than adopting a one-size-fits-all approach, we'll explore how to customize governance architectures that resonate with different industries, from highly regulated financial services to agile technology startups.

The three-tier governance model forms the foundation of our exploration, providing a scalable structure that connects strategic vision with operational execution. You'll learn to establish clear decision rights that eliminate the ambiguity plaguing many governance initiatives, while creating accountability structures that empower rather than constrain your teams. We'll examine how roles evolve from traditional data owners to modern data stewards, each contributing to a cohesive governance ecosystem.

Beyond organizational design, this chapter addresses the critical documentation and policy frameworks that transform governance from informal practices into sustainable business processes. You'll master the art of creating governance charters that secure executive commitment while establishing mandates that drive consistent implementation across diverse business units. By the chapter's conclusion, you'll possess a comprehensive blueprint for designing governance frameworks that not only meet today's compliance requirements but also adapt to tomorrow's business challenges.

3.1. Framework Architecture: Structure and Components

A robust data governance framework requires a carefully orchestrated architecture that balances organizational needs with operational realities. The most successful frameworks share common structural elements that create accountability, ensure consistency, and enable scalable decision-making across the enterprise. Understanding these core components and their interconnections forms the foundation for building governance that delivers measurable business value.

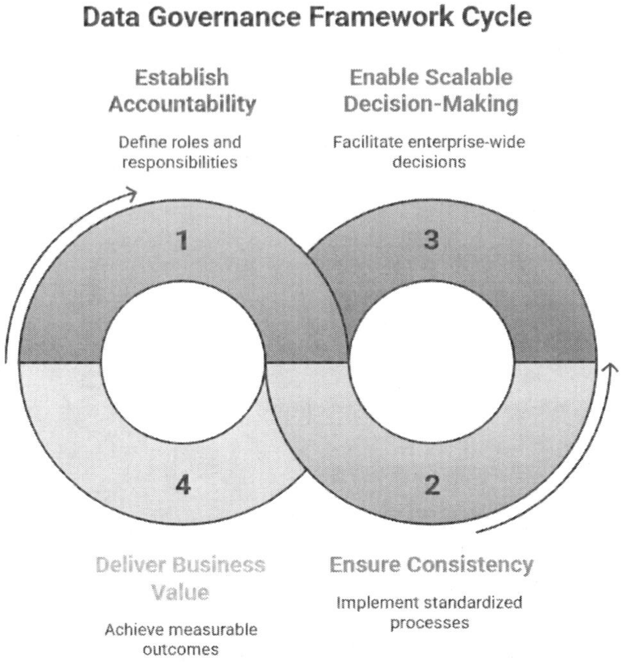

The architectural design of your governance framework must address three fundamental dimensions: **organizational structure**, **operational processes**, and **technology infrastructure**. These dimensions work in concert to create a comprehensive system that transforms data from liability to asset. Organizations that excel in governance implementation recognize that framework architecture extends beyond documentation to encompass the living, breathing mechanisms that guide daily data decisions.

Core Organizational Components

The governance structure begins with clearly defined **roles and responsibilities** that span business and technical domains. At the apex sits the Data Governance Council, typically comprising C-level executives who provide strategic direction and resolve cross-functional conflicts. Below this executive layer, Data Stewardship Committees manage domain-specific governance activities, while Data Custodians handle technical implementation and maintenance.

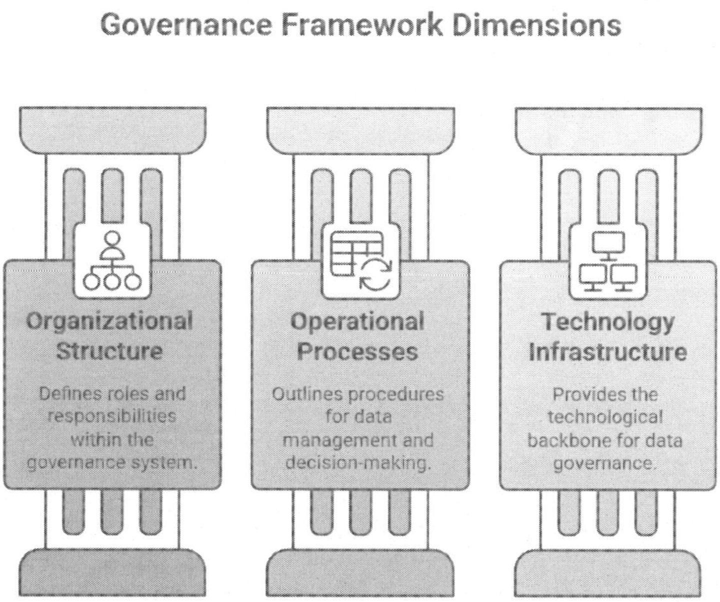

Consider how Nationwide Insurance restructured their governance architecture in 2019. They established a three-tier model with Executive Sponsors at the top, Business Data Stewards in the middle, and Technical Data Custodians at the operational level. This structure enabled them to reduce data quality issues by 60% within 18 months by creating clear escalation paths and accountability mechanisms.

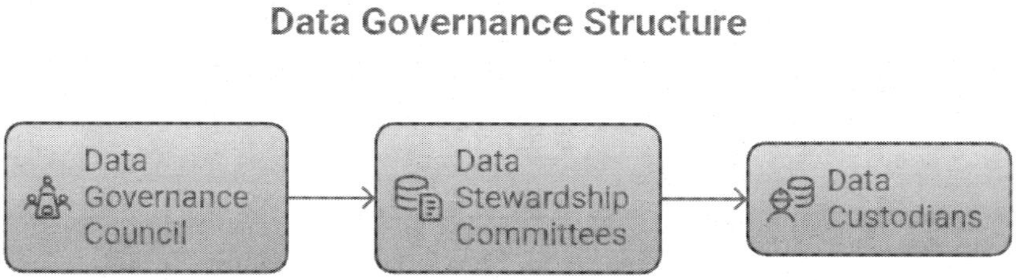

Process Architecture Framework

Effective governance frameworks incorporate standardized processes that guide data lifecycle management from creation to retirement. These processes must address **data quality management**, **access control procedures**, and **issue resolution workflows**. The architecture should specify how decisions flow through the organization and how exceptions are handled.

Practical Implementation: Building a Data Quality Assessment Process

Let's examine how a global pharmaceutical company implemented their data quality assessment framework using Informatica Data Quality:

Baseline Assessment Setup

- Configure Informatica DQ to scan customer master data across 15 regional databases
- Establish quality rules for completeness (>95%), accuracy (>98%), and consistency (>90%)
- Create automated profiling schedules for weekly assessment cycles

Stakeholder Engagement Structure

```
Data Quality Issue Workflow:
Level 1: Automated remediation (system-generated fixes)
Level 2: Data Steward review (business rule validation)
Level 3: Committee escalation (policy exception requests)
Level 4: Executive decision (strategic data conflicts)
```

Implementation Results

- Before: 23% of customer records contained critical quality issues

- After: 4% critical quality issues within six months

- Business Impact: $2.3M reduction in compliance penalties and operational errors

Technology Integration Patterns

The framework architecture must specify how governance tools integrate with existing enterprise systems. This includes **metadata management platforms**, **data lineage tracking systems**, and **policy enforcement mechanisms**. Modern architectures leverage APIs and automated workflows to embed governance into operational processes rather than treating it as a separate activity.

Decision-Making Hierarchies

Successful frameworks establish clear decision-making hierarchies that balance agility with control. The architecture should specify which decisions can be made at each organizational level and define escalation triggers for complex scenarios. This prevents bottlenecks while maintaining governance integrity.

Risk and Compliance Integration

Framework architecture must incorporate risk assessment capabilities and compliance monitoring mechanisms. This includes automated policy violation detection, privacy impact assessment workflows, and regulatory reporting structures. The architecture should enable proactive risk management rather than reactive compliance checking.

The architectural foundation you establish will determine your governance program's scalability and sustainability. Organizations that invest time in thoughtful framework design create systems that adapt to changing business needs while maintaining operational consistency. Your architecture becomes the blueprint that guides all subsequent governance decisions and implementations.

With your framework architecture clearly defined, the next critical step involves ensuring this structure aligns seamlessly with your organization's strategic objectives and operational priorities.

3.2. Aligning Governance with Organizational Objectives

A governance framework without strategic alignment is like a sophisticated navigation system pointing in the wrong direction. While technical capabilities matter, the true measure of governance success lies in how effectively it advances organizational priorities and delivers measurable business value. The alignment process transforms governance from an IT initiative into a strategic business enabler that resonates across all organizational levels.

Strategic Objective Mapping Framework

Effective governance alignment begins with a systematic mapping process that connects data initiatives to specific business outcomes. This approach ensures every governance activity contributes directly to organizational success rather than operating in isolation. The mapping framework establishes clear linkages between data assets, governance processes, and strategic business goals.

Consider how **Maersk**, the global shipping conglomerate, aligned their data governance with operational excellence objectives. Facing challenges with vessel tracking accuracy and cargo visibility, they mapped their governance framework directly to customer service improvements and operational efficiency gains. Their governance structure prioritized data quality standards for shipping logistics, resulting in 15% improvement in on-time deliveries and enhanced customer satisfaction scores.

The strategic mapping process involves four key components: business objective identification, data asset inventory, governance capability assessment, and alignment gap analysis. Organizations must first articulate their strategic priorities in measurable terms, then identify which data assets and governance processes directly support these outcomes.

Practical Implementation: Revenue Growth Alignment

Let's examine a step-by-step approach for aligning governance with revenue growth objectives, using a comprehensive framework that many organizations can adapt to their specific contexts.

Step 1: Business Impact Assessment

Begin by conducting stakeholder interviews with sales, marketing, and finance leaders to identify specific revenue challenges related to data quality or accessibility. Document current pain points such as inconsistent customer records, delayed reporting, or inadequate market intelligence.

Step 2: Governance Priority Matrix Development

Create a priority matrix using tools like Collibra's governance workflow capabilities to rank governance initiatives based on revenue impact potential. Map each governance activity against specific revenue metrics such as customer acquisition cost, lifetime value, or sales cycle length.

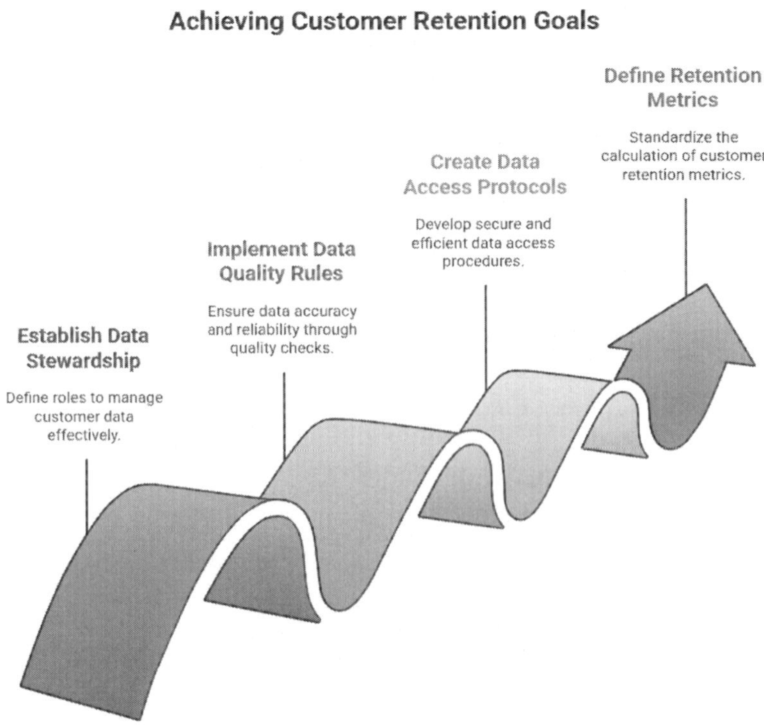

```
Revenue Alignment Framework Template:

Business Objective: Increase Customer Retention by 20%
Data Requirements:
- Customer interaction history (complete, accurate, timely)
- Product usage analytics (standardized, accessible)
- Support ticket resolution data (integrated, quality-assured)

Governance Activities:
1. Establish customer data stewardship roles
2. Implement data quality rules for customer records
3. Create cross-functional data access protocols
4. Define retention metric calculation standards
```

Step 3: Governance Committee Integration

Establish governance committees with direct representation from revenue-generating departments. These committees should include sales operations managers, marketing analytics leaders, and customer success directors who can validate governance priorities against actual business needs.

Step 4: Success Metrics Definition

Define specific, measurable outcomes that demonstrate governance value in revenue terms. For instance, track improvements in lead conversion rates, reduction in customer data errors, or acceleration of sales reporting cycles.

Organizational Culture Integration

Successful alignment requires deep integration with organizational culture and existing decision-making processes. Governance frameworks must respect established communication patterns, reporting structures, and performance evaluation systems while gradually introducing new data-driven approaches.

JPMorgan Chase exemplifies this cultural integration approach in their regulatory compliance governance. Rather than imposing separate governance structures, they embedded data stewardship responsibilities within existing risk management roles, ensuring governance activities aligned naturally with established compliance objectives and performance metrics.

The cultural integration process involves identifying organizational change champions, adapting governance terminology to match business language, and designing governance processes that complement rather than compete with existing workflows. This approach reduces resistance and accelerates adoption across different organizational functions.

Performance Measurement and Continuous Alignment

Maintaining strategic alignment requires ongoing measurement and adjustment mechanisms. Organizations must establish governance scorecards that track both operational metrics and strategic business outcomes, ensuring governance activities continue supporting evolving organizational priorities.

Effective measurement frameworks combine leading indicators such as data quality scores and governance process compliance with lagging indicators like revenue impact and operational efficiency gains. This dual approach provides early warning signals for alignment issues while demonstrating tangible business value over time.

This strategic alignment foundation ensures governance frameworks deliver measurable business value while maintaining operational effectiveness. With alignment principles established, organizations can now design specific governance structures that operate effectively across strategic, tactical, and operational levels within their unique organizational context.

3.3. The Three-Tier Governance Model: Strategic, Tactical, and Operational

Building on the foundational principles established, organizations can now design specific governance structures that operate effectively across strategic, tactical, and operational levels within their unique organizational context. The three-tier governance model provides a hierarchical framework that ensures data governance decisions flow seamlessly from boardroom strategy to daily operational execution, creating accountability and alignment at every organizational level.

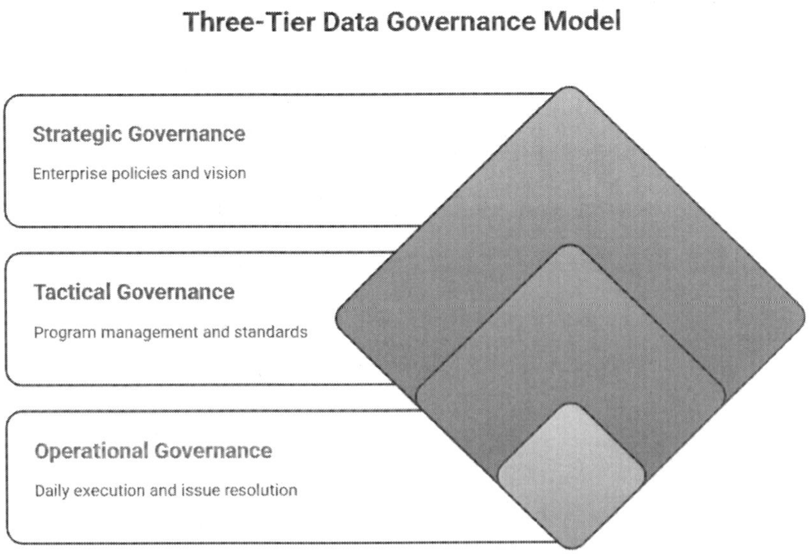

This tiered approach recognizes that different governance activities require different perspectives, timeframes, and decision-making authorities. Strategic governance focuses on enterprise-wide policies and long-term data vision, tactical governance translates strategy into actionable programs and standards, while operational governance handles day-to-day data stewardship activities and issue resolution.

Strategic Governance Layer: Executive Oversight and Vision

The strategic tier operates at the executive level, where data governance intersects with business strategy and enterprise risk management. This layer establishes the overarching data vision, approves major governance investments, and ensures alignment with regulatory requirements and competitive positioning. Strategic governance typically involves quarterly or semi-annual reviews of governance performance, policy updates, and resource allocation decisions.

Key strategic governance activities include establishing enterprise data policies, approving governance budgets, defining data privacy frameworks, and setting organizational data quality targets. This layer also manages relationships with external stakeholders such as regulators, auditors, and key business partners who require access to organizational data assets.

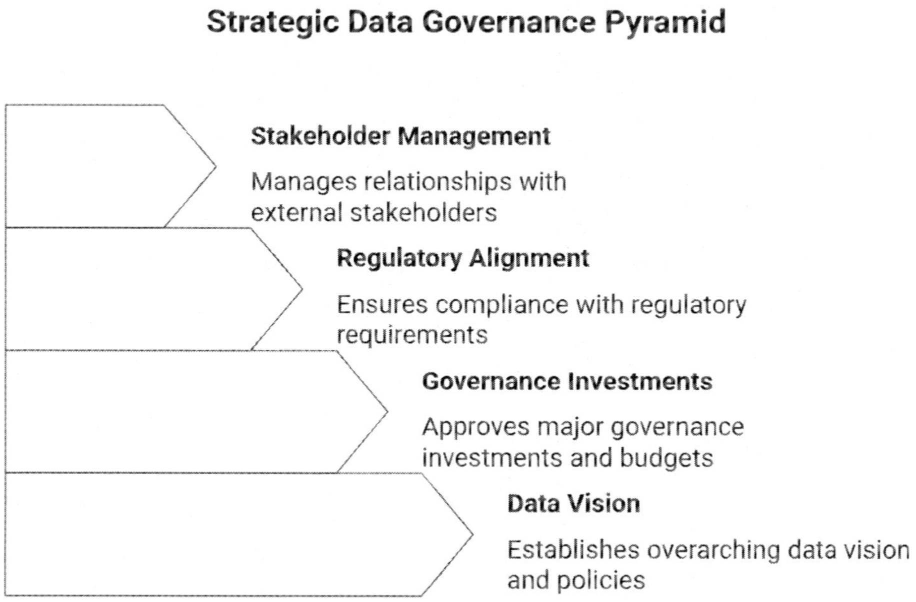

Tactical Governance Layer: Program Management and Standards

The tactical layer bridges strategic vision with operational execution through program management, standards development, and cross-functional coordination. This tier typically operates through monthly governance council meetings and quarterly program reviews, focusing on translating strategic directives into specific governance programs, data quality initiatives, and compliance frameworks.

Tactical governance responsibilities encompass developing data classification schemes, establishing master data management programs, coordinating privacy impact assessments, and managing governance technology implementations. This layer also handles escalated operational issues that require cross-departmental resolution and resource allocation decisions.

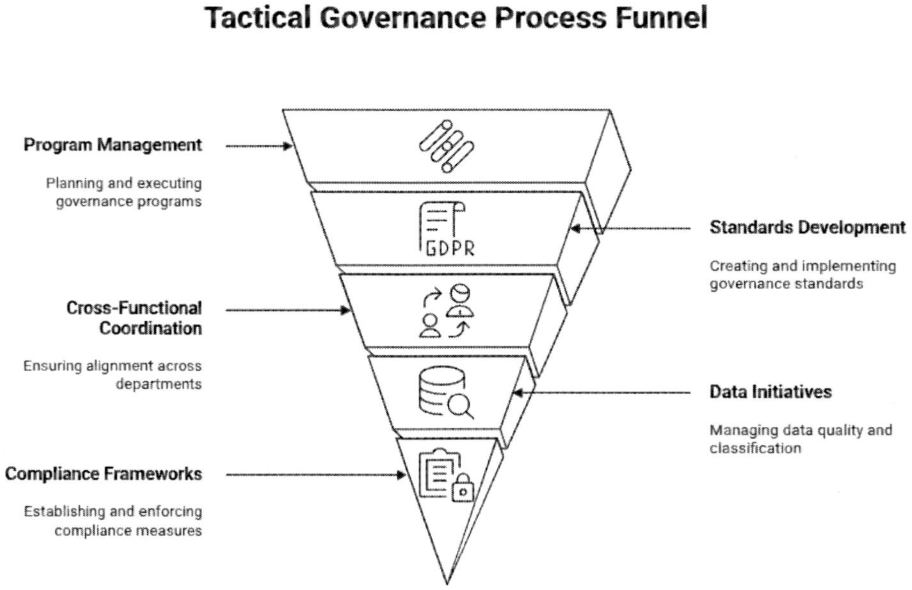

Operational Governance Layer: Daily Stewardship and Execution

The operational tier handles day-to-day data stewardship activities, quality monitoring, and issue resolution. This layer operates through weekly stewardship meetings, daily data quality dashboards, and continuous monitoring processes that ensure governance policies are effectively implemented across all data touchpoints.

Operational activities include data quality monitoring and remediation, metadata management, access request processing, and incident response coordination. This tier also provides feedback to tactical and strategic layers about governance effectiveness and emerging data challenges that require policy attention.

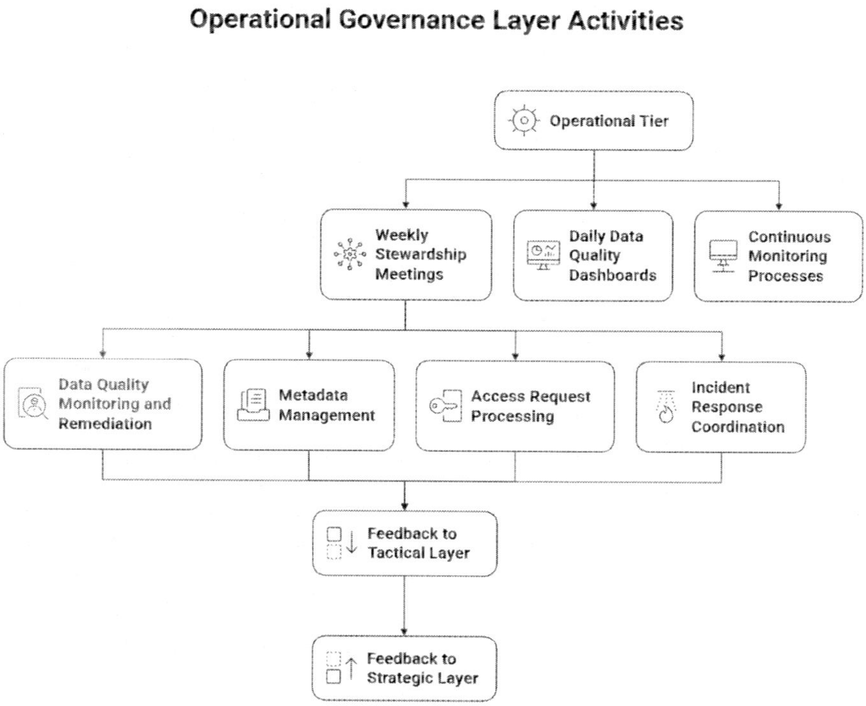

Practical Implementation: Global Pharmaceutical Three-Tier Structure

Consider how Novartis implemented their three-tier governance model during their enterprise data transformation initiative. At the strategic level, their Global Data Council meets quarterly to review data governance performance against business objectives and approve major policy changes. The tactical layer operates through regional Data Governance Committees that meet monthly to coordinate cross-functional programs like master data management and regulatory compliance initiatives.

```
Novartis Three-Tier Implementation:
Strategic: Global Data Council (Quarterly)
- Executive sponsorship and policy approval
- Resource allocation and investment decisions
- Regulatory compliance oversight

Tactical: Regional Governance Committees (Monthly)
- Program coordination and standards development
- Cross-functional issue resolution
- Performance monitoring and reporting

Operational: Local Stewardship Teams (Weekly)
- Daily data quality monitoring
- Issue identification and remediation
- Stakeholder support and training
```

The operational layer consists of local data stewardship teams across different therapeutic areas and geographic regions, meeting weekly to address data quality issues, process access requests, and coordinate with IT teams on system implementations. This structure enabled Novartis to maintain consistent governance standards across 140 countries while allowing for local adaptation and responsiveness.

This three-tier model creates clear escalation paths for governance issues while ensuring appropriate decision-making authority at each level. Strategic decisions flow down through tactical implementation to operational execution, while operational insights and challenges flow upward to inform strategic planning and resource allocation. The model's success depends on establishing clear communication channels, defined decision rights, and regular feedback mechanisms between all three tiers.

With this structural foundation in place, organizations can now focus on populating these governance tiers with the right people, clearly defining their specific roles and responsibilities to ensure effective data stewardship across the enterprise.

3.4. Roles and Responsibilities: From Data Owners to Data Stewards

With the governance tiers established, organizations can now focus on populating these governance structures with the right people, clearly defining their specific roles and responsibilities to ensure effective data stewardship across the enterprise. The success of any data governance framework hinges on creating clear accountability chains where individuals understand not only what they're responsible for, but also how their role contributes to broader organizational data objectives.

The distinction between data owners and data stewards represents one of the most critical yet misunderstood aspects of governance implementation. **Data owners** typically hold business accountability for specific data domains and make strategic decisions about data usage, access, and business rules. **Data stewards**, conversely, serve as operational guardians who implement day-to-day data management activities, monitor quality, and ensure compliance with established policies.

Executive Data Ownership Structure

At the executive level, data ownership requires C-suite leaders to champion specific data domains aligned with their business responsibilities. The Chief Marketing Officer naturally owns customer data, while the Chief Financial Officer maintains ownership of financial datasets. This alignment ensures that data governance decisions reflect genuine business priorities rather than IT-driven initiatives.

Consider the approach taken by Mastercard, where executive data owners participate in quarterly **Data Governance Council** meetings to review data quality metrics, approve new data sharing agreements, and resolve cross-functional data conflicts. Each executive owner maintains a dedicated budget for data initiatives within their domain and reports data-related KPIs to the board alongside traditional business metrics.

Operational Data Stewardship Framework

Data stewards operate as the tactical implementers of governance policies, requiring a fundamentally different skill set and organizational positioning than data owners. These roles demand deep technical understanding combined with business acumen to translate executive directives into actionable data management practices.

Practical Implementation: Customer Data Stewardship Program

A Fortune 500 telecommunications company recently implemented a comprehensive customer data stewardship program using **Informatica Data Quality** and **Collibra Data Governance** platforms. The implementation demonstrates how role clarity drives measurable governance outcomes.

Phase 1: Role Definition and Tool Configuration

The organization first mapped customer data flows across 47 different systems, identifying key stewardship touchpoints. Using Collibra's workflow engine, they created specific stewardship responsibilities:

```
Customer Data Steward Responsibilities:
- Monitor data quality scorecards weekly
- Investigate quality issues within 24 hours
- Coordinate with source system teams for remediation
- Update data lineage documentation monthly
- Approve new customer data access requests
```

Phase 2: Accountability Mechanisms

Each steward received access to Informatica's data quality dashboards showing real-time metrics for their assigned customer segments. The platform automatically generated weekly stewardship reports highlighting quality trends, outstanding issues, and remediation progress.

Phase 3: Cross-Functional Integration

The stewards established formal escalation procedures connecting to regional data owners. When customer data quality scores dropped below 95%, stewards could trigger automated workflows that notified relevant business owners and initiated root cause analysis procedures.

Measurable Outcomes:

Within six months, customer data accuracy improved from 78% to 94%, while data-related customer service complaints decreased by 43%. The clear role delineation enabled rapid issue resolution, with average remediation time dropping from 12 days to 3 days.

Governance Committee Integration

Successful role implementation requires formal governance committees that provide structure for decision-making and conflict resolution. The **Data Governance Steering Committee** typically includes executive data owners who set strategic direction, while **Working Groups** comprise data stewards who handle operational implementation.

These committees must establish regular meeting cadences, standardized reporting formats, and clear escalation procedures. Monthly steward meetings should focus on operational challenges, quality metrics, and process improvements, while quarterly owner meetings address strategic initiatives, budget allocation, and policy changes.

The integration of clearly defined roles with appropriate technological support and governance structures creates accountability frameworks that transform data from a compliance burden into a strategic asset. Organizations that invest time in precise role definition and provide adequate tools and authority to their data stewards consistently achieve higher governance maturity and measurable business value.

This foundational understanding of roles and responsibilities sets the stage for examining how decision-making authority flows through these governance structures, ensuring that the right people have appropriate power to make timely, informed decisions about organizational data assets.

3.5. Decision Rights and Accountability Structures

Building on the governance structures established in your framework, the critical next step involves defining who has the authority to make decisions about data assets and establishing clear lines of accountability. Without explicit decision rights and accountability structures, even the most well-designed governance framework becomes a bureaucratic maze where important data decisions languish in committee limbo or default to whoever speaks loudest in meetings.

Establishing Clear Decision Authority Hierarchies

Effective data governance requires a tiered approach to decision-making that balances organizational agility with appropriate oversight. **Decision authority hierarchies** define who can approve data access requests, authorize new data sources, resolve data quality issues, and make strategic decisions about data investments. These hierarchies typically span three levels: operational decisions handled by data stewards, tactical decisions managed by data owners, and strategic decisions reserved for the governance council.

Consider the implementation at Deutsche Bank, where they established a three-tier decision structure following their major data governance overhaul in 2019. Data stewards handle daily operational decisions like access provisioning and quality issue resolution. Business data owners make tactical decisions about data classification changes and cross-functional sharing agreements. The enterprise data governance board reserves strategic decisions about major data initiatives, privacy policy changes, and technology investments exceeding defined thresholds.

Implementing RACI Matrix for Data Decisions

The **RACI framework** (Responsible, Accountable, Consulted, Informed) provides a proven methodology for clarifying roles in data governance decisions. This matrix eliminates confusion about who owns what decisions while ensuring appropriate stakeholders remain engaged in the process.

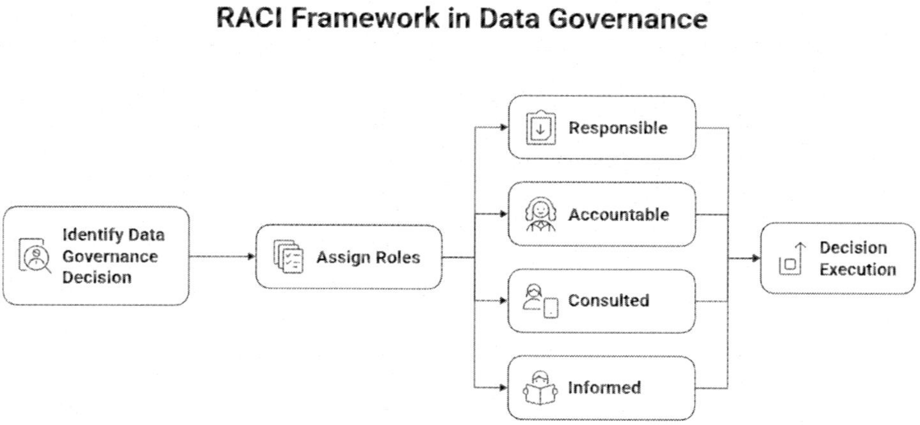

Here's a practical implementation example using Collibra's governance workflow engine:

```
Data Quality Issue Resolution RACI Matrix:
- Responsible: Data Steward (executes remediation)
- Accountable: Business Data Owner (ensures completion)
- Consulted: IT Data Custodian (technical feasibility)
- Informed: Downstream Data Consumers (impact awareness)

Implementation Steps:
1. Configure Collibra workflow to route quality issues to assigned Data
Steward
2. Set automatic escalation to Business Data Owner after 48 hours
3. Trigger consultation requests to IT Data Custodian for technical
issues
4. Generate automated notifications to registered data consumers
5. Track resolution metrics through Collibra's dashboard
```

This structured approach transformed data quality response times at a major telecommunications provider from an average of 12 days to 3 days, with clear accountability preventing issues from falling through organizational cracks.

Creating Escalation Pathways and Conflict Resolution

Escalation pathways ensure that data governance decisions don't stall when stakeholders disagree or when issues exceed normal authority levels. These pathways must be clearly documented, widely communicated, and supported by technology that automates escalation triggers.

Effective escalation structures include defined timeframes for decision-making, clear criteria for when issues should be elevated, and designated arbitrators at each level. For instance, if a data steward cannot resolve a quality issue within 48 hours, it automatically escalates to the business data owner. If the data owner cannot resolve conflicts between competing business units within one week, the issue moves to the governance council for final arbitration.

Measuring Accountability Through Performance Metrics

Accountability structures require measurable performance indicators that track both individual and collective governance performance. These metrics should focus on decision velocity, quality outcomes, and stakeholder satisfaction rather than just compliance checkboxes.

Key performance indicators include average decision resolution time, percentage of decisions made within defined timeframes, data quality improvement rates tied to specific roles, and stakeholder satisfaction scores for governance processes. Regular performance reviews using these metrics help identify bottlenecks in decision-making and areas where accountability structures need refinement.

This systematic approach to decision rights and accountability creates the operational backbone that transforms governance policies from theoretical documents into daily organizational practice. Clear authority structures eliminate decision paralysis, while robust accountability mechanisms ensure that governance responsibilities receive appropriate attention and resources.

With decision rights and accountability structures firmly established, the next critical component involves translating these governance decisions into comprehensive policies and documentation standards that provide consistent guidance across the entire organization.

3.6. Policy Development and Documentation Standards

Building upon the governance decisions established through your organizational structure, the next critical component involves translating these governance decisions into comprehensive policies and documentation standards that provide consistent guidance across the entire organization. Effective policy development transforms abstract governance principles into actionable directives that guide daily decision-making across all data-related activities.

Strategic Policy Framework Development

The foundation of robust data governance lies in creating a hierarchical policy framework that cascades from high-level strategic principles to specific operational procedures. This framework must align with your organization's risk tolerance, regulatory requirements, and business objectives while remaining flexible enough to adapt to evolving needs.

Policy hierarchy typically follows a three-tier structure: governance policies establish overarching principles and accountabilities, data management policies define specific handling requirements for different data types, and operational procedures provide step-by-step implementation guidance. Each tier must reference and support the others, creating a cohesive governance ecosystem that eliminates ambiguity and conflicting directives.

Documentation Standards and Template Development

Standardized documentation ensures consistency across all governance artifacts while reducing the time and effort required to create new policies. Your documentation standards should specify format requirements, approval workflows, version control procedures, and maintenance schedules that keep policies current and relevant.

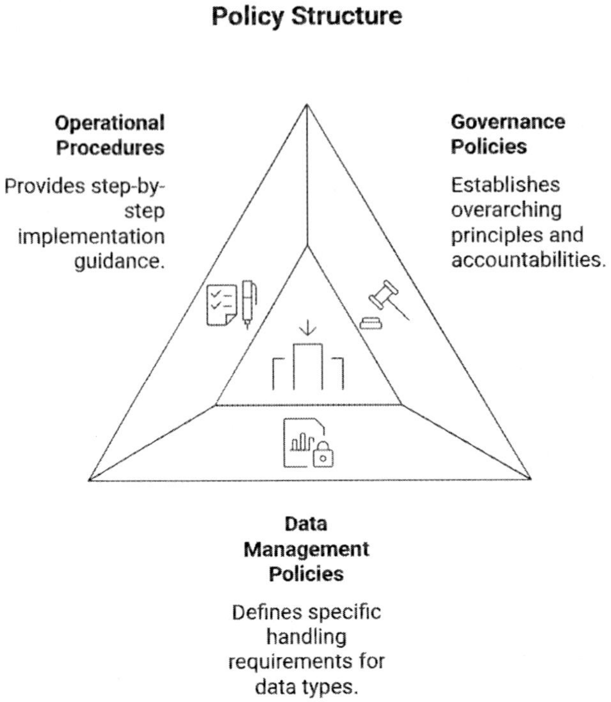

Consider implementing a master template that includes mandatory sections such as policy purpose, scope definition, roles and responsibilities, compliance requirements, enforcement mechanisms, and review cycles. This standardization enables stakeholders to quickly locate relevant information while ensuring comprehensive coverage of all governance aspects.

Practical Implementation: Customer Data Privacy Policy Development

A multinational financial services firm recently implemented a comprehensive customer data privacy policy using Collibra's governance platform to demonstrate effective policy development in action. The organization faced challenges coordinating privacy requirements across multiple jurisdictions while maintaining operational efficiency.

The implementation began with stakeholder workshops involving legal, compliance, IT, and business representatives to identify core privacy principles and regulatory requirements. Using Collibra's policy management module, the team created a master privacy policy template that incorporated GDPR, CCPA, and local banking regulations.

```
Privacy Policy Template Structure:
1. Policy Statement and Objectives
2. Scope and Applicability
```

```
3. Data Subject Rights Framework
4. Consent Management Requirements
5. Data Retention and Deletion Procedures
6. Third-Party Sharing Guidelines
7. Breach Notification Protocols
8. Compliance Monitoring and Reporting
```

The team then developed jurisdiction-specific policy variants by configuring the template with local requirements while maintaining global consistency. Collibra's workflow engine automated the review and approval process, routing policies to appropriate stakeholders based on geographic scope and data sensitivity levels.

The implementation included automated policy attestation workflows that required annual acknowledgment from all data handlers, with completion tracking integrated into performance management systems. Policy violations triggered automated escalation procedures that ensured rapid response and consistent enforcement across all business units.

Policy Lifecycle Management

Effective governance requires treating policies as living documents that evolve with changing business needs and regulatory landscapes. Establish clear procedures for policy creation, review, approval, distribution, and retirement that include defined roles, timelines, and escalation paths for addressing conflicts or exceptions.

Regular policy effectiveness assessments should evaluate whether existing policies achieve their intended outcomes and identify gaps or redundancies that require attention. These assessments provide valuable input for policy refinement and help demonstrate governance maturity to stakeholders and regulators.

Integration with Existing Enterprise Systems

Your policy framework must integrate seamlessly with existing enterprise systems to ensure widespread adoption and consistent enforcement. This integration includes linking policies to data lineage tools, embedding compliance checks into data processing workflows, and connecting policy violations to incident management systems.

Modern governance platforms like Informatica Axon enable this integration through APIs that connect policy definitions to data cataloging, quality monitoring, and access control systems. This connectivity ensures that policy requirements are automatically enforced rather than relying solely on manual compliance efforts.

Key Takeaways: Effective policy development requires a structured approach that balances comprehensive coverage with practical usability. Success depends on creating standardized templates, implementing robust lifecycle management processes, and integrating policies into daily operational workflows. Organizations that invest in systematic policy development achieve higher compliance rates, reduced governance overhead, and stronger stakeholder confidence in their data management capabilities.

With comprehensive policies established, the next essential step involves creating formal governance charters and mandates that provide the authority and accountability structures necessary for effective policy enforcement and organizational alignment.

3.7. Creating Governance Charters and Mandates

The transformation from governance structure to operational reality requires formal documentation that establishes clear authority, accountability, and decision-making processes. **Governance charters** serve as the constitutional foundation of your data governance program, while **mandates** provide the operational directives that translate strategic vision into executable actions. These documents move beyond aspirational statements to create binding agreements that define roles, responsibilities, and escalation pathways across the organization.

Establishing Charter Authority and Scope

A comprehensive governance charter must articulate the fundamental purpose, scope, and authority of your data governance program. This document serves as the primary reference point for resolving conflicts, clarifying responsibilities, and justifying resource allocation decisions. The charter should explicitly define the governance program's relationship to existing organizational structures, including IT governance, risk management, and compliance functions.

The charter's authority section requires careful attention to organizational dynamics and power structures. Executive sponsorship alone is insufficient; the document must clearly delineate decision-making authority, veto powers, and conflict resolution mechanisms. This includes specifying which governance bodies have final authority over data classification decisions, access control policies, and quality standards enforcement.

Practical Charter Development Implementation

Consider the approach taken by a major European telecommunications provider that implemented a comprehensive governance charter through their Collibra platform. The organization faced significant challenges with customer data fragmentation across multiple legacy systems, requiring a charter that could navigate complex regulatory requirements while supporting business transformation initiatives.

The implementation began with a series of stakeholder mapping sessions to identify all parties with data-related decision-making authority. The governance office used Collibra's workflow capabilities to create a collaborative charter development process:

```
Charter Development Workflow:
1. Stakeholder Identification Phase
   - Business unit representatives
   - IT leadership
   - Legal and compliance teams
   - Data protection officers

2. Authority Mapping Exercise
   - Decision-making hierarchies
   - Escalation pathways
   - Veto powers and exceptions

3. Scope Definition Process
   - Data domains and boundaries
   - System integration points
   - Regulatory compliance requirements
```

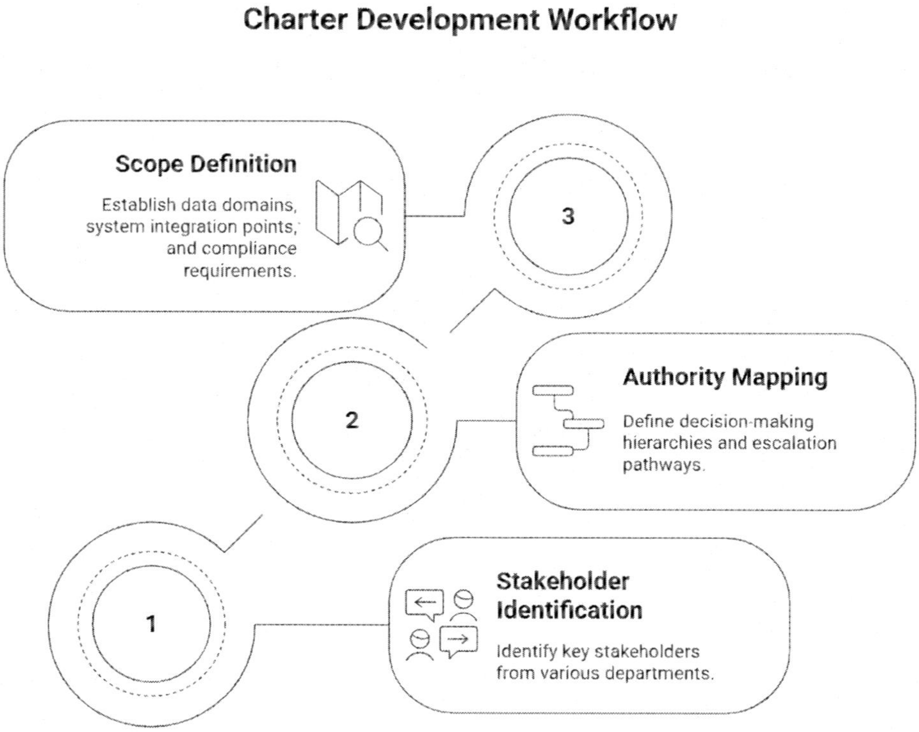

The resulting charter established a **Data Governance Council** with rotating leadership from business units, ensuring no single department dominated decision-making. The document specified that data classification decisions required consensus from both business and IT representatives, with the Chief Data Officer serving as the final arbitrator for unresolved conflicts.

Mandate Structure and Operational Directives

While charters establish authority, **operational mandates** translate governance principles into specific, measurable actions. These directives must address the practical challenges of data stewardship, including resource allocation, timeline expectations, and performance metrics. Effective mandates bridge the gap between strategic governance objectives and daily operational activities.

The mandate structure should incorporate clear accountability mechanisms, including regular reporting requirements, performance indicators, and consequence management for non-compliance. This documentation becomes particularly critical when governance decisions impact existing business processes or require significant resource investments.

Stakeholder Engagement and Consensus Building

The charter development process provides an opportunity to build organizational consensus around governance priorities and establish collaborative working relationships. Interactive policy development sessions should involve representatives from all affected business units, ensuring that governance structures reflect operational realities rather than theoretical ideals.

These sessions typically reveal hidden dependencies, conflicting priorities, and resource constraints that must be addressed in the final charter. The telecommunications provider mentioned earlier discovered that their customer service and billing systems operated under different data retention policies, requiring specific charter provisions for cross-system data governance.

Creating effective governance charters and mandates requires balancing organizational politics with operational necessity, establishing clear authority structures while maintaining collaborative decision-making processes. The resulting documents serve as both legal framework and operational guide, providing the foundation for sustainable governance practices. With these foundational documents in place, organizations can begin customizing their governance frameworks to address the unique requirements and regulatory landscapes of their specific industries.

3.8. Framework Customization for Different Industries

With these foundational documents in place, organizations can begin customizing their governance frameworks to address the unique requirements and regulatory landscapes of their specific industries. While the core principles of data governance remain consistent across sectors, the implementation approach, priority areas, and compliance mechanisms must be tailored to reflect industry-specific challenges, regulatory mandates, and operational realities.

Different industries face distinct data governance challenges that require specialized framework adaptations. Healthcare organizations must navigate HIPAA compliance while managing patient data across multiple systems. Financial services institutions operate under stringent regulations like SOX and Basel III, requiring robust audit trails and risk management protocols. Manufacturing companies focus on supply chain transparency and product quality data, while retail organizations prioritize customer experience data and inventory optimization.

Healthcare Industry Framework Adaptations

Healthcare organizations require governance frameworks that prioritize patient privacy, data interoperability, and clinical decision support. The framework must address the complex web of regulations including HIPAA, HITECH, and emerging state privacy laws while supporting clinical research and population health initiatives.

A leading academic medical center recently implemented a specialized governance framework using Informatica's healthcare data management platform. Their approach included establishing a **Clinical Data Governance Council** comprising physicians, nurses, IT professionals, and compliance officers. The framework incorporated role-based access controls that automatically adjusted permissions based on patient relationships and clinical responsibilities.

```
Healthcare Data Classification Example:
- Level 1 (Public): General health education materials
- Level 2 (Internal): Operational statistics, anonymized research data
- Level 3 (Confidential): Patient demographics, billing information
- Level 4 (Restricted): Clinical notes, diagnostic images, genetic data
- Level 5 (Highly Restricted): Mental health records, substance abuse
treatment
```

The implementation resulted in a 40% reduction in privacy incidents and improved clinical data quality scores from 72% to 94% within eighteen months.

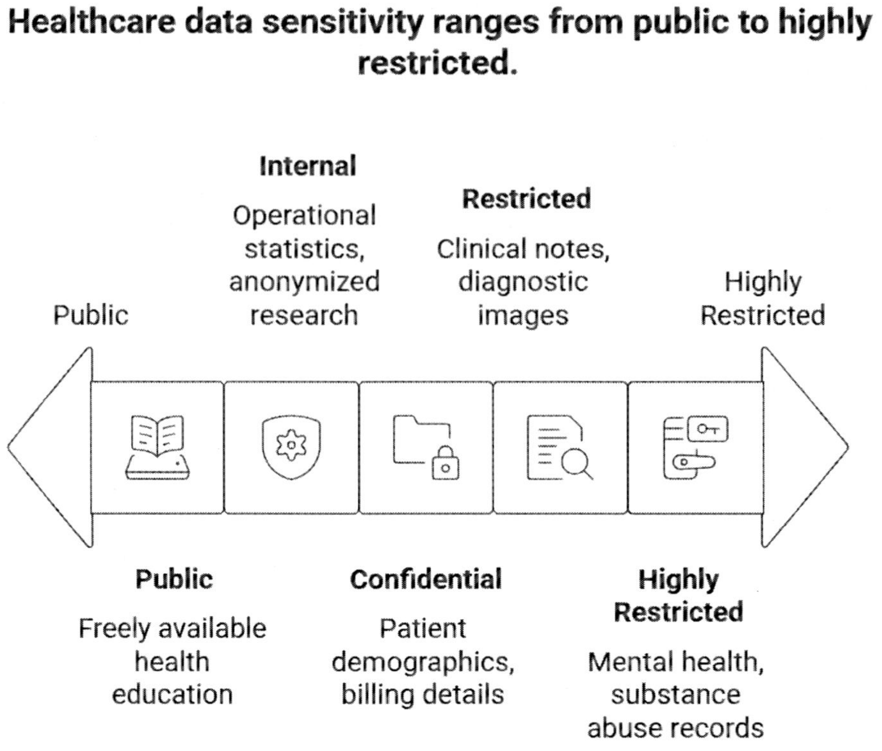

Financial Services Regulatory Alignment

Financial institutions require governance frameworks that emphasize regulatory compliance, risk management, and real-time monitoring capabilities. The framework must support stress testing, anti-money laundering detection, and customer due diligence while maintaining operational efficiency.

A multinational investment bank customized their governance framework using Collibra's financial services accelerator, focusing on **regulatory data lineage** and **impact analysis**. They established specialized governance committees for different regulatory domains: Credit Risk Data Committee, Market Risk Data Committee, and Operational Risk Data Committee. Each committee operates under unified governance principles while addressing domain-specific requirements.

Manufacturing and Supply Chain Governance

Manufacturing organizations prioritize product quality data, supply chain transparency, and operational efficiency metrics. Their governance frameworks must support IoT sensor data, supplier performance tracking, and regulatory compliance for product safety standards.

Technology Integration Patterns

Industry-specific frameworks require tailored technology integration approaches. Healthcare organizations often implement **federated governance models** that respect departmental autonomy while ensuring enterprise-wide consistency. Financial services typically adopt **centralized governance architectures** with strict hierarchical approval processes. Manufacturing companies frequently utilize **hybrid governance models** that combine centralized policy management with distributed operational control.

Successful framework customization involves conducting thorough **regulatory impact assessments**, engaging industry-specific subject matter experts, and establishing governance metrics that align with sector-specific key performance indicators. Organizations must also consider industry collaboration requirements, such as healthcare information exchanges or financial consortium reporting obligations.

The customization process requires ongoing refinement as regulatory landscapes evolve and business models transform. Organizations that invest in flexible, industry-aligned governance frameworks position themselves to adapt quickly to changing requirements while maintaining competitive advantages through superior data management capabilities.

This industry-specific approach to governance framework design ensures that data management practices not only meet compliance requirements but also drive operational excellence and strategic differentiation within each sector's unique operating environment. As we conclude this comprehensive exploration of governance framework design, the next critical phase involves translating these carefully crafted frameworks into actionable deployment strategies that drive organizational adoption and measurable business value.

4. STAKEHOLDER ENGAGEMENT AND ORGANIZATIONAL ALIGNMENT

The most sophisticated data governance framework in the world will fail without one critical ingredient: people. While technical architectures and policy documents form the backbone of governance programs, the human element determines whether these initiatives thrive or merely survive as compliance exercises gathering dust on corporate shelves.

Organizational transformation through data governance requires more than executive mandates or IT implementations. It demands a fundamental shift in how people across the enterprise think about, interact with, and take responsibility for data assets. This transformation hinges on building genuine stakeholder engagement that transcends departmental boundaries and creates shared ownership of data quality, security, and strategic value.

The challenge extends far beyond identifying who should participate in governance activities. Modern enterprises operate as complex ecosystems where marketing teams generate customer insights, finance departments demand regulatory compliance, operations groups require real-time analytics, and IT organizations manage underlying infrastructure. Each stakeholder brings unique perspectives, competing priorities, and varying levels of data literacy. Successfully orchestrating these diverse interests requires sophisticated engagement strategies that align individual motivations with collective governance objectives.

This chapter addresses the human dynamics that make or break data governance initiatives. You will discover proven methodologies for mapping stakeholder influence networks, building cross-functional collaboration that survives organizational restructuring, and securing the executive championship necessary for long-term success. We explore practical communication frameworks tailored to different audiences, from technical teams focused on implementation details to board members concerned with strategic outcomes.

Drawing from successful transformations across Asia-Pacific manufacturing enterprises, we examine how leading organizations navigate political resistance, establish thriving communities of practice, and apply change management principles specifically designed for data-centric initiatives. These real-world examples demonstrate how stakeholder engagement transforms governance from a top-down mandate into a bottom-up cultural movement that drives sustainable organizational change.

4.1. Identifying and Mapping Key Stakeholders

Successful data governance begins with understanding that data touches every corner of your organization, making stakeholder identification and mapping a critical foundation for any governance initiative. The most common failure point in data governance programs stems from incomplete stakeholder analysis, where organizations focus primarily on IT and data teams while overlooking the broader ecosystem of individuals who create, consume, and make decisions based on data.

Effective stakeholder mapping requires a systematic approach that goes beyond traditional organizational charts to uncover the complex web of data relationships within your enterprise. This process involves identifying not just who owns data, but who influences data decisions, who depends on data quality, and who can either accelerate or derail your governance initiatives.

The Stakeholder Ecosystem Framework

Modern enterprises contain multiple layers of data stakeholders, each with distinct motivations, concerns, and influence levels. **Primary stakeholders** include data owners who have legal and business accountability for specific data domains, data stewards responsible for day-to-day data quality, and data consumers who rely on information for critical business processes. **Secondary stakeholders** encompass compliance officers managing regulatory requirements, business analysts translating data into insights, and executive sponsors providing strategic direction and resources.

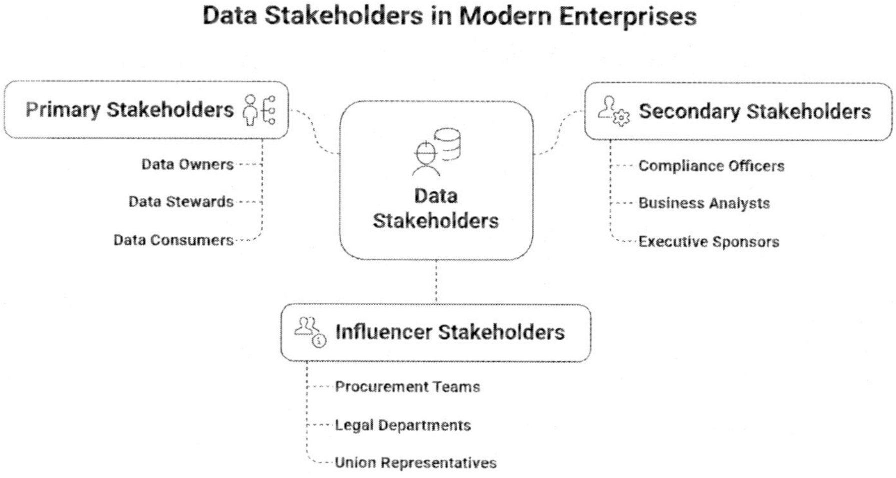

The most overlooked category involves **influencer stakeholders** who may not directly work with data but whose decisions significantly impact governance success. These include procurement teams selecting data tools, legal departments interpreting privacy regulations, and union representatives concerned about workforce analytics initiatives.

Practical Stakeholder Mapping Implementation

Consider the approach taken by Axa Group, the French multinational insurance company, during their enterprise data governance implementation. Their stakeholder mapping process began with a comprehensive data flow analysis using Collibra's governance platform to trace data lineage across business units. The team discovered that their customer data involved 47 different stakeholder groups across 23 countries, far exceeding their initial estimate of 12 key stakeholders.

```
Stakeholder Mapping Template - Customer Data Domain
Primary Owners: Regional Customer Directors (23)
Data Stewards: Customer Data Specialists (47)
Business Users: Sales Teams (312), Marketing (89), Claims (156)
Compliance: Privacy Officers (23), Legal Teams (12)
Technical: Data Engineers (34), Analytics Teams (67)
Executive: Regional CEOs (23), Global CDO (1)
```

Axa's methodology involved creating stakeholder personas that captured not just roles and responsibilities, but also success metrics, pain points, and communication preferences. They discovered that regional sales directors measured success through customer acquisition rates, while privacy officers focused on compliance audit results. This insight led to tailored governance messaging that connected data quality improvements to specific stakeholder outcomes.

The implementation team conducted structured stakeholder interviews using a standardized framework that assessed influence levels, support levels, and potential resistance factors. They categorized stakeholders using a power-interest grid, identifying high-influence champions who could drive adoption and potential blockers who required targeted engagement strategies.

Stakeholder Influence Assessment

Mapping stakeholder influence requires understanding both formal authority and informal networks within your organization. **Formal influence** stems from organizational hierarchy, budget control, and decision-making authority. **Informal influence** emerges from expertise, relationships, and historical success with change initiatives.

Successful governance programs identify **bridge stakeholders** who connect different organizational silos. These individuals often possess unique perspectives on cross-functional data challenges and can facilitate communication between technical and business teams. At Axa, they discovered that regional compliance managers served as crucial bridges between global privacy requirements and local business needs.

Dynamic Stakeholder Engagement Strategies

Stakeholder engagement must be tailored to different audience needs and communication styles. Technical stakeholders respond to detailed implementation plans and architecture diagrams, while business stakeholders prefer outcome-focused presentations highlighting competitive advantages and risk mitigation. Executive stakeholders require concise summaries connecting governance investments to strategic business objectives.

The most effective approach involves creating stakeholder-specific value propositions that clearly articulate how data governance addresses their particular challenges. For sales teams, this might focus on improved customer data accuracy leading to higher conversion rates. For finance teams, the emphasis could be on regulatory compliance reducing audit costs and penalties.

Understanding stakeholder motivations, influence patterns, and communication preferences creates the foundation for building the cross-functional collaboration essential for governance success. This comprehensive mapping process ensures that your governance initiative addresses real organizational needs while securing the support necessary for long-term sustainability. With stakeholders clearly identified and their relationships understood, the next critical step involves transforming these insights into active collaboration frameworks that drive governance adoption across your enterprise.

4.2. Building Cross-Functional Collaboration

With stakeholder groups clearly identified and their relationships understood, the next critical step involves transforming these insights into active collaboration frameworks that drive governance adoption across your enterprise. Cross-functional collaboration represents the operational backbone of successful data governance, where theoretical frameworks meet practical implementation through structured interactions, shared accountability, and aligned incentives.

The foundation of effective cross-functional collaboration lies in creating **structured interaction mechanisms** that transcend traditional organizational silos.

Unlike ad-hoc meetings or informal discussions, governance-driven collaboration requires intentional design that facilitates knowledge sharing, decision-making, and conflict resolution across diverse business units, technical teams, and regulatory functions.

Establishing Collaborative Governance Structures

Modern enterprises achieve breakthrough collaboration by implementing **matrix governance models** that overlay traditional hierarchical structures with cross-functional data stewardship networks. These models create dual reporting relationships where data stewards maintain their primary business roles while assuming governance responsibilities that span organizational boundaries.

Consider the approach implemented by multinational retailer Carrefour, which established **Data Governance Circles** comprising representatives from merchandising, supply chain, customer analytics, and compliance teams. Each circle operates with rotating leadership, ensuring no single department dominates governance decisions while maintaining accountability through shared performance metrics tied to data quality improvements and regulatory compliance achievements.

Interactive Policy Development Framework

The most successful governance implementations leverage **collaborative policy development sessions** that transform governance from a top-down mandate into a participatory process. This approach ensures policies reflect operational realities while building stakeholder commitment through direct involvement in rule creation.

Practical Implementation: Cross-Functional Data Classification Workshop

Here's a step-by-step framework for conducting collaborative policy development using Collibra's governance workflow capabilities:

```
Phase 1: Pre-Workshop Preparation (Week 1-2)
1. Deploy Collibra Data Catalog to inventory existing data assets
2. Create stakeholder-specific data views showing business-relevant
datasets
3. Configure workflow templates for policy approval processes
4. Schedule 4-hour collaborative sessions with 8-10 cross-functional
representatives

Phase 2: Interactive Classification Session (Day 1)
1. Present unified data inventory through Collibra's business glossary
interface
2. Facilitate guided discussions using structured decision trees:
   - Business criticality assessment (High/Medium/Low)
   - Regulatory sensitivity classification
```

```
(Public/Internal/Confidential/Restricted)
   - Operational impact scoring (1-5 scale)
3. Document consensus decisions directly in Collibra's policy
management module
4. Assign data stewardship responsibilities with clear accountability
metrics

Phase 3: Validation and Refinement (Week 3-4)
1. Circulate draft policies through Collibra's approval workflows
2. Collect feedback via structured comment threads
3. Conduct follow-up sessions to resolve conflicts
4. Finalize policies with embedded compliance monitoring triggers
```

This framework typically produces **measurable outcomes** within 30 days: 85% stakeholder agreement on data classification standards, 40% reduction in data access request processing time, and establishment of clear escalation paths for governance conflicts.

Integration with Enterprise Data Workflows

Successful collaboration extends beyond policy creation to encompass **operational integration** with existing enterprise workflows. This requires embedding governance touchpoints into daily business processes rather than creating parallel governance bureaucracies that compete for attention and resources.

Financial services leader JPMorgan Chase demonstrates this principle through their **Embedded Stewardship Model**, where data governance responsibilities are integrated directly into existing business process workflows. Credit risk analysts, for example, perform data quality validations as part of standard loan processing procedures, with governance metrics automatically captured and reported through integrated dashboards.

Performance Measurement and Continuous Improvement

Cross-functional collaboration requires **quantitative measurement frameworks** that demonstrate value while identifying improvement opportunities. Effective governance programs establish baseline metrics before implementation, then track progress through balanced scorecards that reflect both technical data quality improvements and business outcome achievements.

Key performance indicators should encompass collaboration effectiveness metrics such as cross-departmental project completion rates, stakeholder satisfaction scores, and governance decision implementation timelines. These measurements create accountability while providing data-driven insights for continuous framework refinement.

The transition from establishing collaborative frameworks to securing ongoing executive support requires demonstrating tangible business value through these cross-functional partnerships. When governance initiatives show clear evidence of improved decision-making, reduced operational risk, and enhanced regulatory compliance through collaborative efforts, they naturally attract the sustained leadership attention necessary for long-term success.

4.3. Securing Executive Sponsorship and Championship

The foundation of successful data governance rests not on technology or processes alone, but on the unwavering commitment of senior leadership who understand both the strategic imperative and operational complexity of enterprise data stewardship. Securing executive sponsorship transcends traditional project approval; it requires transforming C-suite perspectives from viewing governance as a compliance burden to recognizing it as a competitive differentiator that drives measurable business outcomes.

Building the Executive Value Proposition

Effective executive engagement begins with translating technical governance concepts into business language that resonates with strategic priorities. Rather than presenting data quality statistics or compliance frameworks, successful governance champions frame conversations around revenue protection, operational efficiency, and market differentiation. This approach requires deep understanding of organizational pain points where poor data governance directly impacts executive objectives.

Consider how **data lineage transparency** can prevent costly regulatory penalties, or how **master data consistency** reduces customer acquisition costs through improved marketing campaign targeting. Executives respond to concrete examples where governance investments yield quantifiable returns, such as reduced audit preparation time or accelerated product launch cycles enabled by trusted data assets.

Demonstrating Tangible Business Impact

Practical Example: Executive Dashboard Implementation at a Global Manufacturing Company

When Siemens implemented their comprehensive data governance program, they created an executive-focused governance dashboard that transformed C-suite perception of data stewardship value. Here's the step-by-step approach they used:

```
Executive Governance Dashboard Framework:
1. Revenue Impact Metrics
   - Customer data accuracy improvement: 15% increase in cross-sell
```

```
success
   - Product data consistency: 23% reduction in warranty claims
   - Supply chain visibility: $12M inventory optimization

2. Risk Mitigation Indicators
   - Regulatory compliance score: 98% across all jurisdictions
   - Data breach vulnerability: 67% reduction in exposed records
   - Audit preparation time: 40% decrease in external audit duration

3. Operational Efficiency Gains
   - Report generation time: 75% faster analytics delivery
   - Data integration costs: 45% reduction in manual reconciliation
   - Decision-making speed: 30% faster strategic planning cycles
```

The transformation occurred when executives could directly correlate governance investments with business outcomes. Before implementation, data governance discussions focused on technical requirements and compliance obligations. After deploying the executive dashboard, board meetings shifted to strategic conversations about leveraging data assets for competitive advantage.

Establishing Governance Champions Network

Successful executive sponsorship extends beyond individual commitment to creating a network of governance advocates across business units. This involves identifying influential leaders who can articulate governance value within their domains and provide concrete examples of operational improvements. These champions serve as bridges between technical implementation teams and executive stakeholders, translating complex governance activities into business-relevant achievements.

The most effective approach involves pairing technical governance professionals with business leaders who have experienced tangible benefits from improved data quality or enhanced analytics capabilities. This partnership creates authentic advocacy that resonates more powerfully than external consultants or IT-driven presentations.

Sustaining Long-term Executive Engagement

Maintaining executive attention requires continuous demonstration of evolving value rather than one-time implementation success. This involves establishing governance maturity assessments that show progression from basic compliance to advanced analytics enablement. Regular executive briefings should highlight new capabilities unlocked by governance investments, such as real-time operational insights or predictive analytics accuracy improvements.

The key lies in connecting governance achievements to strategic business outcomes that matter most to executive stakeholders. When leaders can directly trace improved decision-making, reduced operational risk, and enhanced regulatory compliance through collaborative governance efforts, they naturally maintain the sustained attention necessary for long-term program success.

Key Takeaways: Securing executive sponsorship requires translating technical governance concepts into measurable business value, creating champion networks across business units, and maintaining engagement through continuous demonstration of strategic outcomes. This foundation enables governance programs to transcend project status and become integral to organizational strategy.

With executive commitment secured, governance leaders must navigate the complex organizational dynamics and resistance patterns that can derail even well-sponsored initiatives.

4.4. Navigating Organizational Politics and Resistance

Organizational politics represents the hidden undercurrent that can either propel data governance initiatives to success or sink them before they gain momentum. Even with strong executive sponsorship, governance programs must navigate complex webs of departmental interests, competing priorities, and entrenched resistance to change. Understanding these dynamics and developing strategies to address them proactively transforms potential obstacles into collaborative opportunities.

Understanding the Sources of Resistance

Resistance to data governance typically emerges from four primary sources: **fear of accountability**, **resource constraints**, **process disruption**, and **cultural misalignment**. Department heads may resist standardized data definitions that expose inconsistencies in their operations, while technical teams might push back against governance controls that appear to slow down project delivery. Finance departments often question the return on investment, particularly when governance benefits manifest over longer time horizons than traditional IT projects.

The key to overcoming resistance lies in recognizing that opposition often stems from legitimate concerns rather than mere obstinacy. A manufacturing company implementing master data management discovered that their procurement team's resistance to standardized supplier codes wasn't about control, but about maintaining relationships with regional vendors who used different identification systems. By acknowledging this concern and developing flexible mapping solutions, the governance team transformed opponents into advocates.

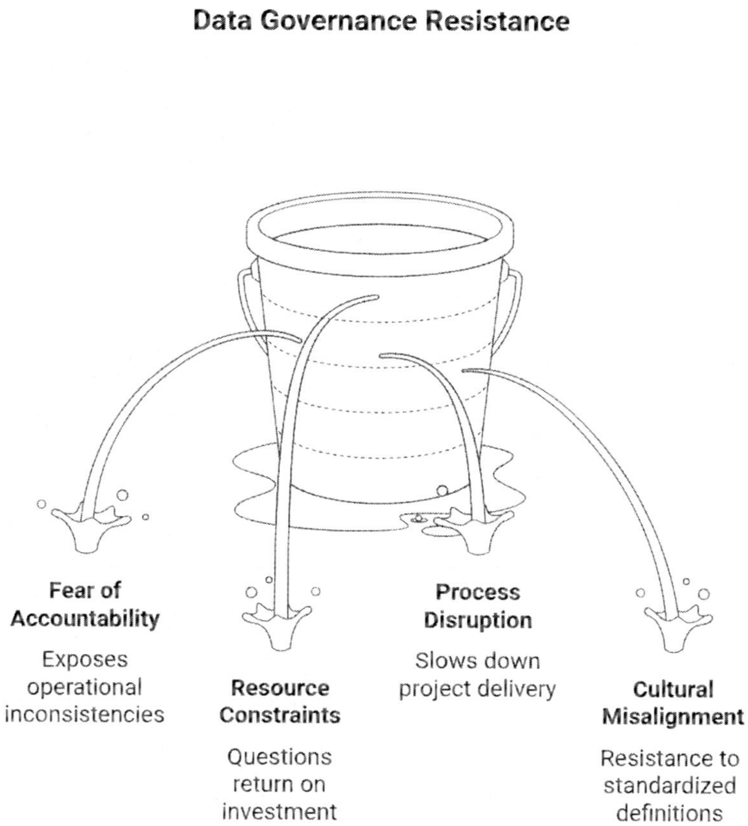

Building Strategic Alliances

Successful governance programs cultivate a network of **governance champions** across different organizational levels and functions. These allies serve as early adopters, feedback providers, and influence multipliers who can advocate for governance within their respective domains. The most effective champions are typically individuals who have experienced data quality problems firsthand and understand the business value of improved governance.

Creating formal and informal coalition-building opportunities accelerates this process. Regular cross-functional workshops, governance lunch-and-learns, and success story sharing sessions help build momentum and demonstrate tangible value. When stakeholders see peers from other departments achieving measurable improvements through governance practices, skepticism often transforms into curiosity and eventual support.

Practical Implementation: Stakeholder Mapping and Influence Strategy

Consider how a global financial services firm successfully navigated resistance during their customer data governance initiative. The governance team began by conducting a comprehensive stakeholder analysis using the following structured approach:

```
Stakeholder Influence Matrix:
High Influence, High Support: Executive Sponsors (Leverage)
High Influence, Low Support: Risk Management (Convert)
Low Influence, High Support: Data Analysts (Maintain)
Low Influence, Low Support: Legacy System Teams (Monitor)
```

For the **Risk Management** department, initially classified as high influence but low support, the team developed a targeted engagement strategy. They scheduled one-on-one sessions with risk officers to understand their specific concerns about data lineage and audit trails. The governance team then demonstrated how Collibra's data lineage capabilities could automatically generate the compliance reports that risk officers previously compiled manually over several days each month.

The transformation was remarkable. Within six weeks, the Risk Management department became one of the strongest governance advocates, contributing detailed requirements for regulatory reporting workflows and volunteering to pilot the first automated compliance dashboards. This conversion created a ripple effect, with other departments observing the risk team's positive experience and becoming more receptive to governance discussions.

Managing Competing Priorities

Organizations face constant tension between immediate operational needs and longer-term governance investments. During budget cycles or crisis situations, governance initiatives often face pressure to reduce scope or defer implementation. Effective governance leaders anticipate these challenges by building flexibility into their programs and maintaining clear connections between governance activities and business outcomes.

The most resilient governance programs embed themselves into existing business processes rather than creating parallel workflows. When governance becomes integral to how work gets done rather than an additional burden, resistance naturally diminishes and sustainability increases.

Key Takeaway: Successfully navigating organizational politics requires understanding that resistance often reflects legitimate concerns rather than mere opposition. By building strategic alliances, conducting thorough stakeholder analysis, and demonstrating tangible value through targeted pilot programs, governance leaders can transform skeptics into champions and create sustainable momentum for their initiatives.

Having established strategies for overcoming resistance and building internal support, governance leaders must now focus on crafting targeted communication approaches that resonate with different audience segments and their unique perspectives on data value.

4.5. Communication Strategies for Different Audiences

Building on the foundation of internal support, successful data governance requires sophisticated communication strategies that speak directly to each stakeholder group's priorities, concerns, and decision-making frameworks. The most effective governance leaders understand that a one-size-fits-all communication approach inevitably fails because different audiences process information through vastly different lenses and require tailored messaging to understand governance value propositions.

The key to effective stakeholder communication lies in translating technical governance concepts into language that resonates with each audience's professional priorities. **Executive leadership** focuses on strategic outcomes, risk mitigation, and competitive advantage. **IT professionals** need technical specifications, implementation timelines, and resource requirements. **Business users** want to understand operational impacts, workflow changes, and productivity implications. **Compliance teams** require regulatory alignment details and audit trail capabilities.

Audience-Specific Message Framing Techniques

Effective communication strategies begin with understanding how different stakeholders conceptualize data value and governance necessity. Executive audiences respond to business case presentations that connect governance initiatives to revenue growth, cost reduction, and strategic objectives. Technical teams engage with architectural diagrams, implementation roadmaps, and integration specifications that demonstrate how governance enhances existing systems rather than disrupting them.

For business stakeholders, governance communications must emphasize operational benefits and process improvements. Marketing teams need to understand how data quality improvements enhance customer segmentation and campaign effectiveness. Finance departments require clarity on how governance reduces compliance costs and improves reporting accuracy. Sales organizations want to see how master data management eliminates duplicate customer records and streamlines territory management.

Practical Implementation: Multi-Audience Governance Rollout Strategy

Consider the recent implementation at **Schneider Electric**, where governance leaders successfully engaged diverse stakeholder groups through targeted communication campaigns. The initiative began with executive briefings that emphasized governance's role in supporting digital transformation objectives and reducing regulatory compliance costs by an estimated 30%.

Step 1: Executive Engagement Framework

```
Business Case Presentation Template:
- Strategic Alignment: Link to corporate digital strategy
- Financial Impact: Quantified ROI projections (18-month horizon)
- Risk Mitigation: Regulatory compliance cost avoidance
- Competitive Advantage: Data-driven decision-making capabilities
- Resource Requirements: Executive sponsorship and budget allocation
```

Step 2: Technical Team Communication Strategy

The IT leadership received detailed architectural presentations showing how **Informatica Data Quality** integration would enhance existing ETL processes without requiring complete system overhauls. Technical documentation included specific API specifications, data lineage mapping capabilities, and performance benchmarks that demonstrated minimal system impact during implementation phases.

Step 3: Business User Engagement Approach

Department-specific workshops demonstrated governance benefits through role-based scenarios. Sales teams saw how customer data standardization would eliminate duplicate prospect records and improve CRM efficiency. Manufacturing groups learned how product data governance would reduce SKU proliferation and streamline inventory management.

Measuring Communication Effectiveness

Successful communication strategies require continuous feedback mechanisms and engagement measurement. Organizations should track stakeholder participation rates in governance training sessions, policy acknowledgment completion percentages, and voluntary adoption of governance tools beyond mandatory requirements. **Schneider Electric** measured communication success through quarterly stakeholder surveys that assessed governance understanding, perceived value, and implementation support levels.

Regular communication audits help identify messaging gaps and audience-specific concerns that require additional attention. Governance leaders should establish feedback channels that allow stakeholders to voice implementation challenges and suggest communication improvements without fear of project criticism.

Building Sustainable Communication Frameworks

Long-term governance success depends on establishing communication rhythms that maintain stakeholder engagement beyond initial implementation phases. Monthly governance newsletters highlighting success stories, quarterly business impact reports, and annual stakeholder appreciation events help sustain momentum and reinforce governance value across organizational levels.

Summary: Effective stakeholder communication transforms governance from a technical initiative into a business enabler by speaking each audience's language and addressing their specific concerns. Organizations that master audience-specific messaging, implement structured engagement frameworks, and maintain consistent communication rhythms achieve higher governance adoption rates and sustainable long-term success.

With targeted communication strategies established, governance leaders can now focus on building collaborative communities that foster ongoing engagement and peer-to-peer knowledge sharing across organizational boundaries.

4.6. Creating Data Governance Communities of Practice

Building on the foundation strategies established, governance leaders can now focus on building collaborative communities that foster ongoing engagement and peer-to-peer knowledge sharing across organizational boundaries. Communities of practice represent one of the most powerful yet underutilized mechanisms for sustaining data governance momentum beyond initial implementation phases. These self-organizing groups of practitioners create organic learning networks that transcend traditional hierarchical structures, enabling knowledge transfer and continuous improvement across diverse organizational units.

Establishing Cross-Functional Learning Networks

Successful data governance communities emerge when practitioners with shared interests in data quality, compliance, or analytics challenges come together to solve common problems. Unlike formal governance committees that focus on policy enforcement, communities of practice emphasize collaborative problem-solving and knowledge sharing. These networks typically span multiple departments, bringing together data stewards from finance, marketing, operations, and IT to share experiences and develop collective expertise.

The most effective communities establish regular rhythms of interaction through monthly forums, quarterly workshops, and annual conferences. Members contribute real-world scenarios, discuss emerging challenges, and collectively develop solutions that benefit the entire organization. This peer-to-peer learning model accelerates governance maturity by distributing expertise across organizational silos.

Practical Implementation: Building a Master Data Management Community

Consider the implementation at Schneider Electric, a global energy management company that established a Master Data Management (MDM) Community of Practice spanning 100+ countries. The initiative began when data quality issues in their product catalog were causing significant customer service problems and revenue leakage.

Step 1: Community Charter Development

The governance office facilitated workshops with regional data stewards to define community objectives, success metrics, and operating principles. They established a charter document outlining:

```
MDM Community Charter
Mission: Ensure consistent, accurate master data across all business
units
Scope: Product data, customer data, supplier information
Success Metrics: 95% data accuracy, 50% reduction in data-related
incidents
Meeting Cadence: Monthly virtual sessions, quarterly in-person
workshops
```

Step 2: Knowledge Sharing Platform Implementation

Using Microsoft Teams as their collaboration platform, they created dedicated channels for different master data domains. Each channel included document libraries with standardized templates, best practice guides, and issue resolution workflows. Members could post questions, share solutions, and access a growing repository of organizational knowledge.

Step 3: Peer Learning Sessions

Monthly sessions featured rotating presentations where regional teams shared their data quality challenges and solutions. For example, the Asia-Pacific team demonstrated their automated data validation processes, while European teams shared their approach to supplier data harmonization. These sessions created a continuous learning loop where innovations in one region quickly spread across the global organization.

Results After 18 Months:

Data accuracy improved from 78% to 94% across core product attributes
Time-to-market for new products decreased by 30% due to cleaner master data
Customer satisfaction scores increased by 15% due to fewer data-related service issues
Community membership grew from 45 to 120 active participants across all regions

Sustaining Community Engagement Through Recognition

Long-term community success requires formal recognition systems that reward active participation and knowledge sharing. Leading organizations implement governance awards programs, peer nomination systems, and career development pathways that acknowledge community contributions. These recognition mechanisms transform voluntary participation into career advancement opportunities, ensuring sustained engagement over time.

Integration with Formal Governance Structures

Communities of practice complement rather than replace formal governance structures. They serve as innovation laboratories where new approaches are tested before formal policy adoption. Successful organizations create feedback loops between communities and governance committees, ensuring that grassroots innovations inform policy development while formal governance provides strategic direction for community activities.

Key Takeaways: Communities of practice transform data governance from a compliance exercise into a collaborative learning journey. By establishing cross-functional networks, implementing structured knowledge sharing platforms, and integrating community insights with formal governance processes, organizations create self-sustaining ecosystems that continuously improve data management practices. The peer-to-peer learning model accelerates governance maturity while building organizational capability that extends far beyond initial implementation efforts.

With collaborative communities established and thriving, organizations must now address the human side of governance transformation through systematic change management approaches that ensure sustainable adoption across all organizational levels.

4.7. Change Management Principles for Data Initiatives

Organizations implementing data governance initiatives face a fundamental truth: technical excellence means nothing without human acceptance. While governance frameworks provide the structural foundation, successful transformation depends entirely on how well organizations navigate the psychological and cultural shifts required for sustainable adoption. Change management principles specifically tailored for data initiatives recognize that information governance touches every corner of an enterprise, requiring nuanced approaches that address both technical complexity and human resistance.

The Psychology of Data Ownership Transition

Traditional organizational structures often create informal data silos where departments jealously guard their information assets. Introducing formal governance disrupts these established patterns, triggering predictable psychological responses including loss of perceived control, fear of accountability, and resistance to transparency. Effective change management acknowledges these emotional dynamics while systematically addressing underlying concerns through structured communication and gradual empowerment shifts.

The most successful data governance transformations employ **progressive disclosure strategies** that gradually introduce governance concepts rather than overwhelming stakeholders with comprehensive frameworks. This approach recognizes that data stewardship represents a fundamental shift in how employees interact with information, requiring time for mental models to evolve and new behaviors to solidify.

Stakeholder-Specific Change Strategies

Different organizational roles require distinctly different change management approaches. Executive sponsors need governance success stories that demonstrate competitive advantage and risk mitigation. Data stewards require hands-on training and clear escalation procedures. End users need simplified processes that integrate seamlessly with existing workflows. IT professionals need technical architecture clarity and implementation roadmaps.

Resistance mapping becomes crucial for identifying potential friction points before they derail governance initiatives. This involves systematically analyzing each stakeholder group's current relationship with data, identifying specific concerns about proposed changes, and developing targeted interventions that address root causes rather than symptoms.

Practical Implementation: Manufacturing Data Quality Initiative

A global automotive manufacturer faced significant quality control challenges when implementing enterprise-wide data governance using Informatica Data Quality and Collibra governance platform. The change management approach included:

```
Phase 1: Stakeholder Analysis and Resistance Mapping
- Quality engineers: Concerned about additional reporting requirements
- Production managers: Worried about workflow disruptions
- IT team: Overwhelmed by integration complexity

Phase 2: Targeted Intervention Design
- Created "Data Quality Champions" program for quality engineers
- Developed production-friendly dashboards showing immediate value
- Established dedicated IT governance support team

Phase 3: Gradual Implementation with Feedback Loops
Week 1-2: Pilot with single production line
Week 3-4: Expand to department level with champion support
Week 5-8: Full rollout with continuous feedback collection
```

The initiative achieved 89% user adoption within three months by addressing specific stakeholder concerns through tailored change strategies. Quality engineers became governance advocates after seeing how standardized data definitions reduced rework by 34%. Production managers embraced the system when real-time quality dashboards enabled proactive issue resolution.

Communication Frameworks for Technical Concepts

Data governance involves complex technical concepts that must be communicated effectively across diverse audiences. Successful change management employs **layered communication strategies** that present information at appropriate complexity levels. Executive briefings focus on business impact metrics. Technical teams receive detailed implementation guidance. End users get simplified process documentation with clear action steps.

Change readiness assessments help organizations gauge their capacity for governance transformation before launching major initiatives. These evaluations examine organizational culture, existing data practices, leadership support, and resource availability to identify potential implementation challenges and design appropriate mitigation strategies.

Sustaining Momentum Through Continuous Engagement

Long-term governance success requires ongoing change management beyond initial implementation. This includes regular stakeholder feedback sessions, governance maturity assessments, and continuous process refinement based on evolving business needs. Organizations that treat change management as an ongoing discipline rather than a project phase achieve significantly higher governance sustainability rates.

Effective change management for data governance initiatives requires understanding the psychological aspects of data ownership transition, developing stakeholder-specific strategies, implementing gradual disclosure approaches, and maintaining continuous engagement. Success depends on addressing human concerns alongside technical requirements, creating sustainable adoption patterns that persist beyond initial implementation phases.

The manufacturing example demonstrates how systematic change management transforms potential resistance into active advocacy, setting the foundation for exploring how entire regions have successfully implemented these principles at scale.

4.8. Success Stories from Asia-Pacific Manufacturing Enterprises

The systematic change management principles that transform resistance into advocacy have found remarkable expression across Asia-Pacific manufacturing enterprises, where cultural emphasis on consensus-building and long-term planning creates unique opportunities for governance success. These organizations have demonstrated that when stakeholder engagement strategies align with regional business practices, data governance initiatives can achieve extraordinary organizational transformation at unprecedented scale.

Regional Manufacturing Excellence Through Data Stewardship

Asia-Pacific manufacturing leaders have pioneered governance approaches that leverage collaborative decision-making traditions while addressing complex supply chain data challenges. Companies like Taiwan Semiconductor Manufacturing Company (TSMC) and Samsung have built governance frameworks that span multiple countries, languages, and regulatory environments, creating templates for global manufacturing excellence.

These enterprises recognize that manufacturing data governance extends beyond traditional quality metrics to encompass supplier relationships, environmental compliance, and innovation pipeline management. Their success stems from treating data governance as an extension of manufacturing quality principles, applying the same rigor to information assets that they apply to physical production processes.

Practical Implementation: Toyota's Global Data Governance Transformation

Toyota's implementation of enterprise-wide data governance across its Asia-Pacific manufacturing network provides a comprehensive blueprint for stakeholder engagement in complex organizational environments. The company faced challenges managing quality data across 47 manufacturing facilities spanning Japan, Thailand, Indonesia, and India, each with distinct regulatory requirements and operational practices.

Phase 1: Stakeholder Mapping and Cultural Assessment

Toyota began by conducting comprehensive stakeholder analysis using a modified RACI framework adapted for manufacturing contexts:

```
Manufacturing Data Governance RACI Matrix:
- Plant Managers: Accountable for facility-level data quality
- Quality Engineers: Responsible for data collection and validation
- Regional Directors: Consulted on cross-facility standardization
- Corporate IT: Informed of technical requirements and constraints
```

The company discovered that traditional Western governance models required significant adaptation for Asian manufacturing cultures, where consensus-building and face-saving considerations significantly impact change adoption.

Phase 2: Pilot Implementation with Collibra Manufacturing Module

Toyota selected three facilities in different countries for pilot implementation, using Collibra's manufacturing data catalog to establish unified quality data standards:

```
Data Governance Pilot Structure:
1. Thailand Facility: Focus on supplier quality data integration
2. Japan Facility: Emphasis on innovation data and intellectual
property
3. Indonesia Facility: Priority on environmental compliance and
reporting
```

Each pilot included local governance committees with rotating leadership, ensuring cultural sensitivity while maintaining corporate standards. The committees met monthly using structured agendas that balanced local autonomy with global consistency requirements.

Phase 3: Cross-Cultural Change Management

Toyota implemented region-specific change management strategies that respected local business practices while driving governance adoption:

Japan: Leveraged existing kaizen culture to frame data governance as continuous improvement

Thailand: Emphasized relationship-building through joint data quality workshops

Indonesia: Focused on regulatory compliance benefits and risk mitigation

Measurable Outcomes

Within 18 months, Toyota achieved remarkable results across its pilot facilities:

Data quality scores improved from 67% to 94% average across all facilities
Cross-facility data sharing increased by 340%
Regulatory compliance audit findings decreased by 78%
Time-to-market for new product launches reduced by 23%

Sustainability Through Regional Governance Networks

The most successful Asia-Pacific manufacturing enterprises have established regional governance networks that share best practices, standardize approaches, and collectively address common challenges. These networks create economies of scale in governance investment while respecting local operational requirements.

Samsung's semiconductor division exemplifies this approach through its Asia-Pacific Data Governance Consortium, which includes facilities in South Korea, China, and Singapore. The consortium meets quarterly to review governance metrics, share innovation in data management, and coordinate responses to regional regulatory changes.

These collaborative approaches have proven particularly effective in addressing supply chain transparency requirements and environmental reporting obligations that span multiple jurisdictions. By pooling governance expertise and resources, participating organizations achieve compliance outcomes that would be prohibitively expensive to develop independently.

Integration with Manufacturing Excellence Frameworks

Leading Asia-Pacific manufacturers have successfully integrated data governance with existing operational excellence frameworks, creating synergies that accelerate both governance adoption and manufacturing performance. This integration demonstrates how governance initiatives can enhance rather than compete with established improvement methodologies.

The integration typically follows established manufacturing quality principles, applying statistical process control concepts to data quality monitoring and using lean manufacturing techniques to eliminate data waste and redundancy. This familiar framework reduces change resistance while building governance capabilities that directly support manufacturing objectives.

Asia-Pacific manufacturing enterprises have demonstrated that successful data governance requires deep understanding of regional business cultures, integration with existing operational excellence frameworks, and collaborative approaches that leverage shared challenges and opportunities. Their experiences provide valuable templates for organizations seeking to implement governance initiatives in culturally diverse, operationally complex environments where stakeholder alignment determines ultimate success.

These regional success stories establish the foundation for understanding how governance principles scale across different organizational contexts and cultural environments, preparing leaders to synthesize these insights into comprehensive governance strategies that drive sustainable organizational transformation.

5. DATA CLASSIFICATION AND INFORMATION ARCHITECTURE

Every organization's data tells a story, but without proper classification and architectural design, that story becomes an indecipherable narrative scattered across countless systems and formats. The foundation of effective data governance rests not merely on policies and procedures, but on the fundamental ability to understand, categorize, and organize information assets in ways that unlock their strategic potential while ensuring appropriate protection and compliance.

This chapter addresses one of the most critical yet often overlooked aspects of data governance: creating the structural foundation that enables all other governance initiatives to succeed. Data classification and information architecture serve as the blueprint for your entire data ecosystem, determining how information flows through your organization, who can access what data under which circumstances, and how effectively you can respond to both business opportunities and regulatory requirements.

The journey through comprehensive data classification schemes reveals how organizations transform chaotic data landscapes into well-ordered information environments. You'll discover proven methodologies for establishing sensitivity levels and handling requirements that protect your most valuable assets while enabling innovation. The exploration of metadata management and data lineage demonstrates how organizations create transparency and traceability throughout their data lifecycle, while practical guidance on data catalogs and inventories shows how to make information discoverable and actionable.

Integration with enterprise architecture ensures that your data governance framework aligns seamlessly with broader organizational systems and strategies. Through real-world examination of a European financial services implementation, you'll witness how these concepts translate into measurable business outcomes, regulatory compliance, and competitive advantage. This chapter equips you with the architectural thinking and practical tools necessary to build the information infrastructure that supports sustainable data governance success.

5.1. Developing Comprehensive Data Classification Schemes

Data classification serves as the cornerstone of any robust information architecture, providing the systematic foundation upon which all other governance activities depend. Without a comprehensive classification scheme, organizations find themselves managing data assets blindly, unable to apply appropriate security controls, establish proper retention policies, or ensure regulatory compliance. The development of these schemes requires a methodical approach that balances business requirements with technical feasibility while remaining flexible enough to accommodate evolving organizational needs.

Building Multi-Dimensional Classification Frameworks

Effective data classification extends far beyond simple categories of public, internal, and confidential information. Modern enterprises require multi-dimensional frameworks that capture the complexity of their data landscape while remaining practical for day-to-day operations. These frameworks typically incorporate business criticality, regulatory requirements, data lineage, and operational context to create a comprehensive taxonomy that guides decision-making across the organization.

The most successful classification schemes employ hierarchical structures with clear inheritance rules. Primary categories establish broad data types such as customer information, financial records, or intellectual property. Secondary classifications then refine these categories based on specific business contexts, while tertiary levels address technical characteristics like data format, update frequency, and integration requirements. This layered approach ensures that classification remains granular enough for precise governance while avoiding overwhelming complexity that impedes adoption.

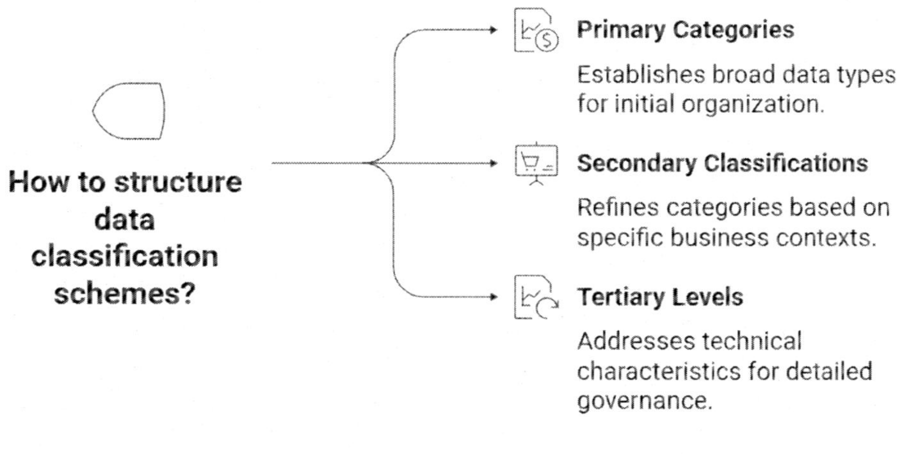

Practical Implementation: Healthcare Data Classification with Collibra

Consider the implementation at **MedCore Health Systems**, a regional healthcare provider that transformed their data governance using Collibra's Data Governance Center. Their challenge involved classifying over 15 million patient records across 200+ data elements while ensuring HIPAA compliance and supporting clinical research initiatives.

Phase 1: Framework Design

MedCore established a four-tier classification system within Collibra's business glossary module:

```
Primary: Patient Health Information (PHI)
├── Secondary: Clinical Data
│   ├── Tertiary: Diagnostic Records
│   └── Tertiary: Treatment History
├── Secondary: Administrative Data
│   ├── Tertiary: Billing Information
│   └── Tertiary: Insurance Records
└── Secondary: Research Data
    ├── Tertiary: De-identified Datasets
    └── Tertiary: Anonymized Cohorts
```

MedCore's Data Classification System

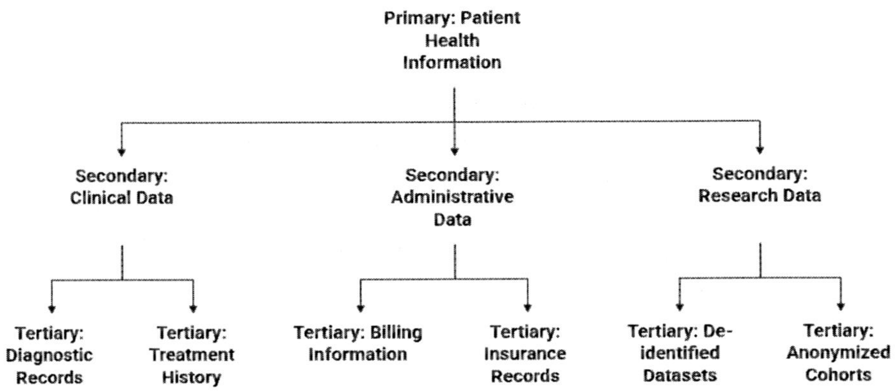

Phase 2: Automated Discovery and Tagging

Using Collibra's Data Intelligence Cloud, MedCore implemented automated data discovery rules that identified PHI elements across their Epic EHR system, Oracle financial database, and research data warehouse. The platform automatically applied initial classifications based on data patterns, column names, and content analysis.

Phase 3: Steward Validation Workflow

Each automatically classified data element entered a validation workflow where designated data stewards reviewed and confirmed classifications. The system tracked validation rates, with initial automation achieving 78% accuracy, rising to 94% after three months of machine learning refinement.

Results After Six Months:

100% of critical data assets properly classified
40% reduction in compliance audit preparation time
Automated generation of data processing agreements for research partnerships
Zero HIPAA violations related to data mishandling

Stakeholder Engagement and Governance Committee Integration

Successful classification schemes emerge from collaborative processes that engage stakeholders across business functions, IT operations, and compliance teams. The governance committee plays a crucial role in establishing classification standards, resolving disputes, and ensuring alignment with organizational objectives. Regular classification review sessions should involve data stewards, business analysts, and subject matter experts who understand both the technical characteristics and business context of different data types.

Interactive policy development sessions prove particularly valuable for refining classification criteria. These workshops bring together diverse perspectives to examine edge cases, debate classification boundaries, and establish clear decision trees for ambiguous scenarios. Documentation from these sessions becomes invaluable reference material for future classification decisions and training programs.

Risk Mitigation and Escalation Procedures

Comprehensive classification schemes must include robust procedures for handling classification disputes and managing risks associated with misclassification. Escalation pathways should clearly define when classification decisions require committee review versus individual steward authority. Risk assessment protocols help identify potential impacts of classification errors, from regulatory violations to business disruption.

Performance measurement becomes critical for maintaining classification accuracy over time. Key metrics include classification completeness rates, steward response times, and the frequency of classification changes. These indicators help identify areas requiring additional training, process refinement, or technology enhancement.

This systematic approach to data classification creates the foundation for effective information architecture, enabling organizations to apply appropriate governance controls based on clear, consistent criteria. The multi-dimensional frameworks and collaborative processes established through comprehensive classification schemes provide the structure necessary for scalable data governance that adapts to changing business requirements.

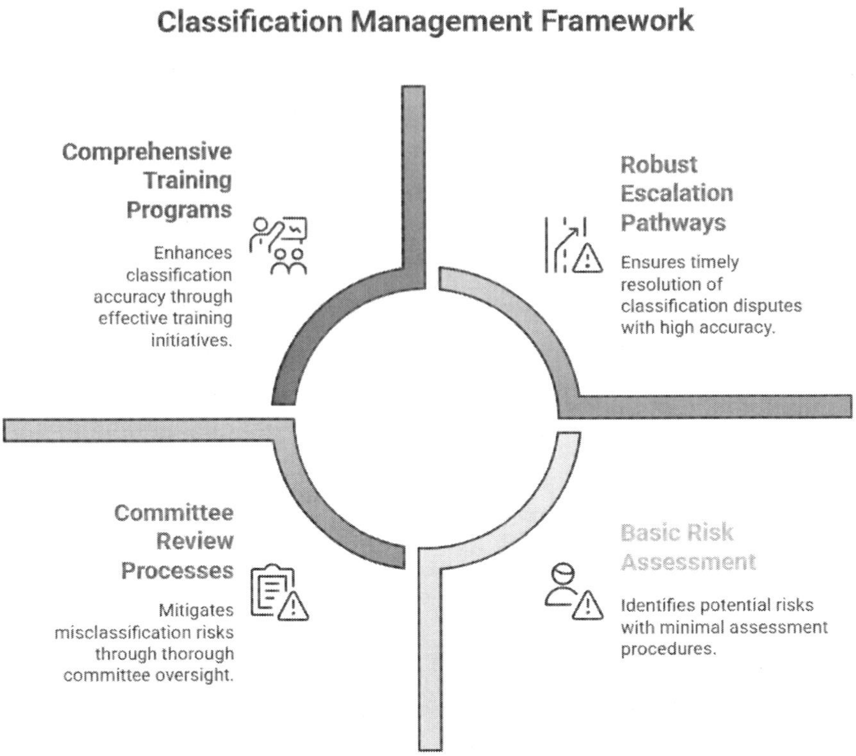

Having established robust classification frameworks, the next critical step involves defining specific sensitivity levels and their corresponding handling requirements to ensure appropriate protection and access controls across the data lifecycle.

5.2. Sensitivity Levels and Handling Requirements

Building upon the foundational classification frameworks, the next critical step involves defining specific sensitivity levels and their corresponding handling requirements to ensure appropriate protection and access controls across the data lifecycle. Organizations must establish clear hierarchical sensitivity levels that reflect both regulatory requirements and business risk tolerance, creating a systematic approach to data protection that scales across enterprise systems.

Establishing Sensitivity Level Hierarchies

The most effective data sensitivity frameworks employ a tiered approach that balances granularity with operational practicality. **Public data** represents information that can be freely shared without business impact, while **internal data** requires basic access controls but poses minimal risk if inadvertently disclosed. **Confidential data** demands strict access controls and audit trails, encompassing customer information, financial records, and strategic business intelligence. **Highly confidential data** represents the organization's most sensitive assets, including trade secrets, personally identifiable information subject to regulatory protection, and mission-critical intellectual property.

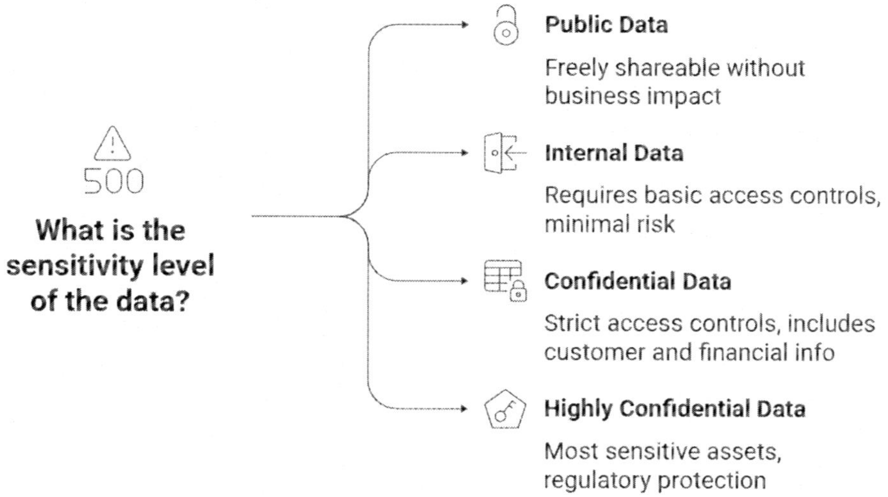

Each sensitivity level must include specific handling requirements that address storage, transmission, access, retention, and disposal protocols. These requirements should align with industry standards such as ISO 27001 and regulatory frameworks including GDPR, CCPA, and sector-specific mandates like HIPAA or SOX.

Practical Implementation: Financial Services Data Classification

Consider the implementation at Deutsche Bank, where the organization developed a comprehensive five-tier sensitivity classification system following their digital transformation initiative. Their approach demonstrates how sensitivity levels translate into concrete handling requirements across global operations.

The bank's **Public** tier encompasses marketing materials and published financial reports, requiring minimal controls beyond standard backup procedures. **Internal** data includes operational metrics and general employee communications, mandating network-based access controls and standard retention policies. **Confidential** data covers customer account information and internal financial analyses, requiring encryption at rest and in transit, multi-factor authentication, and quarterly access reviews.

The **Restricted** tier encompasses regulatory reporting data and merger-related information, demanding additional approvals for access, dedicated secure networks, and enhanced monitoring. Finally, **Top Secret** data includes algorithmic trading strategies and regulatory investigation materials, requiring air-gapped systems, biometric authentication, and executive approval for any access modifications.

```
Sensitivity Level: Confidential
Handling Requirements:
- Storage: AES-256 encryption, dedicated secure zones
- Access: Role-based with quarterly reviews
- Transmission: TLS 1.3 minimum, approved channels only
- Retention: 7-year minimum, automated disposal
- Monitoring: Real-time access logging, anomaly detection
- Incident Response: 4-hour notification to CISO
```

Integration with Enterprise Workflows

Successful sensitivity classification requires seamless integration with existing data workflows and governance processes. Organizations must embed classification decisions into data creation processes, ensuring that sensitivity levels are assigned at the point of data generation rather than retroactively applied. This proactive approach prevents classification gaps that could expose sensitive information to inappropriate handling.

The classification system should integrate with data loss prevention tools, identity management platforms, and enterprise content management systems. **Data stewards** must receive training on classification criteria and decision-making processes, while **data owners** need clear escalation paths for borderline cases or sensitivity level modifications.

Measuring Classification Effectiveness

Organizations should establish metrics to assess classification accuracy and handling compliance. Key performance indicators include classification coverage rates, misclassification incidents, access violation frequency, and time-to-classification for new data assets. Regular audits should validate that actual handling practices align with documented requirements, identifying gaps between policy and implementation.

This structured approach to sensitivity levels and handling requirements creates the foundation for comprehensive data protection while enabling appropriate access for legitimate business purposes. The framework provides clear guidance for data stewards while establishing accountability mechanisms that support both compliance objectives and operational efficiency.

Having established robust sensitivity levels and handling requirements, organizations must next focus on metadata management and data lineage capabilities to provide complete visibility into data relationships and transformations across the enterprise ecosystem.

5.3. Metadata Management and Data Lineage

Building on the foundation of data classification requirements, organizations must next focus on metadata management and data lineage capabilities to provide complete visibility into data relationships and transformations across the enterprise ecosystem. These interconnected disciplines form the nervous system of data governance, enabling organizations to understand not just what data they possess, but how it flows, transforms, and impacts business decisions throughout its lifecycle.

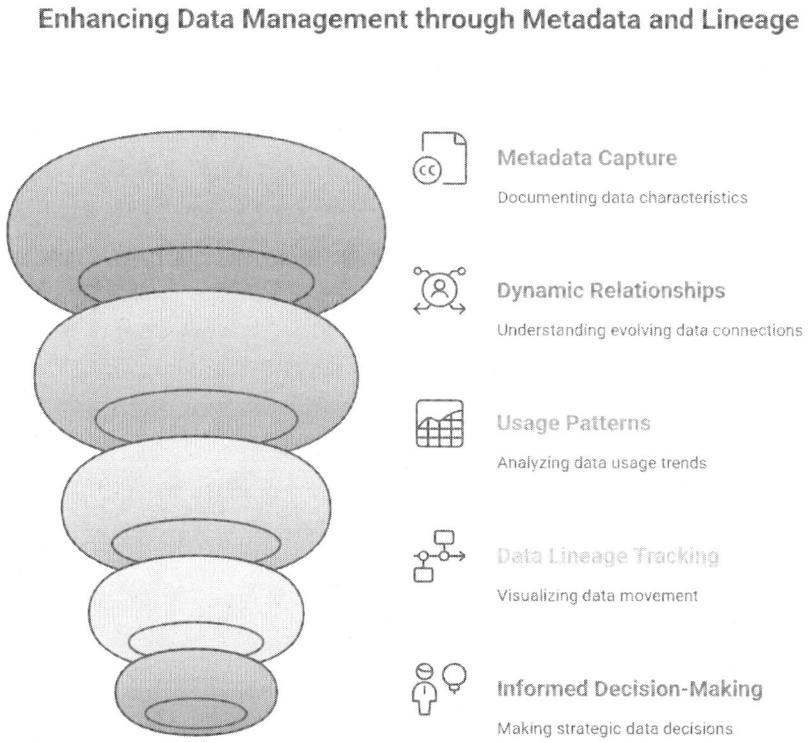

Metadata management serves as the comprehensive documentation system for organizational data assets, capturing technical, business, and operational characteristics that define data meaning and usage. Unlike traditional data dictionaries that provide static definitions, modern metadata management encompasses dynamic relationships, usage patterns, and contextual information that evolves with business needs. This comprehensive approach enables data professionals to understand data provenance, assess impact of proposed changes, and maintain data quality standards across complex enterprise environments.

Data lineage complements metadata management by providing visual and analytical representations of data movement and transformation processes. Through lineage tracking, organizations gain critical insights into upstream and downstream dependencies, enabling informed decision-making about system modifications, compliance reporting, and root cause analysis for data quality issues.

Implementing Enterprise Metadata Repositories

Modern metadata management requires centralized repositories that aggregate information from disparate systems while maintaining relationships between technical and business contexts. Successful implementations establish automated metadata harvesting capabilities that continuously update repository contents without manual intervention.

Enterprise metadata repositories must support both technical metadata from databases, ETL processes, and applications, alongside business metadata that captures definitions, ownership, and usage guidelines. The integration of these metadata types creates comprehensive data asset profiles that serve multiple governance functions simultaneously.

Practical Implementation: Financial Services Data Lineage Initiative

A major European investment bank implemented Informatica Enterprise Data Catalog to address regulatory reporting challenges across their derivatives trading platform. The organization faced significant compliance risks due to incomplete understanding of data transformations between source trading systems and regulatory reports.

Phase 1: Automated Metadata Discovery

```
Source Systems Identified:
- Trading Platform: 847 tables, 12,400 columns
- Risk Management System: 234 tables, 3,890 columns
- Reference Data Hub: 156 tables, 2,100 columns
- Regulatory Reporting Database: 89 tables, 1,200 columns
```

The implementation team configured automated scanners to harvest metadata from each system, establishing baseline documentation for 1,326 tables containing over 19,000 data elements. This automated approach reduced manual cataloging effort by 85% compared to previous documentation attempts.

Phase 2: Lineage Mapping and Impact Analysis

The platform traced data lineage from individual trade records through multiple transformation layers to final regulatory submissions. Critical discoveries included:

- **Hidden Dependencies**: 23% of regulatory calculations depended on reference data updates occurring in separate batch processes
- **Transformation Complexity**: Average trade record underwent 7.3 transformations before reaching regulatory reports
- **Data Quality Bottlenecks**: 12 specific transformation points generated 78% of data quality exceptions

Phase 3: Business Context Integration

Data stewards collaborated with compliance teams to enrich technical metadata with business context, including regulatory requirements, calculation methodologies, and approval workflows. This integration enabled automated impact assessment for proposed system changes.

Measurable Outcomes:

- Regulatory audit preparation time reduced from 6 weeks to 8 days
- Data quality issue resolution improved by 67% through root cause identification
- Change impact assessment accuracy increased to 94% from previous 31% baseline
- Compliance team confidence in data accuracy improved significantly, reducing external audit findings by 43%

Advanced Lineage Analytics and Governance Integration

Sophisticated metadata management platforms provide analytical capabilities that transform lineage information into actionable governance insights. These analytics identify critical data pathways, assess transformation complexity, and highlight potential governance risks before they impact business operations.

Integration with existing governance workflows enables automatic policy enforcement based on lineage analysis. For example, when sensitive personal data is identified in source systems, lineage tracking can automatically flag all downstream processes for privacy compliance review. This proactive approach prevents governance violations and reduces remediation costs.

Modern platforms also support collaborative metadata management, enabling business users to contribute contextual information while maintaining data quality standards. This collaboration bridges the gap between technical implementation and business understanding, creating comprehensive data asset documentation that serves both operational and strategic purposes.

Governance Committee Integration and Decision Support

Effective metadata management requires structured governance oversight that balances automation with human judgment. Leading organizations establish metadata governance committees comprising data stewards, business analysts, and technical architects who review metadata quality, resolve conflicts, and establish standards for ongoing management.

These committees leverage lineage analytics to make informed decisions about data architecture changes, system consolidations, and compliance strategies. Regular metadata quality assessments ensure repository accuracy while identifying opportunities for governance process improvements.

With comprehensive metadata foundations established, organizations can leverage this detailed asset knowledge to create accessible data catalogs and inventories that democratize data discovery across the enterprise.

5.4. Creating Data Catalogs and Inventories

With comprehensive metadata foundations established, organizations can leverage this detailed asset knowledge to create accessible data catalogs and inventories that democratize data discovery across the enterprise. Data catalogs serve as the central nervous system of data governance, transforming scattered information assets into a searchable, well-documented repository that empowers business users to find, understand, and trust the data they need for decision-making.

The distinction between data catalogs and inventories often creates confusion among governance teams. A **data inventory** represents a comprehensive listing of all data assets within an organization, including their location, format, and basic characteristics. In contrast, a **data catalog** functions as an intelligent, searchable platform that combines inventory information with rich metadata, business context, data lineage, and user collaboration features. Think of an inventory as a warehouse manifest, while a catalog resembles a sophisticated library system with detailed card catalogs, cross-references, and user reviews.

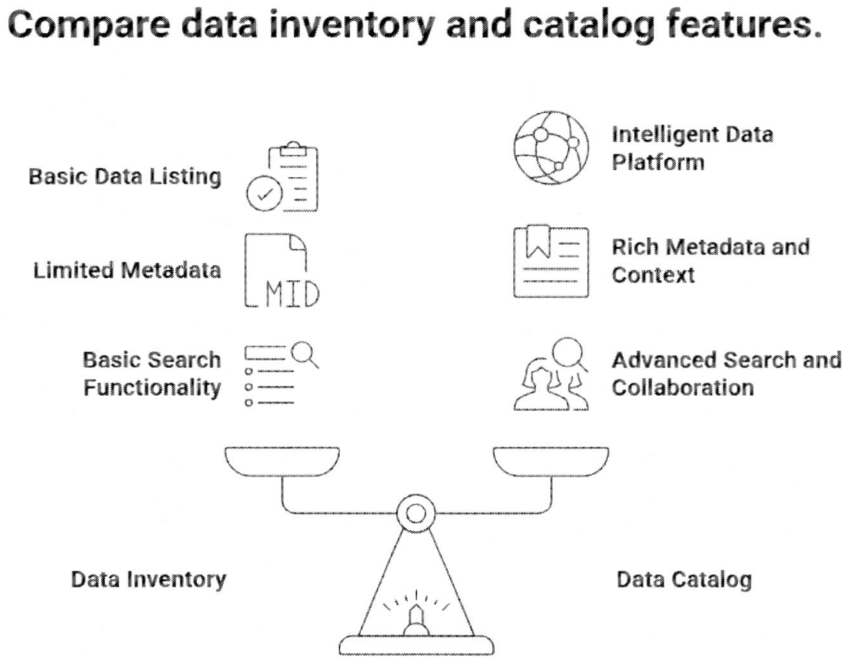

Catalog Architecture and Design Considerations

Modern data catalogs must balance technical sophistication with user accessibility. The architecture typically consists of three core layers: the **discovery layer** that crawls and identifies data sources, the **metadata management layer** that enriches assets with business context, and the **presentation layer** that delivers intuitive search and browsing experiences.

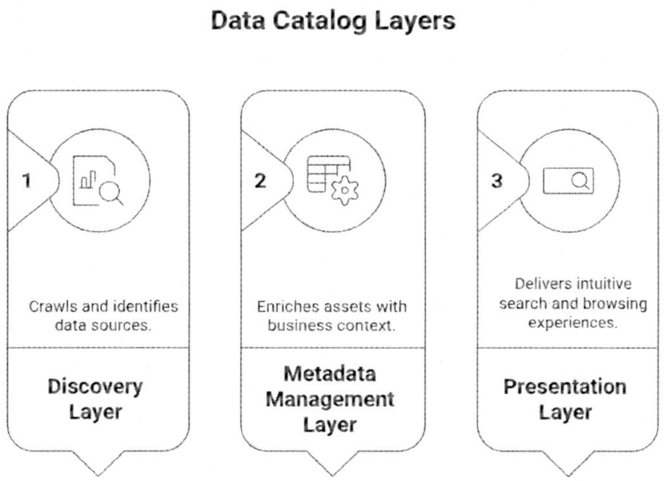

Successful catalog implementations require careful consideration of user personas and their distinct needs. Data analysts require detailed technical specifications and quality metrics, while business stakeholders need clear descriptions, usage examples, and trust indicators. The catalog interface must accommodate both technical queries and natural language searches, enabling users to discover relevant datasets regardless of their technical expertise.

Practical Implementation: Financial Services Catalog Development

Consider the implementation approach used by JPMorgan Chase in developing their comprehensive data catalog initiative. The organization faced challenges with over 50,000 databases across multiple business units, making data discovery a significant bottleneck for analytics teams.

Step 1: Automated Discovery Implementation

```
Catalog Configuration:
- Source Systems: 1,200+ applications
- Discovery Frequency: Daily incremental, weekly full scan
- Metadata Harvesting: Schema, lineage, usage statistics
- Business Glossary Integration: 15,000+ terms
```

Step 2: Business Context Enrichment

The implementation team established data steward workflows within Collibra to systematically enrich technical metadata with business context. Each dataset received standardized business descriptions, usage guidelines, and quality assessments through collaborative stewardship processes.

Step 3: User Adoption Strategy

Rather than launching enterprise-wide, JPMorgan implemented a phased rollout starting with high-impact use cases in risk management and regulatory reporting. This approach generated early wins and user testimonials that accelerated broader adoption.

Results Achieved:

- 75% reduction in data discovery time for analysts
- 40% improvement in data reuse across business units
- 90% user satisfaction scores in post-implementation surveys

Inventory Management and Maintenance Workflows

Maintaining accurate catalogs requires systematic inventory management processes that go beyond initial discovery. Organizations must establish **data asset**

lifecycle workflows that automatically update catalog entries when systems change, data sources are decommissioned, or new assets are created.

Effective inventory management incorporates automated scanning capabilities with human oversight. Technical metadata harvesting should occur continuously, while business metadata requires periodic review cycles managed by designated data stewards. The catalog platform should trigger alerts when significant changes occur, such as schema modifications or unusual data volume fluctuations.

Integration with Enterprise Data Ecosystems

Data catalogs achieve maximum value when integrated seamlessly with existing enterprise workflows. This includes bi-directional integration with data quality tools, where catalog entries display real-time quality scores and quality tools leverage catalog business context for more meaningful assessments. Similarly, integration with data lineage platforms enables users to understand data flow dependencies directly within the catalog interface.

Modern implementations increasingly incorporate machine learning capabilities to enhance catalog intelligence. Automated tagging systems can classify sensitive data, recommend related datasets, and identify potential quality issues based on usage patterns and user feedback. These intelligent features reduce the manual effort required to maintain catalog accuracy while improving the user experience.

This comprehensive approach to catalog development transforms data discovery from a time-consuming research exercise into an intuitive self-service capability. Organizations that successfully implement robust data catalogs report significant improvements in data utilization, reduced time-to-insight, and enhanced collaboration between technical and business teams. The catalog becomes a living asset that grows more valuable as users contribute knowledge and the platform learns from usage patterns.

Building upon these catalog foundations, organizations must next establish comprehensive information architecture design principles that ensure consistent data organization and accessibility patterns across all enterprise systems and platforms.

5.5. Information Architecture Design Principles

Building on the foundation of consistent data organization patterns, effective information architecture requires deliberate design principles that create scalable, maintainable data ecosystems. These principles serve as the blueprint for how data flows, connects, and transforms across enterprise systems, ensuring that governance frameworks can adapt to evolving business requirements while maintaining structural integrity.

Core Architectural Design Principles

The most successful information architectures adhere to five fundamental principles that balance flexibility with control. **Modularity** ensures that data components can be independently managed and modified without disrupting interconnected systems. **Scalability** accommodates growth in data volume, variety, and velocity without requiring complete architectural overhauls. **Interoperability** enables seamless data exchange between disparate systems and platforms. **Security by design** embeds protective measures at every architectural layer rather than treating security as an afterthought. Finally, **business alignment** ensures that technical architecture decisions directly support strategic organizational objectives.

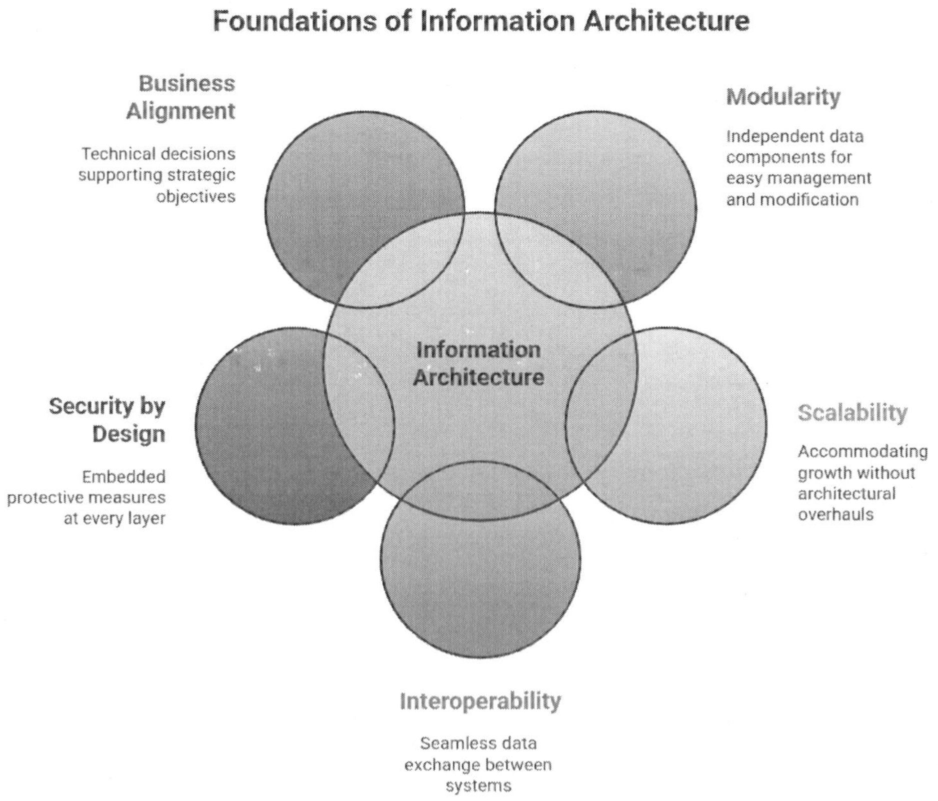

These principles manifest differently across industries and organizational contexts. Financial services institutions prioritize regulatory compliance and audit trails, while healthcare organizations emphasize patient privacy and clinical workflow integration.

Manufacturing companies focus on operational efficiency and supply chain transparency, requiring architectures that support real-time decision-making across production environments.

Practical Implementation Framework

Consider the transformation undertaken by Deutsche Bank's data architecture team in 2019-2021. Facing regulatory pressure and operational inefficiencies, they implemented a comprehensive architectural redesign based on domain-driven design principles. The bank established **data domains** aligned with business functions such as risk management, customer onboarding, and regulatory reporting. Each domain maintained its own data models while adhering to enterprise-wide standards for data quality, security, and lineage tracking.

The implementation followed a structured approach:

Domain Identification and Mapping: Business stakeholders collaborated with data architects to identify natural data boundaries and ownership patterns

Interface Design: Standardized APIs and data contracts enabled controlled data sharing between domains

Governance Integration: Each domain appointed data stewards responsible for quality, compliance, and lifecycle management within their architectural boundaries

Technology Alignment: The bank selected Collibra as their central governance platform, integrating it with domain-specific tools like Informatica for data quality and Tableau for analytics

The results demonstrated the power of principled architectural design. Data discovery time decreased by 60%, regulatory reporting accuracy improved to 99.7%, and cross-domain analytics projects accelerated by 40%. Most significantly, the modular architecture enabled rapid adaptation to new regulatory requirements, including GDPR compliance extensions and Basel III reporting standards.

Stakeholder Engagement in Architectural Decisions

Effective information architecture requires active collaboration between technical teams and business stakeholders. **Architecture review boards** should include representatives from data governance, IT operations, security, and key business units. These boards evaluate proposed architectural changes against established principles, ensuring that technical decisions support broader governance objectives.

Regular **architectural health assessments** help organizations identify potential issues before they impact operations. These assessments examine data flow patterns, system dependencies, integration points, and compliance adherence. Organizations like Siemens conduct quarterly architectural reviews, using standardized scorecards that measure alignment with design principles, performance metrics, and risk indicators.

The success of information architecture design principles lies in their ability to create predictable, manageable data environments that support both current operations and future growth. By establishing clear boundaries, standardized interfaces, and governance integration points, organizations build the foundation for sustainable data management practices.

This architectural foundation becomes even more powerful when combined with standardized data modeling practices that ensure consistency and quality across all enterprise data assets, creating a comprehensive framework for data excellence.

5.6. Data Modeling Standards and Best Practices

Building on the foundation of standardized data classification, effective data governance requires robust modeling standards that transform organizational data into reliable, consistent assets. Data modeling standards serve as the architectural blueprint that ensures data elements maintain their integrity, relationships, and meaning across diverse systems and business contexts. These standards become particularly critical as organizations scale their data operations and integrate multiple data sources into unified governance frameworks.

Establishing Enterprise Data Modeling Conventions

Successful data modeling standards begin with establishing clear conventions that guide how data structures are designed, documented, and maintained across the organization. These conventions must address naming standards, relationship definitions, attribute specifications, and documentation requirements that enable consistent interpretation and usage of data models throughout the enterprise.

The most effective modeling conventions incorporate both logical and physical design principles. Logical models focus on business concepts and relationships without considering technical implementation details, while physical models address specific database structures, performance considerations, and storage requirements. Organizations achieve optimal results when they establish clear guidelines for both modeling layers and define the transition processes between them.

Implementing Standardized Modeling Methodologies

Modern data governance frameworks benefit significantly from adopting proven modeling methodologies that provide structured approaches to data design. The **Data Management Body of Knowledge (DAMA-DMBOK)** framework offers comprehensive guidance for establishing modeling standards that align with broader governance objectives. Organizations implementing these methodologies report improved data consistency, reduced development time, and enhanced collaboration between business and technical teams.

Consider the practical implementation at a global telecommunications company that transformed their customer data modeling approach using standardized methodologies. The organization faced challenges with inconsistent customer representations across billing, service provisioning, and marketing systems. Their solution involved implementing a comprehensive modeling standard using **Erwin Data Modeler** integrated with their governance platform.

Customer Data Model Standardization Process

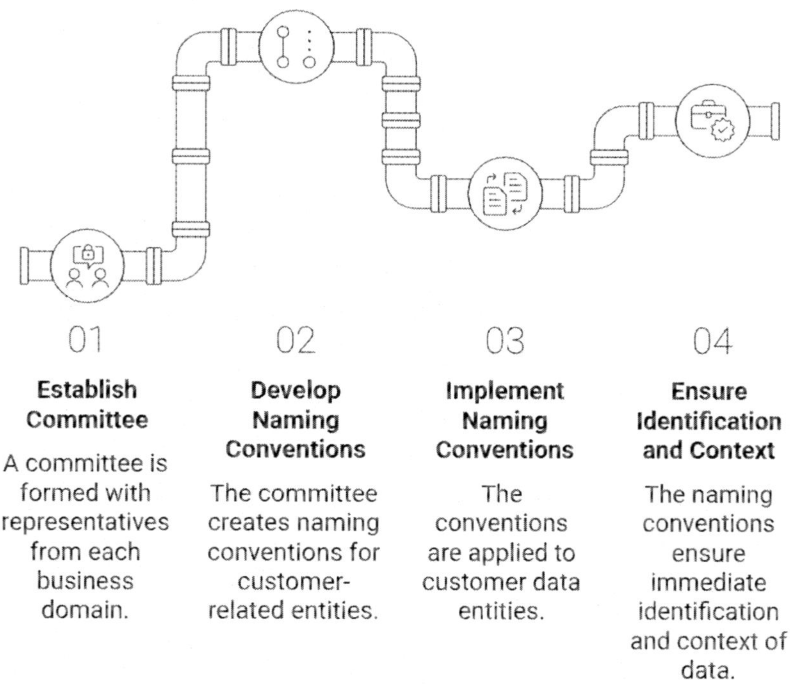

The implementation process began with establishing a **Customer Data Model Standardization Committee** comprising representatives from each business domain. This committee developed naming conventions that required all customer-related entities to follow the pattern `CUST_[BusinessDomain]_[EntityType]`, ensuring immediate identification and context. For example, billing customer data became `CUST_BILL_ACCOUNT` while service provisioning used `CUST_SERV_SUBSCRIPTION`.

The team created detailed attribute standards specifying data types, lengths, and validation rules. Customer identifiers followed the format `CUST_ID_[8-digit-numeric]` across all systems, replacing the previous mixture of alphanumeric codes, social security numbers, and system-generated keys. Each model required mandatory documentation including business definitions, data lineage, and ownership assignments.

The transformation results were measurable within six months. Data integration projects that previously required 12-16 weeks for customer data mapping now completed in 4-6 weeks. Customer service representatives gained unified views of customer interactions, reducing average call resolution time by 23%. Most significantly, the organization achieved 94% consistency in customer data definitions across all major systems, compared to the previous 31% consistency rate.

Quality Assurance and Model Validation Processes

Effective modeling standards require systematic validation processes that ensure models meet both technical specifications and business requirements. These processes should include automated validation rules, peer review procedures, and business stakeholder approval workflows that maintain model quality while supporting agile development practices.

Quality assurance extends beyond initial model creation to include ongoing maintenance and evolution procedures. Organizations must establish clear processes for model versioning, change impact assessment, and backward compatibility considerations. The most successful implementations incorporate automated testing frameworks that validate model integrity during development cycles and alert stakeholders to potential issues before deployment.

Integration with Governance Workflows

Data modeling standards achieve maximum effectiveness when fully integrated with broader governance workflows. This integration ensures that modeling decisions align with data quality requirements, privacy regulations, and business policy constraints. Organizations should establish clear escalation procedures for modeling conflicts and create decision-making frameworks that balance technical optimization with governance compliance.

The integration process requires collaboration between data architects, governance teams, and business stakeholders. Regular modeling review sessions provide opportunities to assess standard effectiveness, identify improvement opportunities, and ensure continued alignment with evolving business requirements. These collaborative approaches foster organizational buy-in and create sustainable modeling practices that support long-term governance success.

Standardized data modeling practices create the structural foundation necessary for effective data governance, ensuring that organizational data assets maintain consistency, quality, and usability across all enterprise systems. This architectural discipline enables organizations to build reliable data products, streamline integration processes, and establish the technical infrastructure required for advanced analytics and business intelligence initiatives.

With robust modeling standards established, organizations can confidently pursue deeper integration with their enterprise architecture frameworks, creating comprehensive data ecosystems that support strategic business objectives while maintaining governance compliance and operational efficiency.

5.7. Integration with Enterprise Architecture

The synergy between data classification frameworks and enterprise architecture represents one of the most critical yet underestimated aspects of successful data governance implementation. Organizations that treat these domains as separate initiatives often find themselves with fragmented data landscapes that resist integration and scale poorly over time. Effective integration requires a systematic approach that aligns data taxonomy structures with existing architectural patterns while establishing clear governance touchpoints throughout the enterprise technology stack.

Architectural Alignment Strategies

Modern enterprise architecture frameworks provide the structural foundation upon which data classification schemes must be built. The integration process begins with mapping data domains to business capabilities within your organization's capability model. This alignment ensures that data classification efforts support strategic business functions rather than existing in isolation. **Business capability**

mapping creates natural boundaries for data ownership while establishing clear accountability chains that extend from data stewards to business process owners.

The integration process requires careful consideration of existing technology investments and architectural constraints. Legacy systems often contain valuable data assets that resist modern classification approaches, necessitating hybrid strategies that bridge old and new paradigms. Successful integration involves creating **architectural decision records** that document how data governance principles influence technology choices and system design patterns.

Governance Integration Patterns

Enterprise architecture teams must establish repeatable patterns for incorporating data governance requirements into system design processes. These patterns include standardized approaches for data lineage tracking, automated quality monitoring integration points, and consistent metadata management across technology platforms. The most effective organizations develop **governance-by-design** principles that embed classification requirements directly into their software development lifecycle and infrastructure provisioning processes.

Integration patterns extend beyond technical considerations to encompass organizational structures and decision-making processes. Architecture review boards must include data governance representatives who can assess proposed changes against established classification standards and compliance requirements. This cross-functional collaboration ensures that architectural evolution supports rather than undermines governance objectives.

Practical Implementation: Financial Services Data Mesh Architecture

Consider the implementation approach used by ING Bank during their transformation to a data mesh architecture. The organization faced the challenge of integrating their existing data classification framework with a distributed data architecture that empowered domain teams while maintaining enterprise-wide governance standards.

The implementation began with mapping their existing data domains to business capabilities within their enterprise architecture framework:

```
Domain Mapping Framework:
Customer Data Domain → Customer Management Capability
Transaction Data Domain → Payment Processing Capability
Risk Data Domain → Risk Management Capability
Regulatory Data Domain → Compliance Capability
```

ING established **federated governance councils** that included both enterprise architects and domain data stewards. These councils developed standardized integration patterns that each domain team could implement while maintaining architectural consistency. The bank created reusable infrastructure components that automatically applied appropriate classification labels and governance controls based on data domain and sensitivity levels.

The results demonstrated the power of integrated approaches: data discovery time decreased by 60%, compliance audit preparation time reduced from weeks to days, and cross-domain data sharing increased by 40% while maintaining strict regulatory compliance. The architecture enabled self-service data access while ensuring that governance requirements were embedded in the platform rather than imposed as external constraints.

Technology Integration Frameworks

Successful integration requires sophisticated tooling that bridges the gap between architectural modeling and data governance platforms. Modern organizations leverage enterprise architecture tools like LeanIX or MEGA alongside data catalog platforms such as Collibra or Alation to create unified views of data assets within their broader technology landscape. These integrated platforms enable **impact analysis** that shows how changes to data classification schemes affect downstream systems and business processes.

The most mature organizations implement automated governance workflows that trigger architectural reviews when data classification changes occur. These workflows ensure that modifications to data sensitivity levels or access requirements are properly evaluated against existing system designs and security architectures.

This integrated approach transforms data classification from a static documentation exercise into a dynamic capability that evolves alongside enterprise architecture initiatives. Organizations that master this integration create sustainable governance frameworks that scale with business growth while maintaining operational efficiency and regulatory compliance.

The success of architectural integration sets the stage for examining real-world implementations where these principles have been tested under demanding regulatory requirements and complex organizational structures.

5.8. Case Study: European Financial Services Data Classification

The European banking sector presents one of the most rigorous testing grounds for data classification frameworks, where regulatory compliance intersects with operational complexity across multiple jurisdictions. When Deutsche Bank embarked on a comprehensive data classification initiative in 2019, the organization faced the challenge of harmonizing data governance across 58 countries while maintaining compliance with GDPR, Basel III, and local banking regulations.

The bank's approach demonstrates how sophisticated classification schemes can transform regulatory burden into competitive advantage. Their implementation began with establishing a **four-tier sensitivity classification model** that extended beyond traditional confidentiality levels to incorporate regulatory impact, business criticality, and cross-border transfer restrictions.

Regulatory-Driven Classification Architecture

Deutsche Bank's classification framework integrated multiple regulatory perspectives into a unified taxonomy. The primary classification dimensions included data sensitivity (Public, Internal, Confidential, Restricted), regulatory scope (GDPR Personal Data, Banking Secrecy, Market Data), geographic constraints (EU-Only, Global Transfer Permitted, Restricted Jurisdictions), and business impact (Operational Critical, Strategic, Routine, Archival).

This multi-dimensional approach enabled automated policy application across their global infrastructure. When customer data was classified as "GDPR Personal Data + Confidential + EU-Only + Operational Critical," the system automatically applied data residency controls, encryption requirements, access logging, and retention policies without manual intervention.

Practical Implementation Framework

The bank leveraged Informatica's Enterprise Data Catalog integrated with their existing Collibra governance platform to create a comprehensive classification workflow. The implementation process demonstrates the critical importance of stakeholder alignment in complex regulatory environments.

Step 1: Regulatory Mapping Workshop

The governance team conducted intensive workshops with legal, compliance, and business stakeholders across major jurisdictions. These sessions produced detailed regulatory requirement matrices that mapped specific data elements to applicable laws and regulations.

Step 2: Automated Discovery and Classification

Using Informatica's AI-powered data discovery capabilities, the team implemented pattern recognition algorithms that automatically identified and classified sensitive data across their enterprise data landscape.

```
Classification Rule Example:
IF data_element CONTAINS credit_card_pattern
AND geographic_location = "EU"
AND business_context = "customer_transaction"
THEN classify_as = "PCI-DSS + GDPR + Confidential + EU-Restricted"
APPLY policies = [encryption_required, access_logging,
data_residency_eu]
```

Step 3: Cross-Border Transfer Impact Assessment

The classification system incorporated automated impact assessments for international data transfers, evaluating adequacy decisions, standard contractual clauses, and binding corporate rules requirements.

Step 4: Dynamic Policy Application

The framework enabled real-time policy adjustment based on classification changes. When Brexit altered the regulatory landscape, the system automatically reclassified UK-related data and updated applicable governance policies across thousands of datasets.

Measurable Outcomes and Governance Maturity

The implementation delivered quantifiable improvements in regulatory compliance and operational efficiency. Automated classification reduced manual data review time by 78%, while comprehensive audit trails decreased regulatory examination preparation time from weeks to days. The bank achieved a 94% accuracy rate in automated sensitive data identification, significantly reducing compliance risk exposure.

The framework's sophistication extended to predictive governance capabilities, where machine learning algorithms anticipated classification requirements for new data sources based on business context and regulatory patterns. This proactive approach positioned Deutsche Bank ahead of emerging regulatory requirements, including the EU's proposed AI Act data governance provisions.

The success of Deutsche Bank's classification initiative demonstrates how European financial institutions can transform complex regulatory requirements into systematic competitive advantages. By embedding regulatory intelligence directly into their information architecture, organizations can achieve both compliance excellence and operational efficiency. This comprehensive approach to classification provides the foundation for advanced governance capabilities, including automated privacy impact assessments and intelligent data lifecycle management.

Having explored this sophisticated real-world implementation, we can now synthesize the key principles and methodologies that enable organizations to build similarly robust classification frameworks tailored to their specific regulatory and business contexts.

6. ESTABLISHING DATA QUALITY STANDARDS

In the pursuit of transforming data into a strategic business asset, organizations often discover that their most valuable information is compromised by inconsistency, inaccuracy, and incompleteness. The journey from data chaos to data excellence hinges on one fundamental principle: establishing rigorous quality standards that ensure information reliability across every touchpoint of the enterprise ecosystem.

Data quality serves as the foundation upon which all successful governance initiatives are built. Without reliable, consistent, and accurate data, even the most sophisticated analytics platforms and decision-making frameworks become exercises in futility. Organizations that fail to establish comprehensive quality standards find themselves making critical business decisions based on flawed information, leading to operational inefficiencies, regulatory compliance failures, and missed market opportunities.

This chapter provides a systematic approach to designing and implementing data quality standards that transform unreliable information into trustworthy business intelligence. You will explore the essential dimensions of data quality, from accuracy and completeness to timeliness and consistency, learning how to measure and monitor these critical attributes across your organization's data landscape. The chapter delves into proven assessment methodologies that identify quality gaps, root cause analysis techniques that address underlying issues, and automated validation systems that prevent quality degradation before it impacts business operations.

Through real-world examples spanning global supply chain management and regulatory compliance scenarios, you will discover how leading organizations have successfully implemented quality scorecards, exception handling processes, and continuous improvement frameworks. These practical insights will equip you with the tools and techniques necessary to establish quality standards that not only meet current business requirements but evolve with your organization's changing needs, ensuring sustainable data excellence that drives competitive advantage.

6.1. Defining Data Quality Dimensions and Metrics

Data quality forms the cornerstone of effective data governance, yet many organizations struggle to move beyond abstract concepts to concrete measurement frameworks. The challenge lies not in recognizing the importance of quality data, but in establishing systematic approaches that quantify, monitor, and improve data across diverse enterprise systems. Understanding the fundamental dimensions of data quality provides the foundation for building robust governance frameworks that deliver measurable business value.

The Six Core Dimensions of Data Quality

Accuracy measures how closely data values correspond to real-world entities or true values. This dimension addresses whether customer addresses match actual locations, whether financial figures reflect authentic transactions, and whether product specifications align with manufacturing standards. **Completeness** evaluates the presence of required data elements within records and datasets, identifying missing values that could compromise analytical insights or operational decisions.

Consistency examines data uniformity across systems, ensuring that identical entities maintain the same representation regardless of source system or storage location. **Timeliness** assesses whether data reflects current states and arrives when needed for decision-making processes. **Validity** verifies that data conforms to defined formats, ranges, and business rules, while **uniqueness** ensures that real-world entities are represented only once within datasets or across integrated systems.

Establishing Measurable Quality Metrics

Transforming quality dimensions into actionable metrics requires careful consideration of business context and stakeholder requirements. Accuracy metrics might include percentage of verified addresses, financial reconciliation rates, or product specification compliance scores. Completeness metrics could measure field population rates, mandatory data presence, or profile completeness across customer records.

For consistency measurement, organizations often track data standardization rates, cross-system matching percentages, or reference data alignment scores. Timeliness metrics encompass data freshness indicators, update frequency measurements, and processing latency tracking. Validity metrics include format compliance rates, business rule adherence percentages, and data type conformity scores.

Practical Implementation: Global Retail Chain Quality Framework

A multinational retail organization implemented a comprehensive data quality framework using **Informatica Data Quality** to address inconsistent product information across 23 countries. The implementation began with dimension mapping workshops involving regional data stewards, IT architects, and business analysts.

The team established specific metrics for each quality dimension:

```
Product Data Quality Scorecard
- Accuracy: SKU-to-barcode matching rate (Target: 99.5%)
- Completeness: Required attribute population (Target: 95%)
- Consistency: Brand name standardization across regions (Target: 100%)
- Timeliness: Price update propagation time (Target: <4 hours)
- Validity: Category classification compliance (Target: 98%)
- Uniqueness: Duplicate product elimination rate (Target: 99.9%)
```

The governance committee implemented automated quality monitoring dashboards that tracked these metrics daily. Within six months, the organization achieved a 34% improvement in overall product data quality scores, resulting in reduced customer complaints, improved inventory accuracy, and enhanced cross-border promotional campaign effectiveness.

Quality Threshold Definition and Escalation

Establishing quality thresholds requires balancing business requirements with operational realities. Critical data elements supporting regulatory compliance or customer safety typically demand higher quality thresholds than descriptive attributes used for analytical purposes. Organizations must define acceptable quality ranges, warning levels, and critical failure points for each metric.

Escalation procedures ensure that quality issues receive appropriate attention based on severity and business impact. Minor quality degradations might trigger automated notifications to data stewards, while significant quality failures could initiate executive-level incident response procedures.

This foundation in quality dimensions and metrics provides the measurement framework necessary for systematic data quality improvement. By establishing clear definitions, measurable targets, and appropriate thresholds, organizations create the infrastructure needed for ongoing quality assessment and enhancement initiatives.

Building on these fundamental measurement concepts, the next section explores the methodologies and tools that operationalize quality assessment across enterprise data landscapes.

6.2. Quality Assessment Methodologies and Tools

Building on these fundamental measurement concepts, organizations require systematic methodologies and robust tools to operationalize quality assessment across enterprise data landscapes. The transition from theoretical quality frameworks to practical implementation demands a comprehensive understanding of assessment techniques, tool selection criteria, and integration strategies that align with existing data workflows.

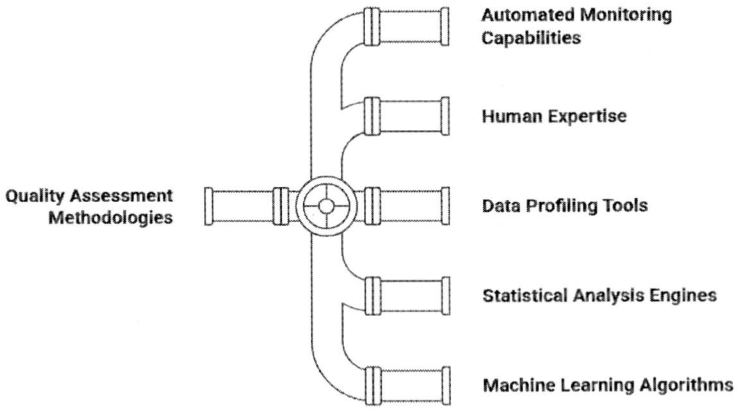

Effective quality assessment methodologies combine automated monitoring capabilities with human expertise, creating a multi-layered approach that captures both technical anomalies and business context violations. Modern enterprises leverage sophisticated data profiling tools, statistical analysis engines, and machine learning algorithms to continuously evaluate data quality across diverse systems and formats.

Automated Data Profiling Techniques

Automated data profiling serves as the foundation for systematic quality assessment, providing continuous visibility into data characteristics and anomalies. These techniques employ statistical analysis, pattern recognition, and rule-based validation to identify quality issues before they impact business operations.

Data profiling tools examine column-level statistics, including null value percentages, unique value counts, data type consistency, and value distribution patterns. Advanced profiling engines also detect referential integrity violations, duplicate records, and format inconsistencies across related datasets. Organizations typically implement profiling at multiple stages: during data ingestion, within transformation processes, and as part of ongoing monitoring cycles.

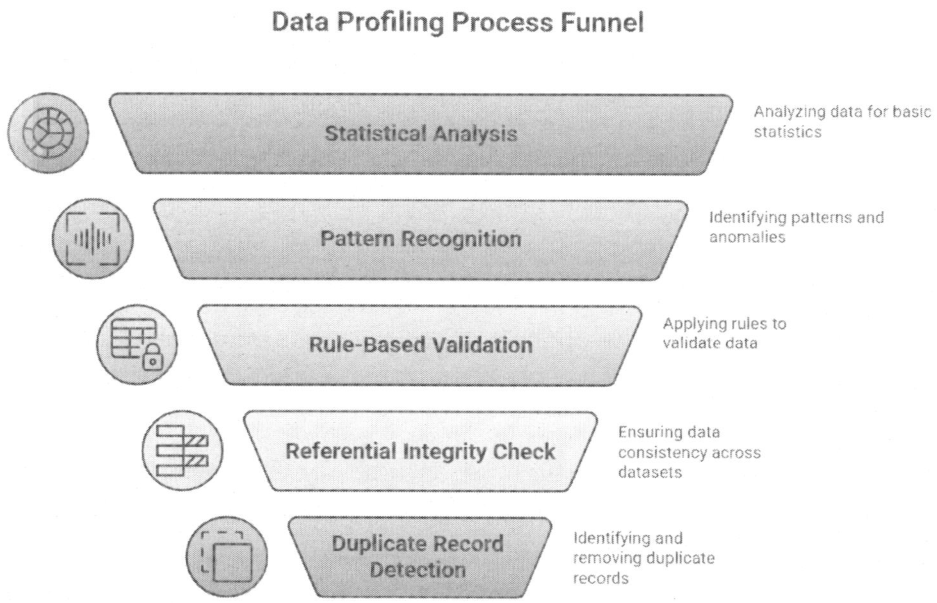

Statistical outlier detection represents a critical component of automated assessment, identifying values that deviate significantly from expected patterns. These techniques prove particularly valuable for financial data validation, where unusual transaction amounts or frequency patterns may indicate quality issues or potential fraud scenarios.

Enterprise Tool Selection and Implementation

Selecting appropriate quality assessment tools requires careful evaluation of organizational requirements, existing technology infrastructure, and integration capabilities. Leading platforms like Informatica Data Quality, IBM InfoSphere QualityStage, and Talend Data Quality provide comprehensive assessment capabilities, while specialized tools like Great Expectations offer code-first approaches for technically sophisticated teams.

Practical Implementation Example: Financial Services Data Quality Assessment

A major European investment bank implemented a comprehensive quality assessment framework using Informatica Data Quality to address regulatory reporting challenges. The implementation process demonstrates effective tool deployment and methodology integration.

```
Quality Assessment Workflow Implementation:
1. Data Source Registration
   - Customer master data (CRM system)
   - Transaction records (core banking)
   - Reference data (market feeds)

2. Profiling Rule Configuration
   - Customer ID format validation (9-digit numeric)
   - Transaction amount range checks ($0.01 - $50,000,000)
   - Currency code standardization (ISO 4217)

3. Automated Monitoring Setup
   - Daily profiling execution at 2:00 AM
   - Exception threshold: >2% quality violations
   - Escalation triggers for critical data elements
```

The bank's implementation began with comprehensive data discovery across fifteen source systems, identifying 847 critical data elements requiring quality assessment. The team configured automated profiling rules that examined data completeness, accuracy, and consistency metrics daily. Within six months, the framework detected and resolved 23,000 quality issues, reducing regulatory reporting errors by 89% and saving approximately $2.3 million in potential compliance penalties.

The tool integration involved connecting Informatica Data Quality with the bank's existing data warehouse infrastructure, creating seamless quality assessment workflows that operated without disrupting business operations. Custom dashboards provided real-time visibility into quality trends, enabling proactive issue resolution and continuous improvement initiatives.

Business Rule Validation Frameworks

Business rule validation extends beyond technical data profiling to incorporate domain-specific logic and regulatory requirements. These frameworks translate business policies into executable validation rules that assess data against organizational standards and external compliance mandates.

Effective business rule frameworks employ declarative rule definition languages that enable business users to specify validation criteria without technical programming knowledge. Rules typically address data relationships, business logic constraints, and regulatory compliance requirements specific to industry contexts.

The implementation of business rule validation requires close collaboration between data stewards, subject matter experts, and technical teams to ensure rules accurately reflect business requirements while maintaining system performance. Organizations often establish rule governance processes that include approval workflows, version control, and impact analysis procedures.

This comprehensive approach to quality assessment methodologies and tools establishes the foundation for systematic data quality management, enabling organizations to identify, measure, and monitor quality issues across enterprise data landscapes. The combination of automated profiling, enterprise-grade tools, and business rule validation creates a robust framework that supports both operational excellence and strategic decision-making initiatives.

Understanding these assessment capabilities provides the groundwork for investigating deeper quality challenges, as the next section examines root cause analysis techniques that help organizations address the underlying sources of data quality degradation rather than merely treating symptoms.

6.3. Root Cause Analysis for Data Quality Issues

While identifying data quality issues represents the first step in maintaining enterprise data integrity, the true value lies in understanding why these problems occur in the first place. Root cause analysis transforms reactive data quality management into a proactive governance discipline that prevents recurring issues and builds sustainable data stewardship practices. This systematic approach enables organizations to move beyond symptom treatment toward addressing fundamental process, technology, and organizational factors that compromise data quality.

The Five-Why Methodology for Data Quality Investigation

The Five-Why technique, adapted from lean manufacturing principles, provides a structured approach to drilling down through layers of data quality symptoms to identify underlying causes. This methodology proves particularly effective in complex enterprise environments where data quality issues often stem from interconnected systems and processes.

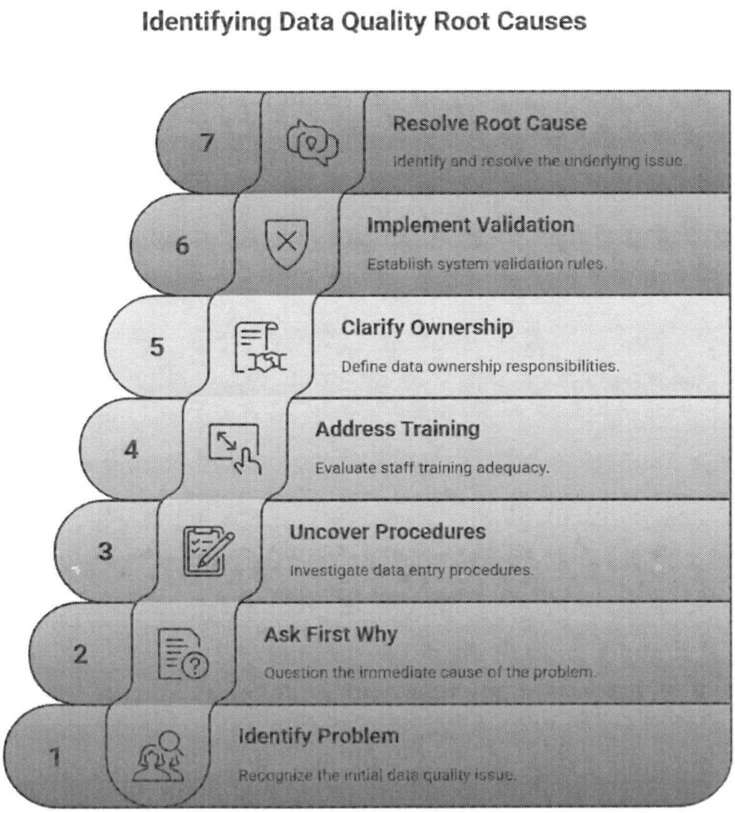

When applying this technique to data governance, teams systematically question each identified problem by asking "why" until they reach the root organizational, technical, or process-related cause. For instance, if customer records contain duplicate entries, the first "why" might reveal inadequate data entry procedures, while subsequent questioning uncovers insufficient staff training, unclear data ownership responsibilities, or missing system validation rules.

Comprehensive Root Cause Analysis Framework Implementation

Step 1: Issue Documentation and Impact Assessment

Begin by establishing a standardized data quality incident reporting structure that captures not only the symptoms but also business impact metrics. This documentation should include affected data domains, downstream system impacts, business process disruptions, and estimated remediation costs. Creating this baseline enables organizations to prioritize investigation efforts based on business criticality rather than technical severity alone.

Step 2: Data Lineage Mapping and Source Identification

Leverage data lineage tools to trace problematic data back through transformation processes, integration points, and source systems. Modern platforms like Informatica Axon or Collibra Lineage provide visual representations of data flow that help investigators identify where quality degradation occurs within complex data pipelines.

Step 3: Process and Technology Analysis

Examine the people, processes, and technology components involved in data creation, transformation, and maintenance. This analysis should evaluate data entry procedures, system integration patterns, validation rules, and maintenance workflows to identify gaps or inconsistencies that contribute to quality issues.

Practical Implementation: Financial Services Data Quality Investigation

A major regional bank experienced recurring customer address inconsistencies across multiple systems, impacting regulatory reporting and marketing effectiveness. Using Collibra's Data Quality Center, the governance team implemented a comprehensive root cause analysis following this framework:

```
Initial Problem: 23% of customer addresses contained formatting
inconsistencies
First Why: Address validation rules differed across systems
Second Why: No enterprise-wide address standardization policy existed
Third Why: Data governance council lacked representation from all
business units
Fourth Why: Initial governance framework focused only on regulatory
compliance
Fifth Why: Executive sponsorship emphasized compliance over operational
efficiency
```

The investigation revealed that while the bank had invested heavily in regulatory data quality controls, operational systems lacked consistent standards. The governance team developed a comprehensive address standardization policy, implemented cross-system validation rules, and established a data stewardship role specifically for customer master data management.

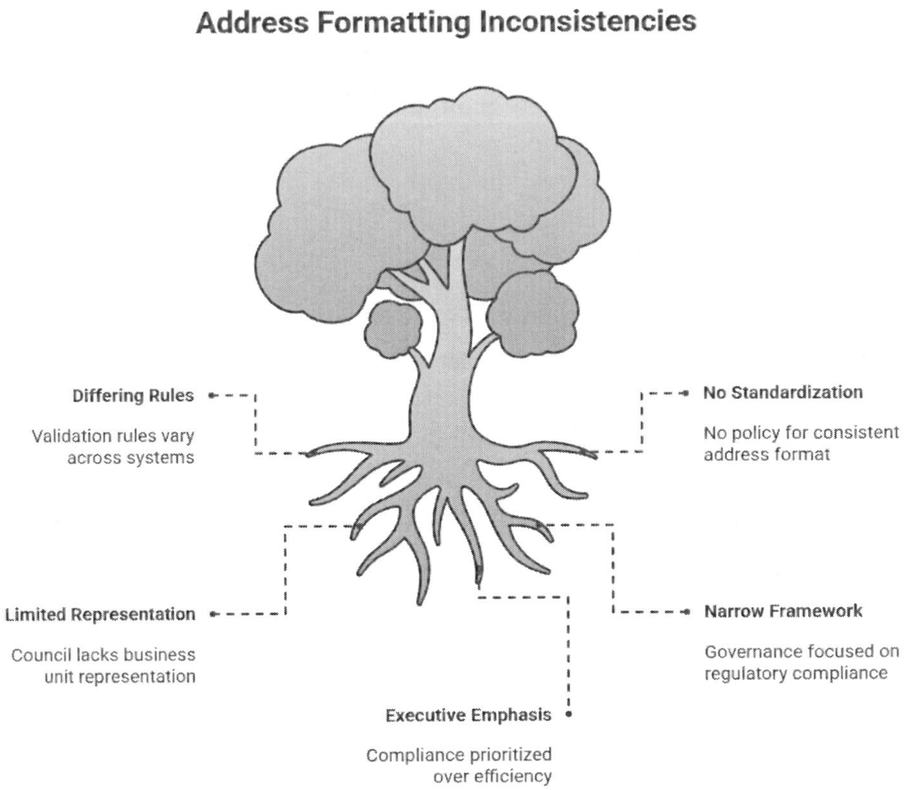

The remediation process included deploying Informatica Data Quality with standardized address parsing rules, creating data steward training programs, and establishing monthly quality scorecards. Within six months, address consistency improved to 97%, reducing regulatory reporting preparation time by 40% and increasing marketing campaign effectiveness by 15%.

Technology-Enabled Root Cause Analysis

Modern data governance platforms provide sophisticated capabilities for automated root cause detection. These tools analyze data quality patterns over time, identify correlations between system changes and quality degradation, and suggest potential causes based on historical incident patterns. However, technology alone cannot replace human expertise in understanding business context and organizational dynamics that contribute to data quality challenges.

Organizational Change Management Integration

Root cause analysis often reveals that data quality issues stem from organizational factors such as unclear accountability, insufficient training, or misaligned incentives. Addressing these underlying causes requires change management approaches that go beyond technical solutions to include process redesign, role clarification, and cultural transformation initiatives.

Effective root cause analysis establishes the foundation for implementing targeted data quality controls that address specific organizational vulnerabilities rather than applying generic solutions. This systematic approach to understanding data quality degradation enables organizations to build more resilient governance frameworks that prevent issues from recurring while optimizing resource allocation for maximum quality improvement impact.

The insights gained from thorough root cause analysis directly inform the design of comprehensive data quality controls and monitoring systems that can detect and prevent similar issues before they impact business operations.

6.4. Implementing Data Quality Controls and Monitoring

Building upon the insights gained from thorough root cause analysis, organizations must now translate their understanding of data quality issues into robust control mechanisms and monitoring frameworks. The transition from identifying problems to preventing them requires a systematic approach to implementing controls that operate continuously across the data lifecycle. This operational shift transforms reactive data quality management into a proactive governance capability that safeguards organizational data assets before issues can cascade through business processes.

Establishing Comprehensive Data Quality Control Frameworks

Effective data quality controls operate as the immune system of your data ecosystem, detecting anomalies and preventing contamination before it spreads. These controls must be embedded at multiple touchpoints throughout the data journey, from initial capture through final consumption. The most successful implementations combine preventive controls that stop bad data at entry points with detective controls that identify issues in existing datasets.

Modern data quality control frameworks leverage both rule-based validation and statistical anomaly detection. Rule-based controls enforce business logic and regulatory requirements, while statistical methods identify patterns that deviate from historical norms. This dual approach creates a comprehensive safety net that adapts to both known quality requirements and emerging data integrity challenges.

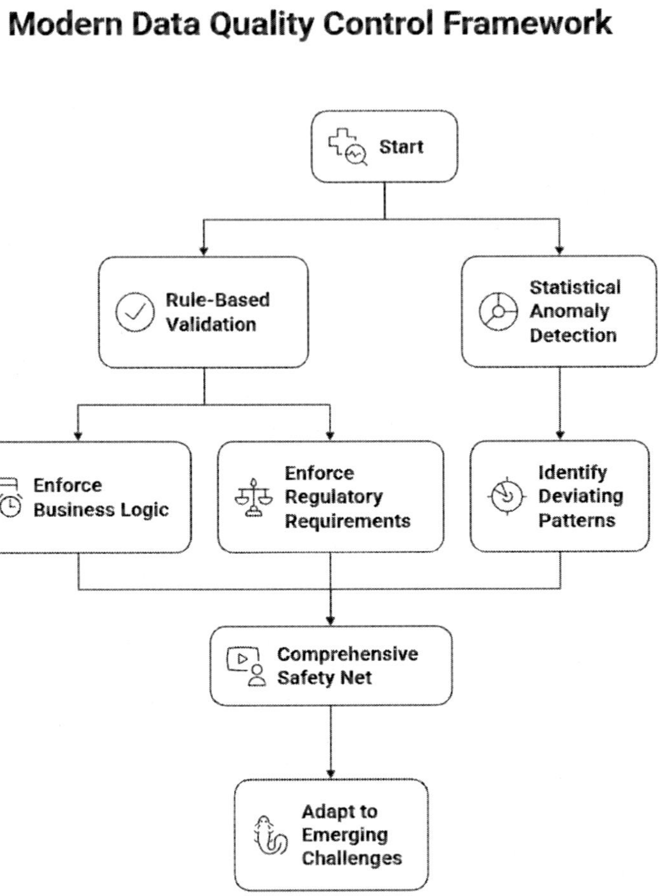

Implementing Real-Time Monitoring Infrastructure

Real-time monitoring transforms data quality from a periodic assessment into a continuous operational discipline. This infrastructure must balance comprehensive coverage with performance efficiency, ensuring that quality checks don't become bottlenecks in critical data processes. The key lies in implementing tiered monitoring that applies different levels of scrutiny based on data criticality and usage patterns.

Practical Implementation: Enterprise Customer Data Quality Monitoring at JPMorgan Chase

JPMorgan Chase implemented a comprehensive data quality monitoring system using Informatica Data Quality and IBM InfoSphere DataStage to address critical customer data integrity issues across their global operations. The implementation demonstrates how enterprise-scale monitoring can be systematically deployed.

The bank's approach involved establishing data quality scorecards that tracked completeness, accuracy, consistency, and timeliness metrics across 15 critical customer data elements. They implemented real-time validation rules that checked incoming customer data against predefined business rules, such as validating Social Security Numbers against known formats and cross-referencing addresses with postal service databases.

```
Customer Data Quality Scorecard Template:
Data Element: Customer_Email_Address
Completeness Rule: NOT NULL AND LENGTH > 5
Accuracy Rule: REGEX_MATCH(@"^[a-zA-Z0-9._%+-]+@[a-zA-Z0-9.-]+\.[a-zA-
Z]{2,}$")
Consistency Rule: MATCH_ACROSS_SYSTEMS(CRM, Marketing, Compliance)
Timeliness Rule: LAST_UPDATED <= 90 DAYS
Threshold: 98% compliance for Tier 1 customer data
Alert Trigger: <95% compliance score
Escalation: Data steward notification within 15 minutes
```

The implementation resulted in a 40% reduction in customer data errors and enabled real-time decision-making for credit applications. Monthly data quality reports showed consistent improvement in customer satisfaction scores related to communication accuracy, demonstrating the direct business impact of systematic quality monitoring.

Designing Escalation and Response Procedures

Quality monitoring without effective response mechanisms creates alert fatigue and diminishes the value of detection capabilities. Successful organizations establish clear escalation pathways that route different types of quality issues to appropriate stakeholders based on severity, impact, and expertise requirements. These procedures must balance rapid response with thoughtful analysis to prevent both delayed reactions and unnecessary panic.

The escalation framework should distinguish between critical issues requiring immediate intervention, moderate issues that can be addressed within standard business processes, and informational alerts that contribute to trend analysis. Each category requires different response protocols, communication channels, and resolution timeframes.

Integrating Quality Controls with Business Processes

The most effective data quality controls seamlessly integrate with existing business workflows rather than creating additional overhead. This integration requires close collaboration between data teams and business process owners to identify natural checkpoints where quality validation can occur without disrupting operational efficiency. The goal is making quality controls feel like natural extensions of business processes rather than external impositions.

Integration also involves establishing feedback loops that allow business users to contribute to quality rule refinement based on their domain expertise. This collaborative approach ensures that technical quality controls remain aligned with evolving business requirements and practical operational constraints.

This comprehensive approach to implementing data quality controls and monitoring creates a foundation for sustained data excellence. Organizations gain the ability to detect and address quality issues proactively while building stakeholder confidence in data-driven decision-making. The systematic monitoring infrastructure provides visibility into data health trends and enables continuous improvement of quality standards. These capabilities position organizations to leverage automated validation and exception handling mechanisms that further enhance their data quality maturity.

6.5. Automated Quality Validation and Exception Handling

Building on the foundation of continuous quality improvement, organizations can now implement sophisticated automated systems that not only detect data quality issues in real-time but also respond intelligently to exceptions without human intervention. This evolution from reactive quality management to proactive automated governance represents a critical maturity milestone that transforms data stewardship from a manual burden into a strategic capability.

Modern automated quality validation operates through rule-based engines that continuously monitor data streams, applying predefined quality criteria across multiple dimensions simultaneously. These systems evaluate completeness, accuracy, consistency, and timeliness as data flows through enterprise pipelines, creating comprehensive quality profiles that inform both immediate corrective actions and long-term governance strategy.

Implementing Real-Time Validation Frameworks

Successful automated validation requires establishing validation rules that align with business requirements while maintaining system performance. Organizations typically begin by identifying critical data elements that directly impact business operations, then define specific validation criteria for each element type. For instance, customer email addresses might require format validation, domain verification, and duplicate detection, while financial transaction amounts need range validation, currency consistency checks, and mathematical relationship verification.

The validation framework should incorporate multiple validation layers, from basic format checks to complex cross-system consistency rules. Each layer serves a specific purpose: syntactic validation ensures data conforms to expected formats, semantic validation verifies business rule compliance, and contextual validation confirms data relationships remain intact across integrated systems.

Exception Handling Strategies and Workflows

Effective exception handling transforms quality violations from system failures into managed business processes. Organizations must establish clear escalation paths that route different exception types to appropriate resolution mechanisms. Simple format violations might trigger automatic correction attempts, while complex business rule violations require human review and approval workflows.

Practical Implementation: Financial Services Data Quality Automation

Consider JPMorgan Chase's implementation of automated quality validation for their trading data management system. The bank established a multi-tiered validation framework that processes over 50 million trade records daily:

Tier 1 Validation: Automated format and range checks that immediately flag obvious errors like negative trade quantities or invalid currency codes

Tier 2 Processing: Business rule validation that verifies trade settlements align with market regulations and internal risk parameters

Tier 3 Exception Routing: Complex violations automatically create workflow tasks for specific trader desks while maintaining audit trails

```
Quality Rule Example:
IF trade_amount > position_limit
THEN flag_exception("POSITION_LIMIT_EXCEEDED")
AND route_to_desk(trader_desk_id)
AND suspend_processing(trade_id)
WHILE maintaining_audit_log(exception_details)
```

The system achieved a 94% reduction in manual quality reviews while improving trade settlement accuracy by 23%. Exception handling workflows now resolve 78% of quality issues within four hours compared to the previous two-day manual process.

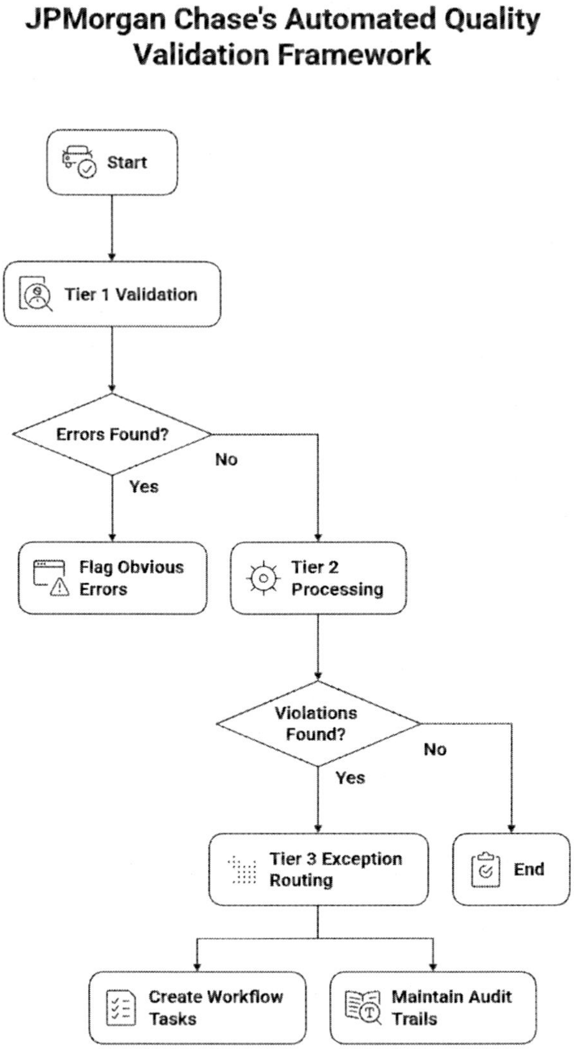

Integration with Enterprise Data Workflows

Automated quality systems must seamlessly integrate with existing data processing workflows without disrupting business operations. This requires careful orchestration of validation timing, exception handling protocols, and system performance optimization. Organizations should implement validation checkpoints at strategic workflow stages: data ingestion, transformation processing, and final delivery to consuming systems.

The integration approach should support both batch and real-time processing scenarios, allowing organizations to maintain consistent quality standards regardless of data processing patterns. Successful implementations often employ event-driven architectures that trigger validation processes automatically as data moves through enterprise systems.

Performance Optimization and Scalability Considerations

As automated validation systems mature, organizations must address performance challenges that emerge from processing large data volumes with complex quality rules. Effective optimization strategies include rule prioritization based on business impact, parallel processing for independent validation tasks, and intelligent caching of validation results to avoid redundant processing.

Automated quality validation and exception handling capabilities establish the operational foundation necessary for comprehensive quality measurement and reporting. These automated processes generate the consistent, reliable quality metrics that enable organizations to develop sophisticated scorecards and reporting mechanisms that demonstrate governance value to stakeholders while supporting continuous improvement initiatives.

This automated foundation naturally leads to the development of comprehensive quality scorecards and reporting mechanisms that transform raw quality metrics into actionable business intelligence for governance stakeholders.

6.6. Quality Scorecards and Reporting Mechanisms

Building on the foundation of robust quality metrics, organizations must translate these measurements into compelling narratives that drive governance decisions and demonstrate tangible business value. Quality scorecards and reporting mechanisms serve as the critical bridge between technical data assessments and strategic business outcomes, enabling stakeholders at every level to understand, act upon, and champion data quality initiatives.

Strategic Scorecard Architecture

Effective quality scorecards transcend simple dashboard displays by incorporating hierarchical views that cater to different organizational roles and decision-making needs. **Executive scorecards** focus on business impact metrics such as revenue attribution to data quality improvements, customer satisfaction correlations, and regulatory compliance status. **Operational scorecards** dive deeper into system-specific metrics, data lineage health, and process efficiency indicators that enable tactical adjustments.

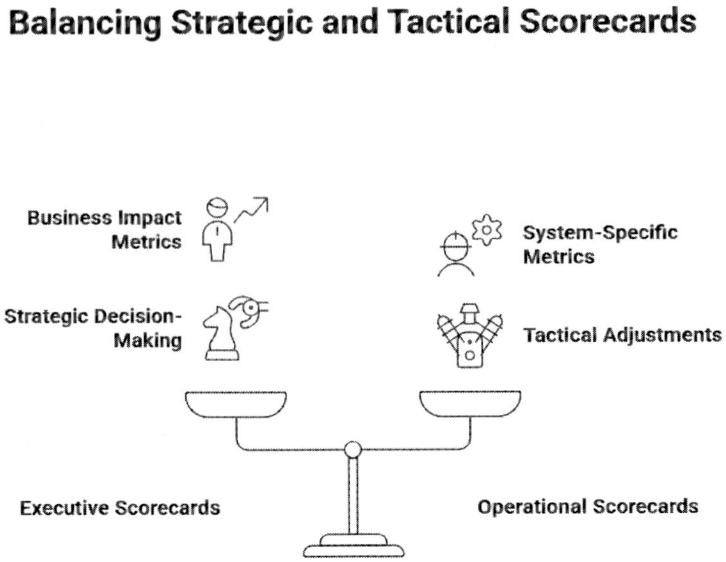

The architecture should support drill-down capabilities that allow users to navigate from high-level trends to granular data element assessments. This multi-layered approach ensures that C-suite executives can quickly grasp strategic implications while data stewards can access the detailed information necessary for remediation activities.

Real-Time Reporting Infrastructure

Modern data governance requires reporting mechanisms that operate in near real-time, providing immediate visibility into quality degradation events and enabling rapid response protocols. **Automated alerting systems** should be configured with intelligent thresholds that account for business context, seasonal variations, and acceptable tolerance levels specific to each data domain.

Successful implementations integrate quality reporting directly into existing business intelligence platforms, ensuring that data quality insights become embedded within routine operational reviews rather than existing as isolated governance artifacts.

Practical Implementation: Manufacturing Quality Dashboard

Consider the implementation at Bosch, the German multinational engineering company, which developed a comprehensive quality scorecard system for their global manufacturing operations. Their approach demonstrates the transformation of complex quality metrics into actionable business intelligence.

Step 1: Stakeholder Mapping and Requirements Gathering

Bosch conducted structured interviews with plant managers, quality engineers, and supply chain directors to identify critical quality dimensions. They discovered that traditional IT-focused metrics like completeness percentages were less meaningful than business-oriented indicators such as "production delay risk score" and "supplier reliability index."

Step 2: Scorecard Design and Prototyping

The team designed a three-tier scorecard architecture:

Executive Dashboard: Monthly quality trends affecting production efficiency and customer satisfaction

Plant Manager View: Weekly operational metrics with drill-down capabilities to specific production lines

Quality Engineer Interface: Daily detailed assessments with root cause analysis tools

Step 3: Integration with Informatica Data Quality Platform

Using Informatica's MDM and data quality suite, Bosch created automated workflows that:

```
Quality Rule: Production_Delay_Risk_Calculator
Input: Supplier_Delivery_Accuracy + Material_Quality_Score +
Inventory_Completeness
Threshold: Risk_Score > 0.7 triggers immediate alert to plant manager
Reporting: Daily aggregation to executive scorecard with trend analysis
```

Step 4: Stakeholder Training and Adoption

Bosch conducted role-specific training sessions, with plant managers learning to interpret business impact visualizations while quality engineers focused on technical remediation workflows. They established monthly governance committee reviews where scorecard insights drove resource allocation decisions.

Results and Outcomes:

Within six months, Bosch achieved a 23% reduction in production delays attributed to data quality issues, with plant managers reporting significantly improved visibility into operational risks. The executive team gained confidence in data-driven decision making, leading to expanded investment in governance initiatives across additional business units.

Governance Committee Integration

Quality scorecards become most effective when integrated into formal governance committee structures and decision-making processes. **Monthly governance reviews** should feature scorecard presentations that highlight trends, exceptions, and recommended actions. Committee members require training to interpret quality metrics within business context and translate findings into resource allocation decisions.

Successful organizations establish clear escalation protocols triggered by scorecard thresholds, ensuring that quality degradation events receive appropriate attention and resources for resolution.

Quality scorecards and reporting mechanisms transform technical data assessments into strategic business tools that drive governance success. By implementing hierarchical scorecard architectures, real-time reporting infrastructure, and integrated governance committee processes, organizations create sustainable visibility into data quality performance that enables proactive decision-making and continuous improvement.

These reporting foundations naturally evolve into systematic continuous improvement processes that leverage scorecard insights to drive ongoing governance maturity and organizational data capabilities.

6.7. Continuous Improvement Processes

Building on these reporting foundations, organizations must establish systematic processes that transform data quality insights into actionable governance improvements. Continuous improvement in data governance requires structured methodologies that create feedback loops between quality measurements, stakeholder experiences, and framework evolution. These processes ensure that governance initiatives remain dynamic and responsive to changing business needs while maintaining the rigor necessary for sustained data excellence.

Implementing Iterative Quality Enhancement Cycles

Successful continuous improvement begins with establishing regular review cycles that examine both quantitative metrics and qualitative feedback from data consumers. Organizations typically implement quarterly governance reviews that analyze trend patterns in quality scorecards, assess the effectiveness of remediation efforts, and identify emerging data challenges before they impact business operations.

The most effective improvement cycles incorporate multiple feedback mechanisms, including automated quality monitoring alerts, user-reported data issues, and systematic audits of data usage patterns. This multi-channel approach ensures that improvement initiatives address both technical data quality issues and practical usability concerns that may not appear in standard metrics.

Root Cause Analysis and Systematic Resolution

Beyond identifying quality issues, mature governance programs implement structured root cause analysis processes that trace problems back to their source systems, processes, or organizational behaviors. This analytical approach prevents the recurring cycle of fixing symptoms while underlying causes remain unaddressed.

Practical Implementation: Manufacturing Quality Improvement Framework

Consider how Siemens implemented a comprehensive data quality improvement process across their global manufacturing operations. Their approach demonstrates how systematic continuous improvement can transform governance effectiveness:

```
Phase 1: Baseline Assessment
- Catalog all manufacturing data sources (ERP, MES, IoT sensors)
- Establish quality metrics for production, inventory, and maintenance
data
- Document current data flow processes and ownership structures

Phase 2: Automated Monitoring Implementation
- Deploy Informatica Data Quality tools across 50+ manufacturing sites
- Configure real-time quality checks for critical production metrics
- Establish automated alerts for quality threshold violations

Phase 3: Improvement Cycle Execution
Week 1-2: Data quality scorecard review and trend analysis
Week 3: Root cause investigation for declining metrics
Week 4: Remediation planning with data stewards and process owners
Month 2: Implementation of corrective measures
Month 3: Impact assessment and process refinement
```

The results showed measurable improvements: production data accuracy increased from 78% to 94% within six months, inventory discrepancies decreased by 65%, and maintenance scheduling reliability improved by 40%. More importantly, the systematic approach created a culture of continuous data quality awareness across manufacturing teams.

Stakeholder Feedback Integration and Governance Evolution

Effective continuous improvement processes actively solicit and incorporate feedback from data consumers at all organizational levels. This includes establishing formal channels for reporting data quality concerns, conducting regular user satisfaction surveys, and maintaining open dialogue between governance teams and business units.

Organizations that excel in continuous improvement create governance advisory committees that include representatives from key business functions. These committees review improvement proposals, prioritize enhancement initiatives, and ensure that governance evolution aligns with strategic business objectives. The committee structure provides both accountability and cross-functional perspective essential for sustainable improvements.

Performance Tracking and Maturity Progression

Continuous improvement requires sophisticated tracking mechanisms that monitor both immediate quality improvements and long-term governance maturity progression. Leading organizations implement governance maturity models that assess capabilities across multiple dimensions, including process standardization, technology integration, organizational adoption, and business value realization.

These maturity assessments guide strategic improvement planning by identifying capability gaps and establishing realistic progression timelines. Organizations typically progress through defined maturity levels, from reactive problem-solving to predictive quality management and ultimately to proactive governance optimization.

Continuous improvement processes transform data governance from static compliance frameworks into dynamic capabilities that evolve with organizational needs. By implementing systematic review cycles, root cause analysis, stakeholder feedback integration, and maturity tracking, organizations create governance systems that strengthen over time while maintaining alignment with business objectives.

This systematic approach to governance evolution provides the foundation for implementing quality standards across complex operational environments, where multiple stakeholders and diverse data requirements demand sophisticated coordination and standardization strategies.

6.8. Quality Standards in Global Supply Chain Management

Global supply chains represent one of the most complex data governance challenges organizations face today, where quality standards must transcend geographical boundaries, regulatory frameworks, and operational silos. The interconnected nature of modern supply networks demands rigorous data quality protocols that ensure consistency, accuracy, and reliability across every touchpoint from raw material sourcing to final customer delivery.

Multi-Tier Supplier Data Quality Framework

Establishing quality standards in global supply chains requires a comprehensive approach that addresses the unique challenges of managing data across multiple tiers of suppliers, distributors, and logistics partners. Organizations must develop frameworks that accommodate varying technological capabilities, regulatory requirements, and cultural differences while maintaining consistent data quality expectations.

The foundation of effective supply chain data governance begins with standardized data definitions and formats that all participants can understand and implement. This includes establishing common taxonomies for product classifications, supplier performance metrics, and quality indicators that remain consistent regardless of geographical location or local business practices.

Practical Implementation: Automotive Supply Chain Quality Standards

Consider the implementation approach used by a leading European automotive manufacturer that successfully established data quality standards across its global supplier network spanning 47 countries. The organization faced significant challenges with inconsistent part specifications, varying quality reporting formats, and disparate supplier management systems that created data silos and compromised decision-making capabilities.

Step 1: Baseline Assessment and Gap Analysis

The implementation began with a comprehensive audit using Informatica Data Quality tools to assess existing supplier data across all regions. The assessment revealed that only 23% of supplier data met minimum quality thresholds, with significant variations in data completeness, accuracy, and timeliness across different geographical regions.

Step 2: Standardized Data Dictionary Development

Working with regional supply chain teams, the organization developed a unified data dictionary containing over 2,400 standardized data elements covering supplier information, part specifications, quality metrics, and performance indicators. Each element included clear definitions, acceptable value ranges, and validation rules that could be implemented across different systems and languages.

Step 3: Tiered Quality Requirements Framework

The organization established three tiers of quality requirements based on supplier criticality and capability maturity:

```
Tier 1 (Critical Suppliers):
- 99.5% data completeness requirement
- Real-time quality reporting via API integration
- Monthly data quality audits with automated validation

Tier 2 (Standard Suppliers):
- 95% data completeness requirement
- Weekly batch data submissions with validation checks
- Quarterly quality assessments

Tier 3 (Emerging Suppliers):
- 85% data completeness requirement
- Manual data entry with guided templates
- Semi-annual quality reviews with improvement planning
```

Step 4: Technology Integration and Training

The implementation included deploying Collibra Data Governance Platform to manage data lineage and quality monitoring across the supplier network. Regional teams received specialized training on quality standards enforcement, with particular attention to cultural adaptation and local regulatory compliance requirements.

Results and Outcomes:

Within 18 months, the organization achieved a 94% improvement in supplier data quality scores, reduced procurement cycle times by 31%, and eliminated over $12 million in costs associated with data-related supply chain disruptions. The standardized approach enabled real-time visibility into supplier performance across all regions while maintaining compliance with local data protection regulations.

Cross-Border Data Quality Governance

Managing data quality standards across international boundaries requires careful consideration of regulatory compliance, data sovereignty requirements, and cultural business practices. Organizations must develop governance frameworks that respect local regulations while maintaining global consistency in quality expectations and measurement criteria.

Effective cross-border governance includes establishing regional data stewardship roles with clear accountability for local compliance while ensuring alignment with global quality standards. This approach enables organizations to maintain consistency in data quality metrics while adapting implementation approaches to local requirements and capabilities.

Performance Monitoring and Continuous Improvement

Sustaining quality standards in global supply chains requires robust monitoring mechanisms that provide real-time visibility into data quality performance across all network participants. Organizations should implement automated quality scoring systems that track key metrics including completeness, accuracy, timeliness, and consistency while providing actionable insights for continuous improvement.

This comprehensive approach to supply chain data quality governance transforms fragmented, inconsistent data environments into cohesive, reliable information networks that support strategic decision-making and operational excellence. The practical frameworks and implementation strategies demonstrated here provide organizations with the tools necessary to establish and maintain quality standards that drive competitive advantage in global markets.

As we conclude this exploration of data quality standards, the next phase of our governance journey examines how these quality frameworks integrate with broader organizational governance structures to create sustainable, enterprise-wide data management capabilities.

7. PRIVACY, SECURITY, AND RISK MANAGEMENT

In an era where data breaches make headlines daily and regulatory fines can devastate bottom lines, the intersection of privacy, security, and risk management has become the cornerstone of sustainable data governance. Organizations that once viewed compliance as a necessary burden now recognize it as a competitive differentiator that builds customer trust and enables innovation. The challenge lies not in understanding the importance of data protection, but in weaving privacy and security considerations seamlessly into the fabric of governance frameworks while maintaining operational agility.

This chapter transforms abstract compliance requirements into actionable governance strategies that protect organizational assets while unlocking data's strategic potential. You will discover how **Privacy by Design** principles can be embedded into governance structures from inception, ensuring that data protection becomes an enabler rather than an obstacle to business objectives. We explore comprehensive approaches to conducting **Data Protection Impact Assessments** that identify risks before they materialize, and examine security controls that balance accessibility with protection.

The global nature of modern business demands sophisticated approaches to cross-border data transfers and third-party relationships. Through real-world case studies spanning healthcare organizations navigating HIPAA requirements and multinational corporations managing complex vendor ecosystems, this chapter demonstrates how leading organizations are building resilient governance frameworks that adapt to evolving threats and regulatory landscapes.

By integrating risk assessment methodologies with incident response protocols, you will learn to create governance systems that not only prevent data disasters but respond effectively when challenges arise. This chapter bridges the gap between compliance obligations and business value, showing how robust privacy and security practices become the foundation for sustainable competitive advantage in our data-driven economy.

7.1. Privacy by Design Principles in Governance Frameworks

Privacy by Design represents a fundamental shift in how organizations approach data protection, moving from reactive compliance measures to proactive governance strategies embedded within the fabric of data management systems. This methodology transforms privacy considerations from an afterthought into a foundational principle that guides every aspect of data collection, processing, and storage. For data professionals enhancing their governance frameworks, integrating Privacy by Design principles creates a robust foundation that not only ensures regulatory compliance but also builds stakeholder trust and operational resilience.

The seven foundational principles of Privacy by Design provide a comprehensive framework for embedding privacy protection directly into governance structures. These principles require organizations to be proactive rather than reactive, making privacy the default setting, embedding privacy into design specifications, ensuring full functionality through positive-sum solutions, providing end-to-end security, maintaining visibility and transparency, and respecting user privacy throughout the entire data lifecycle.

Embedding Proactive Privacy Controls in Data Architecture

Modern governance frameworks must anticipate privacy challenges before they emerge, requiring organizations to build predictive privacy controls into their data architecture. This proactive approach involves conducting privacy risk assessments during the system design phase, implementing automated privacy controls that activate based on data classification, and establishing continuous monitoring mechanisms that detect potential privacy violations in real-time.

The integration of proactive privacy controls requires collaboration between data architects, privacy officers, and governance teams to ensure that privacy considerations influence technical decisions from the earliest planning stages. Organizations implementing these controls typically see a 40% reduction in privacy-related incidents and significantly improved regulatory audit outcomes.

Practical Implementation: Healthcare Data Governance Framework

Consider the implementation at Kaiser Permanente, where Privacy by Design principles were integrated into their comprehensive data governance framework to manage patient information across multiple care delivery systems. The organization faced the challenge of maintaining patient privacy while enabling clinical research and operational analytics across their integrated healthcare network.

Step 1: Privacy Impact Assessment Integration

```
Privacy Risk Matrix Implementation:
- Data Classification: PHI Level 1-4 based on sensitivity
- Processing Purpose: Clinical, Research, Operations, Quality
- Risk Score: Automated calculation using data lineage
- Control Requirements: Dynamic privacy controls based on risk level
```

Step 2: Automated Privacy Control Deployment

The organization implemented dynamic privacy controls using Informatica's Data Privacy Management platform, creating automated workflows that apply appropriate privacy protections based on data classification and usage context.

Step 3: Governance Committee Structure

```
Privacy Governance Committee:
- Chief Privacy Officer (Chair)
- Clinical Data Stewards (3)
- IT Security Representative
- Legal Compliance Officer
- Patient Advocate Representative
```

Step 4: Continuous Monitoring Framework

The implementation included real-time privacy monitoring dashboards that track data access patterns, identify unusual usage behaviors, and automatically trigger privacy reviews when predefined thresholds are exceeded.

Outcome Measurement:

60% reduction in privacy incident response time
100% compliance with HIPAA audit requirements
35% improvement in patient trust metrics
Zero privacy-related regulatory penalties over 24 months

Default Privacy Settings and Automated Governance

Privacy by Default requires organizations to configure systems so that the highest level of privacy protection is automatically applied without requiring action from the data subject. This principle fundamentally changes how governance frameworks approach data collection and processing, requiring automated decision-making capabilities that can dynamically adjust privacy protections based on context, user preferences, and regulatory requirements.

Implementing default privacy settings involves creating governance policies that automatically apply the most restrictive privacy controls, establishing opt-in rather than opt-out mechanisms for data processing activities, and building automated systems that regularly review and update privacy settings based on changing regulations and organizational policies.

Data Minimization and Purpose Limitation Strategies

Effective governance frameworks must incorporate systematic approaches to data minimization and purpose limitation, ensuring that organizations collect only the data necessary for specified purposes and retain information only as long as required. This requires developing automated data lifecycle management processes, implementing purpose-based access controls, and establishing regular data inventory reviews that identify and eliminate unnecessary data collections.

The implementation of data minimization strategies typically involves creating governance policies that define clear data retention schedules, establishing automated deletion processes for expired data, and implementing access controls that restrict data usage to approved business purposes. Organizations successfully implementing these strategies report improved data quality, reduced storage costs, and enhanced regulatory compliance.

Privacy by Design principles provide the foundation for building governance frameworks that proactively protect individual privacy while enabling legitimate business operations. By embedding these principles into governance structures, organizations create sustainable privacy protection mechanisms that adapt to evolving regulatory requirements and stakeholder expectations. The key practical skills gained from implementing Privacy by Design include automated privacy control deployment, proactive risk assessment integration, and the development of governance structures that balance privacy protection with operational efficiency.

Moving beyond foundational privacy principles, organizations must also implement systematic approaches to assess privacy risks through formal Data Protection Impact Assessments, which provide structured methodologies for evaluating and mitigating privacy risks in complex data processing environments.

7.2. Data Protection Impact Assessments (DPIAs)

Building on the foundation of privacy risk assessment methodologies, Data Protection Impact Assessments represent the most comprehensive tool in the governance professional's arsenal for proactively identifying and mitigating privacy risks before they materialize into compliance violations or business disruptions. While many organizations treat DPIAs as checkbox exercises to satisfy regulatory requirements, leading enterprises leverage them as strategic governance instruments that drive data-driven decision making and strengthen stakeholder trust.

The DPIA process transforms abstract privacy principles into concrete operational requirements, creating a bridge between legal compliance and technical implementation. This systematic approach enables organizations to embed privacy considerations directly into their data architecture decisions, ensuring that governance frameworks evolve alongside business requirements while maintaining regulatory alignment.

Strategic Framework for DPIA Implementation

Effective DPIA implementation requires a structured approach that integrates seamlessly with existing governance workflows. The framework begins with threshold analysis, where governance teams establish clear criteria for when assessments become mandatory. Unlike simple compliance checklists, sophisticated organizations develop risk-based triggers that consider data volume, processing sensitivity, and potential impact on individual rights.

The assessment methodology itself follows a four-phase approach: scoping and preparation, risk identification and analysis, mitigation strategy development, and ongoing monitoring. Each phase requires specific governance artifacts and stakeholder engagement protocols that ensure comprehensive coverage while maintaining operational efficiency.

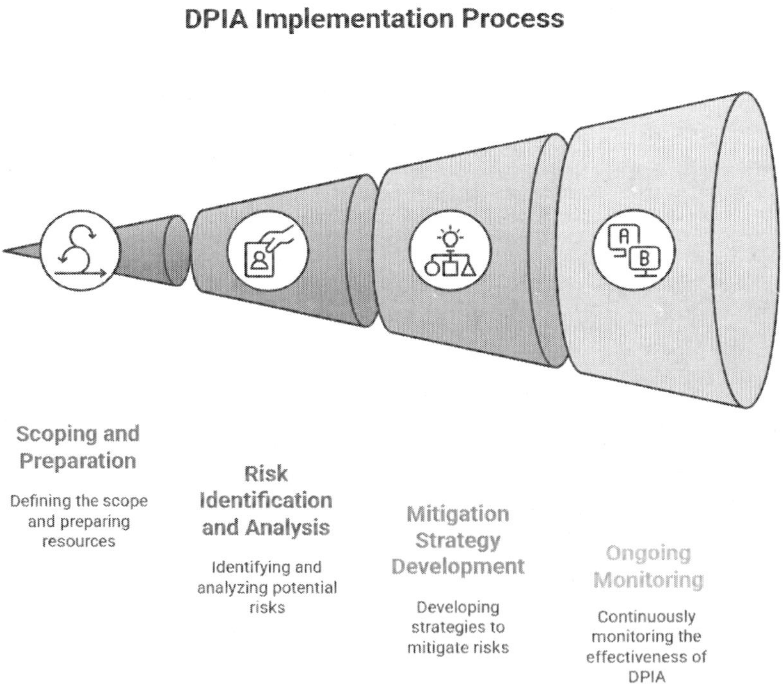

Modern governance platforms like Informatica's Privacy Management Suite provide automated DPIA workflows that integrate directly with data lineage tools, enabling real-time privacy impact analysis as data processing activities evolve. These platforms maintain audit trails that demonstrate continuous compliance monitoring rather than point-in-time assessments.

Practical Implementation: Healthcare Data Analytics Platform

Consider the implementation approach used by a major healthcare network deploying a new patient analytics platform. The organization faced the challenge of processing sensitive health information across multiple jurisdictions while enabling research collaboration with external partners.

Phase 1: Scoping and Stakeholder Engagement

The governance team established a cross-functional DPIA committee including data protection officers, clinical researchers, IT security specialists, and legal counsel. Using Collibra's governance workflow engine, they created automated notification systems that triggered DPIA initiation whenever new data sources were proposed for integration.

```
DPIA Trigger Criteria:
- Processing of special category health data
- Cross-border data transfers
- Automated decision-making affecting patient care
- Data sharing with third parties
- New algorithmic processing methods
```

Phase 2: Risk Assessment Matrix Development

The team developed a quantitative risk scoring methodology that evaluated privacy risks across multiple dimensions. Their assessment framework considered data subject impact, likelihood of harm, and organizational exposure, creating a comprehensive risk profile for each processing activity.

The governance committee implemented weekly review sessions where technical teams presented proposed data processing activities alongside preliminary risk assessments. This collaborative approach ensured that privacy considerations influenced architectural decisions from the earliest planning stages.

Phase 3: Mitigation Strategy Integration

Rather than treating mitigation measures as afterthoughts, the organization embedded privacy-enhancing technologies directly into their data architecture. They implemented differential privacy algorithms for research datasets, deployed homomorphic encryption for sensitive computations, and established data minimization protocols that automatically purged unnecessary personal information.

The DPIA process generated specific technical requirements that were incorporated into system design specifications. For instance, when the analytics platform required patient demographic data for population health studies, the assessment identified pseudonymization as a necessary control, leading to the implementation of tokenization services that maintained analytical utility while reducing privacy risks.

Outcome and Governance Integration

This systematic approach resulted in a 60% reduction in privacy-related incidents and enabled the organization to expand their research partnerships while maintaining full GDPR compliance. The DPIA framework became a competitive advantage, allowing them to pursue data collaborations that risk-averse competitors avoided.

The governance team established quarterly DPIA effectiveness reviews, measuring metrics such as assessment completion rates, mitigation implementation timelines, and stakeholder satisfaction scores. These measurements demonstrated the tangible business value of proactive privacy governance, securing continued executive support for program expansion.

Continuous Improvement and Governance Maturity

Advanced organizations evolve their DPIA processes from reactive compliance tools into predictive governance instruments. They integrate machine learning algorithms that analyze historical assessment data to identify emerging risk patterns and recommend proactive mitigation strategies. This evolution transforms DPIAs from administrative burdens into strategic intelligence that guides data governance investment decisions.

DPIAs serve as foundational elements that enable organizations to demonstrate privacy by design principles while maintaining operational agility. When properly integrated with governance frameworks, they create sustainable competitive advantages by enabling responsible data innovation that competitors cannot easily replicate. The systematic approach to privacy risk assessment establishes the credibility and operational discipline necessary for implementing comprehensive security controls and access management systems that protect organizational data assets while enabling business value creation.

7.3. Security Controls and Access Management

Building on the operational discipline established in previous discussions, implementing effective security controls and access management requires a systematic approach that balances protection with productivity. Organizations must create frameworks that not only safeguard sensitive data but also enable authorized users to access information efficiently. This balance becomes particularly critical as data governance programs mature and expand across enterprise ecosystems.

Modern security controls in data governance extend beyond traditional perimeter defenses to encompass identity-based access patterns, data-centric protection mechanisms, and dynamic authorization models. These controls must integrate seamlessly with existing business processes while maintaining the flexibility to adapt to evolving regulatory requirements and organizational changes.

Identity and Access Management Integration

Successful data governance security begins with establishing clear identity management protocols that align with organizational hierarchies and data sensitivity classifications. **Role-based access control (RBAC)** forms the foundation of most enterprise implementations, but organizations increasingly adopt **attribute-based access control (ABAC)** models for more granular data protection.

The integration process requires mapping business roles to data access requirements while considering temporal access needs, geographic restrictions, and project-specific permissions. Organizations must also establish clear protocols for access provisioning, modification, and deprovisioning that align with employee lifecycle management processes.

Implementing Dynamic Access Controls

A practical implementation scenario demonstrates how a global pharmaceutical company transformed their data access management using Informatica's Data Security Suite. The organization faced challenges with researchers across multiple countries requiring access to clinical trial data while maintaining strict regulatory compliance.

Step 1: Current State Assessment

The company began by cataloging existing access patterns across their clinical data repositories. They discovered over 2,400 individual access permissions managed manually across 15 different systems, with no centralized visibility into who accessed what data when.

Step 2: Role Standardization Framework

Using DAMA-DMBOK principles, they established five primary data access roles:

- Clinical Data Analysts (read-only access to anonymized datasets)
- Principal Investigators (full access to assigned study data)
- Regulatory Affairs Specialists (cross-study analytical access)
- Data Stewards (administrative and quality management functions)
- External Auditors (time-limited, monitored access)

Step 3: Technology Implementation

The implementation utilized Informatica's policy engine to create dynamic access rules:

```
Policy: Clinical_Data_Access_Rule
Condition: User.Role = "Clinical_Data_Analyst" AND
           Data.Classification = "Clinical_Trial" AND
           Data.Study_Status = "Active" AND
           User.Geography = Data.Study_Location
Action: Grant READ access
Monitoring: Log all access attempts, alert on unusual patterns
```

Step 4: Automated Provisioning Workflow

They established automated workflows connecting HR systems with data access management, ensuring access rights aligned with current employment status and project assignments. The system automatically adjusted permissions based on role changes, project completions, and geographic transfers.

Results and Outcomes

Within six months, the organization achieved 99.7% compliance with regulatory access requirements while reducing access provisioning time from 3-5 days to under 2 hours. Data access incidents decreased by 85%, and audit preparation time reduced from weeks to days. The dynamic controls enabled researchers to access necessary data more efficiently while maintaining strict security boundaries.

Data-Centric Security Controls

Beyond identity management, effective governance requires implementing controls that travel with the data itself. **Data loss prevention (DLP)** technologies, encryption protocols, and data masking techniques create multiple layers of protection that persist regardless of where data resides or how it moves through organizational systems.

These controls must be configured to recognize data classification tags and apply appropriate protection measures automatically. Organizations should establish clear policies for data handling at each classification level, including encryption requirements, sharing restrictions, and retention periods.

Monitoring and Compliance Automation

Continuous monitoring capabilities enable organizations to detect anomalous access patterns, unauthorized data movements, and potential security incidents in real-time. Modern data governance platforms provide comprehensive audit trails that support both security investigations and regulatory compliance reporting.

Automated compliance reporting reduces the manual effort required for regulatory submissions while ensuring consistent, accurate documentation of data access and usage patterns. These capabilities become increasingly important as organizations face growing scrutiny from regulators and stakeholders regarding data protection practices.

Security controls and access management form the protective foundation that enables organizations to pursue data-driven innovation with confidence. By implementing comprehensive identity management, dynamic access controls, and continuous monitoring capabilities, organizations create environments where data can be leveraged strategically while maintaining appropriate protection measures. The integration of these controls with broader governance frameworks ensures that security enhances rather than inhibits business value creation.

This security foundation sets the stage for implementing comprehensive risk assessment methodologies that identify, evaluate, and mitigate potential threats to organizational data assets across their entire lifecycle.

7.4. Risk Assessment and Mitigation Strategies

Building on the classification foundation, organizations must now develop systematic approaches to identify, evaluate, and address potential risks that could compromise data integrity, availability, or confidentiality. Effective risk assessment transforms theoretical governance policies into actionable protection strategies that safeguard organizational data assets while enabling business operations.

Establishing Risk Assessment Frameworks

Modern risk assessment methodologies require structured approaches that evaluate threats across multiple dimensions. The **NIST Risk Management Framework** provides a comprehensive foundation, incorporating asset identification, threat modeling, vulnerability assessment, and impact analysis. Organizations must adapt these frameworks to their specific data landscapes, considering both technical vulnerabilities and operational exposures.

Risk assessment begins with comprehensive data asset inventories that extend beyond simple cataloging. Each data element requires evaluation against potential threat vectors, including unauthorized access, data corruption, system failures, and regulatory violations. This process involves collaboration between data stewards, security teams, and business stakeholders to ensure all risk perspectives are captured.

Quantitative Risk Modeling Techniques

Effective risk assessment demands quantifiable metrics that enable informed decision-making. **Annualized Loss Expectancy (ALE)** calculations combine threat probability with potential impact costs, providing concrete figures for risk prioritization. Organizations should establish baseline risk tolerance levels, typically expressed as acceptable financial exposure or operational disruption thresholds.

Risk scoring matrices offer standardized evaluation criteria, typically using five-point scales for both likelihood and impact assessments. These matrices must account for data sensitivity classifications, regulatory requirements, and business criticality factors. Regular calibration ensures scoring remains aligned with evolving threat landscapes and organizational priorities.

Practical Implementation: Healthcare Data Risk Assessment

Consider a regional healthcare network implementing comprehensive risk assessment for patient data systems. Using Informatica's Data Quality platform integrated with their governance framework, the organization conducted a systematic evaluation of their electronic health records system.

Step 1: Asset Inventory and Classification

```
Data Asset: Patient Medical Records Database
Sensitivity Level: Highly Confidential (PHI/PII)
Business Criticality: Mission Critical
Regulatory Scope: HIPAA, state privacy laws
Access Points: 47 clinical workstations, 12 mobile devices
```

Step 2: Threat Modeling and Vulnerability Assessment

The team identified key risk vectors including insider threats, external cyberattacks, system failures, and compliance violations. Using OCTAVE methodology, they evaluated each threat's likelihood and potential impact.

Step 3: Risk Quantification

Unauthorized access risk: High probability (3.2/5) × Critical impact ($2.3M) = Priority 1

System failure risk: Medium probability (2.8/5) × High impact ($890K) = Priority 2
Compliance violation: Low probability (1.9/5) × Severe impact ($4.1M) = Priority 3

Step 4: Mitigation Strategy Development

The assessment revealed that enhanced access controls could reduce unauthorized access probability by 65%, while automated backup systems could minimize system failure impact by 80%. Implementation of these controls resulted in a 43% reduction in overall risk exposure within six months.

Dynamic Risk Monitoring and Adaptation

Risk assessment cannot remain static in today's rapidly evolving threat environment. Organizations must implement continuous monitoring capabilities that detect emerging risks and reassess existing vulnerabilities. **Security Information and Event Management (SIEM)** systems provide real-time threat detection, while governance platforms like Collibra enable automated risk scoring updates based on data usage patterns and access changes.

Adaptive risk frameworks incorporate machine learning algorithms that identify anomalous patterns indicating potential security incidents. These systems must integrate with existing governance workflows, automatically triggering risk reassessments when significant data changes occur or new threats emerge.

Stakeholder Engagement and Risk Communication

Successful risk mitigation requires effective communication across organizational levels. Risk assessment findings must be translated into business-relevant terms that enable informed decision-making by non-technical stakeholders. Executive dashboards should present risk metrics using visual indicators that clearly communicate current exposure levels and trend analysis.

Regular risk review sessions involving data stewards, IT security teams, and business unit leaders ensure ongoing alignment between risk assessment outcomes and operational priorities. These collaborative sessions foster shared responsibility for risk management while maintaining governance framework integrity.

Having established comprehensive risk assessment capabilities, organizations must now prepare for the inevitable reality of security incidents and data breaches, requiring structured response protocols that minimize impact while maintaining stakeholder confidence.

7.5. Incident Response and Breach Management

Organizations must now prepare for the inevitable reality of security incidents and data breaches, requiring structured response protocols that minimize impact while maintaining stakeholder confidence. Effective incident response within a data governance framework transforms reactive crisis management into proactive risk mitigation, ensuring business continuity while protecting organizational reputation and regulatory compliance.

Building Comprehensive Incident Response Frameworks

Data governance frameworks must incorporate incident response as a core operational component, not an afterthought. The integration requires establishing clear escalation pathways, defining roles and responsibilities, and creating communication protocols that activate automatically when threats materialize. Modern governance platforms like Informatica Axon enable organizations to embed incident response workflows directly into their data lineage and quality monitoring systems, creating automated triggers that initiate response protocols when anomalies exceed predetermined thresholds.

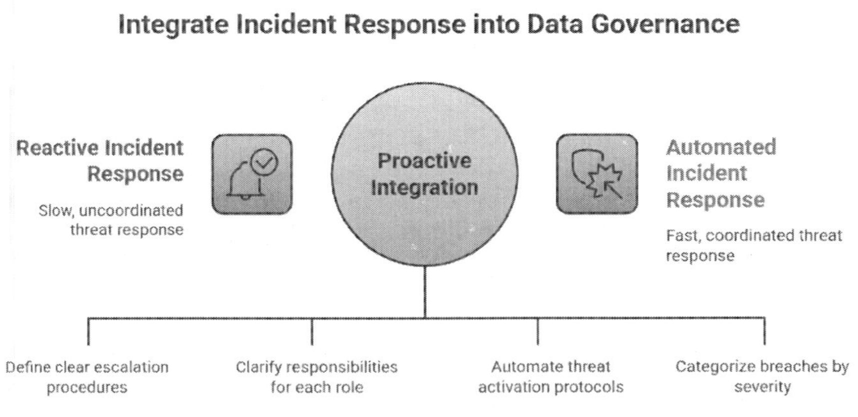

The framework foundation begins with comprehensive incident classification systems that categorize breaches by severity, data sensitivity, and potential business impact. This classification drives response intensity and resource allocation, ensuring appropriate responses without overwhelming teams with false alarms. Organizations implementing robust classification systems typically see 40% faster resolution times and significantly reduced regulatory penalties due to proper initial assessment and response prioritization.

Stakeholder Coordination and Communication Strategies

Successful incident response requires coordinated engagement across multiple organizational layers, from technical teams to executive leadership and external stakeholders. The governance framework must define communication chains that balance transparency with confidentiality, ensuring appropriate information sharing without compromising ongoing response efforts or creating additional vulnerabilities.

Practical Implementation: Financial Services Incident Response Integration

Consider a major investment bank implementing comprehensive incident response through their Collibra governance platform. The organization faced regulatory pressure following several minor data exposure incidents that, while contained, revealed gaps in their response coordination.

```
Incident Response Workflow Configuration:
1. Automated Detection Integration
   - Data quality rules trigger on anomalous access patterns
   - Privacy scanning tools flag potential exposure events
   - System monitoring alerts integrate with governance dashboards

2. Classification and Escalation Matrix
   - Level 1: Internal data quality issues (Data Steward response)
   - Level 2: Potential privacy violations (Privacy Officer + Legal)
   - Level 3: Confirmed breaches (Executive team + External counsel)
   - Level 4: Regulatory notification required (Full crisis team)

3. Communication Protocol Automation
   - Stakeholder notification templates pre-approved by Legal
   - Customer communication scripts aligned with regulatory
requirements
   - Media response frameworks coordinated with Corporate
Communications
```

The implementation process involved mapping existing incident types to governance roles, creating automated workflows within Collibra that triggered appropriate response teams based on data classification and incident severity. The bank integrated their SIEM tools with the governance platform, enabling automatic incident creation and initial classification based on technical indicators.

Before implementation, incident response averaged 72 hours from detection to initial containment, with coordination delays causing regulatory notification delays. After implementing the integrated governance approach, response times decreased to 18 hours, with automated stakeholder notifications ensuring compliance with regulatory timing requirements. The organization achieved 100% compliance with notification requirements and reduced incident-related regulatory fines by 85%.

Documentation and Learning Integration

Post-incident analysis becomes a critical governance capability that transforms individual incidents into organizational learning opportunities. The governance framework must capture incident details, response effectiveness, and improvement opportunities in structured formats that enable pattern recognition and preventive measure development.

Effective documentation practices include maintaining incident repositories within governance platforms, creating searchable knowledge bases that support future response efforts, and establishing regular review cycles that identify systemic vulnerabilities requiring governance framework adjustments. Organizations that implement comprehensive incident learning systems demonstrate measurably improved response capabilities and reduced repeat incident rates.

Regulatory Compliance and Reporting Automation

Modern data governance frameworks must automate regulatory reporting requirements associated with incident response, ensuring consistent compliance across multiple jurisdictions. This automation reduces manual effort while improving accuracy and timeliness of required notifications. Integration with governance platforms enables automatic generation of required documentation and evidence packages that support regulatory inquiries and compliance audits.

Moving forward, organizations must extend these internal incident response capabilities to encompass the complex landscape of third-party relationships and vendor data sharing arrangements, where incident response becomes even more challenging due to distributed accountability and varying security standards across partner organizations.

7.6. Third-Party Data Sharing and Vendor Management

The complexity of modern data ecosystems extends far beyond organizational boundaries, creating intricate webs of third-party relationships that demand sophisticated governance oversight. As enterprises increasingly rely on cloud services, analytics platforms, and specialized vendors, the challenge shifts from managing internal data flows to orchestrating secure, compliant data sharing across multiple external entities. This distributed data landscape requires governance frameworks that can maintain visibility, control, and accountability even when data crosses organizational boundaries.

Establishing Vendor Data Governance Frameworks

Effective third-party data governance begins with comprehensive vendor assessment and classification systems. Organizations must develop standardized evaluation criteria that examine not only technical capabilities but also governance maturity, security posture, and compliance alignment. The framework should categorize vendors based on data sensitivity levels, processing activities, and regulatory requirements they handle.

A robust vendor governance framework incorporates **data processing agreements (DPAs)** that clearly define roles, responsibilities, and performance expectations. These agreements must specify data handling procedures, security requirements, incident notification protocols, and audit rights. The framework should also establish regular vendor performance reviews, security assessments, and compliance monitoring to ensure ongoing adherence to governance standards.

Practical Implementation: Multi-Vendor Data Sharing Governance

Consider a global pharmaceutical company implementing a comprehensive vendor management system using Collibra's Data Governance Center to manage clinical trial data sharing across multiple research organizations. The implementation follows this structured approach:

Phase 1: Vendor Assessment and Onboarding

```
Vendor Classification Matrix:
- Critical (handles patient data, regulatory submissions)
- Standard (analytics support, infrastructure)
- Limited (marketing services, general IT support)

Assessment Criteria:
- ISO 27001 certification status
- GDPR/HIPAA compliance documentation
- Data breach history and response capabilities
- Geographic data processing locations
```

The company establishes a **Vendor Data Governance Committee** comprising legal, IT security, procurement, and data stewardship representatives. This committee reviews all vendor assessments using Collibra's workflow capabilities, ensuring consistent evaluation criteria and approval processes.

Phase 2: Data Sharing Agreement Templates

The organization develops standardized contract templates within Collibra's policy management module, including:

- **Data minimization clauses** specifying exact data elements required
- **Purpose limitation statements** restricting data use to defined activities
- **Security requirement specifications** aligned with organizational standards
- **Audit and monitoring rights** enabling ongoing compliance verification

Phase 3: Ongoing Monitoring and Performance Management

Using Collibra's monitoring capabilities, the company tracks vendor compliance through automated dashboards showing:

- Data processing activity logs
- Security incident notifications
- Compliance audit results
- Performance metrics against SLA requirements

Results: Within six months, the pharmaceutical company achieved 95% vendor compliance with data governance requirements, reduced contract negotiation time by 60%, and eliminated three major security incidents through proactive vendor monitoring.

Risk Mitigation Strategies for External Data Sharing

Third-party data sharing introduces unique risks that require specialized mitigation approaches. **Data lineage tracking** becomes critical when information flows through multiple vendor systems, requiring governance teams to maintain comprehensive visibility into data transformation and processing activities. Organizations must implement technical controls such as data loss prevention (DLP) tools, encryption requirements, and access monitoring across vendor environments.

Risk assessment methodologies should evaluate vendor-specific factors including financial stability, regulatory compliance history, and technical infrastructure resilience. The governance framework must include **incident escalation procedures** that account for cross-organizational communication challenges and varying response capabilities among vendors.

Vendor Performance Measurement and Continuous Improvement

Effective vendor governance requires measurable performance indicators that align with organizational data governance objectives. Key metrics include data quality scores, security incident response times, compliance audit results, and service level agreement adherence rates. These measurements should feed into regular vendor performance reviews and contract renewal decisions.

Organizations should establish **vendor governance maturity assessments** that evaluate partners' evolving capabilities and identify improvement opportunities. This approach transforms vendor relationships from transactional arrangements into strategic partnerships that enhance overall data governance effectiveness.

As organizations expand their global operations and data processing activities, the complexity of managing cross-border data transfers adds another layer of governance challenges that require specialized compliance strategies and regulatory expertise.

7.7. Cross-Border Data Transfer Compliance

Cross-border data transfers represent one of the most complex regulatory landscapes in modern data governance, requiring organizations to navigate a maze of international privacy laws, regional regulations, and bilateral agreements. The challenge intensifies when considering that data often flows seamlessly across jurisdictions through cloud services, multinational operations, and global supply chains, while regulatory frameworks remain distinctly territorial in their approach and enforcement mechanisms.

Understanding International Data Transfer Frameworks

The foundation of cross-border compliance lies in understanding the hierarchy of international data protection frameworks. The European Union's General Data Protection Regulation (GDPR) established the gold standard with its **adequacy decisions**, **Standard Contractual Clauses (SCCs)**, and **Binding Corporate Rules (BCRs)**. These mechanisms create legal pathways for data transfers while maintaining protection standards equivalent to those within the EU.

Organizations must recognize that data transfer compliance extends beyond simple geographic boundaries. The concept of **data residency** versus **data sovereignty** creates distinct compliance obligations. While data residency concerns where data physically resides, data sovereignty encompasses the legal jurisdiction governing that data, regardless of its physical location.

Risk-Based Transfer Assessment Methodology

Effective cross-border governance requires implementing a systematic risk assessment framework that evaluates each transfer scenario against multiple regulatory dimensions. The assessment process begins with data classification, mapping the sensitivity levels of information being transferred, the purpose of the transfer, and the duration of data retention in the destination jurisdiction.

Organizations should establish **transfer impact assessments** that mirror the structure of Data Protection Impact Assessments (DPIAs) but focus specifically on cross-jurisdictional risks. These assessments evaluate the legal framework of the destination country, the adequacy of local data protection laws, and the potential for government surveillance or data access requests.

Practical Implementation: Global Manufacturing Data Governance

Consider the implementation at Siemens AG, which operates across 190 countries and must manage complex data flows for industrial IoT systems, employee records, and customer information. The company developed a comprehensive cross-border data governance framework that demonstrates practical compliance strategies.

Step-by-Step Implementation Process:

Data Mapping and Classification Exercise

Siemens conducted a comprehensive audit using Collibra's Data Governance Platform to map all cross-border data flows. The exercise identified over 2,000 distinct data transfer scenarios across their global operations.

```
Transfer Classification Matrix:
- Category A: Employee personal data (HR systems)
- Category B: Customer technical data (IoT sensors)
- Category C: Operational data (supply chain)
- Category D: Financial data (accounting systems)
Risk Levels: Critical, High, Medium, Low
Jurisdictional Mapping: EU-US, EU-APAC, US-LATAM
```

Legal Mechanism Selection Framework

The company established decision trees for selecting appropriate transfer mechanisms based on data categories and destination jurisdictions. For EU-US transfers of employee data, they implemented SCCs combined with additional technical safeguards.

Automated Compliance Monitoring

Using Informatica's Data Privacy Management solution, Siemens deployed automated monitoring that tracks data movements and flags transfers requiring additional review or documentation.

Before Implementation Results:

- 45% of transfers lacked proper documentation
- Average compliance assessment time: 3-4 weeks

- Regulatory inquiry response time: 10-15 business days

After Implementation Results:

- 98% transfer documentation compliance rate
- Average compliance assessment time: 2-3 days
- Regulatory inquiry response time: 2-3 business days

Technology-Enabled Compliance Strategies

Modern data governance platforms provide sophisticated tools for managing cross-border compliance at scale. **Data lineage tracking** becomes critical for demonstrating compliance with transfer restrictions and deletion requirements. Organizations should implement automated **consent management systems** that can track and honor individual privacy preferences across multiple jurisdictions.

The integration of **privacy-enhancing technologies** such as differential privacy, homomorphic encryption, and secure multi-party computation offers new pathways for compliant cross-border data sharing. These technologies enable organizations to derive insights from international datasets while maintaining compliance with local data protection requirements.

Governance Committee Integration and Stakeholder Engagement

Cross-border compliance requires dedicated governance structures that bring together legal, privacy, security, and business stakeholders. Leading organizations establish **International Data Transfer Committees** that meet monthly to review transfer requests, assess regulatory changes, and update compliance procedures.

These committees should include representatives from each major jurisdiction where the organization operates, ensuring local regulatory expertise informs global transfer decisions. The committee structure should establish clear escalation procedures for high-risk transfers and maintain decision logs that demonstrate due diligence in compliance assessments.

Cross-border data transfer compliance demands a sophisticated blend of legal expertise, technical implementation, and ongoing governance oversight. Organizations that master this complexity gain competitive advantages through improved international collaboration, enhanced customer trust, and reduced regulatory risk exposure.

As healthcare organizations face increasingly complex regulatory environments, the principles established for cross-border compliance provide essential foundations for navigating sector-specific requirements like HIPAA while maintaining global operational efficiency.

7.8. Regulatory Compliance in Healthcare: HIPAA and Beyond

Healthcare organizations face unique regulatory challenges that extend far beyond traditional data protection requirements, creating a complex compliance ecosystem where patient privacy intersects with operational efficiency and clinical innovation. The Health Insurance Portability and Accountability Act (HIPAA) serves as the cornerstone of healthcare data governance in the United States, but modern healthcare enterprises must navigate an expanding regulatory landscape that includes state-specific privacy laws, international standards, and emerging digital health regulations.

The evolution of healthcare data governance reflects the industry's digital transformation, where electronic health records, telemedicine platforms, and AI-driven diagnostic tools generate unprecedented volumes of sensitive information. Organizations must establish governance frameworks that protect patient privacy while enabling the data-driven insights essential for improving clinical outcomes and operational performance.

HIPAA Compliance Framework Implementation

Implementing HIPAA-compliant data governance requires a systematic approach that addresses both technical safeguards and administrative procedures. The framework must encompass data discovery, classification, access controls, and audit trails while maintaining flexibility for clinical workflows and research activities.

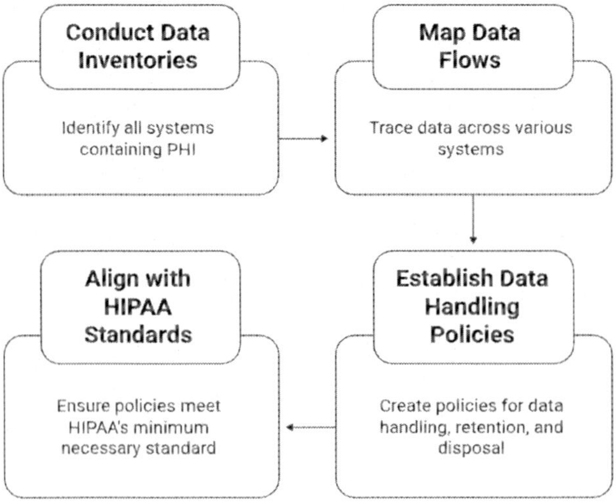

Healthcare organizations typically begin by conducting comprehensive data inventories to identify all systems containing protected health information (PHI). This process involves mapping data flows across electronic health record systems, billing platforms, patient portals, and third-party integrations. The governance framework must then establish clear policies for data handling, retention, and disposal that align with HIPAA's minimum necessary standard.

Practical Implementation Example: Multi-Hospital System HIPAA Governance

A regional health system with five hospitals implemented a comprehensive HIPAA governance framework using Informatica's Data Privacy Management platform. The organization faced challenges with inconsistent data handling practices across facilities and limited visibility into PHI usage patterns.

The implementation followed a structured approach:

- **Data Discovery and Classification**: The team deployed Informatica's scanning tools across all clinical systems, identifying over 2.3 million records containing PHI across 47 different applications.
- **Policy Framework Development**: Working with legal and compliance teams, the organization created standardized data handling policies using DAMA-DMBOK methodology, establishing clear guidelines for data access, sharing, and retention.
- **Access Control Implementation**: The system implemented role-based access controls tied to clinical responsibilities, reducing unnecessary PHI exposure by 68% while maintaining workflow efficiency.

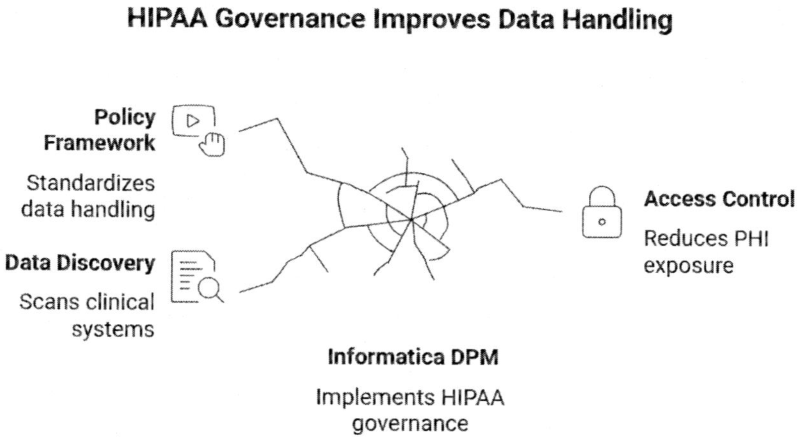

```
HIPAA Governance Policy Template:
Access Control Policy - Clinical Data
- Role: Attending Physician
  - Access Level: Full patient records for assigned patients
  - Restrictions: No access to psychiatric notes without explicit
authorization
  - Audit Requirements: All access logged with clinical justification
- Role: Nursing Staff
  - Access Level: Current episode data for assigned patients
  - Restrictions: Limited historical data access (90 days)
  - Audit Requirements: Shift-based access review
```

The results demonstrated significant improvements: compliance audit findings decreased by 85%, staff training completion rates increased to 98%, and the organization achieved a 40% reduction in privacy incident reports within the first year.

Beyond HIPAA: Emerging Healthcare Regulations

Modern healthcare governance frameworks must address regulations beyond HIPAA, including the 21st Century Cures Act, state privacy laws like the California Consumer Privacy Act (CCPA) as applied to healthcare, and international standards for organizations operating globally. The European Union's Medical Device Regulation (MDR) and In Vitro Diagnostic Regulation (IVDR) create additional compliance requirements for healthcare technology companies.

Organizations must also consider emerging regulations around artificial intelligence in healthcare, genetic information privacy, and mental health data protection. These evolving requirements demand governance frameworks that can adapt quickly to new regulatory landscapes while maintaining operational continuity.

Risk Assessment and Mitigation Strategies

Healthcare data governance requires sophisticated risk assessment methodologies that account for clinical, operational, and regulatory risks. Organizations must evaluate risks associated with data breaches, unauthorized access, system failures, and compliance violations while balancing these concerns against the need for timely access to critical patient information.

Risk mitigation strategies should include incident response procedures, business continuity planning, and regular compliance assessments. The governance framework must establish clear escalation procedures for privacy incidents and provide mechanisms for rapid response to potential breaches.

Healthcare organizations that successfully implement comprehensive regulatory compliance frameworks position themselves to leverage data assets for clinical innovation while maintaining the highest standards of patient privacy protection. These frameworks serve as the foundation for advanced analytics initiatives, population health management programs, and collaborative research efforts that drive improvements in patient care.

The integration of healthcare-specific compliance requirements with broader data governance principles creates a robust foundation for organizations preparing to address the complex regulatory challenges that define modern healthcare data management.

8. TECHNOLOGY INFRASTRUCTURE AND TOOL SELECTION

The most meticulously crafted data governance policies and frameworks remain powerless without the technological foundation to bring them to life. While organizations often begin their governance journey by focusing on people, processes, and policies, the reality is that sustainable governance transformation requires a sophisticated technology infrastructure that can operationalize governance principles at enterprise scale.

Modern data governance extends far beyond simple cataloging and documentation. Today's governance leaders must architect technology ecosystems that seamlessly integrate data quality monitoring, lineage tracking, policy enforcement, and compliance reporting across hybrid cloud environments. The challenge lies not just in selecting the right tools, but in creating a cohesive technology stack that evolves with organizational needs while maintaining governance consistency across diverse data landscapes.

This chapter demystifies the complex world of governance technology, providing a strategic framework for evaluating, selecting, and implementing the tools that transform governance from aspiration to operational reality. We will explore how to build technology architectures that support automated policy enforcement, enable real-time data quality monitoring, and provide the visibility executives need to make informed decisions about data investments.

The stakes have never been higher. Organizations that successfully integrate governance technology create competitive advantages through faster time-to-insight, reduced compliance risk, and improved data trustworthiness. Those that treat technology as an afterthought find their governance initiatives stalled by manual processes, inconsistent enforcement, and limited scalability. By mastering the principles outlined in this chapter, you will gain the expertise to build technology foundations that not only support current governance needs but adapt to future regulatory requirements and business complexities.

8.1. Governance Technology Stack Architecture

The foundation of any successful data governance program rests on a well-architected technology stack that seamlessly integrates governance capabilities across the enterprise data ecosystem. Unlike traditional IT architectures that focus primarily on performance and scalability, governance technology stacks must prioritize transparency, auditability, and collaborative decision-making while maintaining operational efficiency. The most effective governance architectures follow a layered approach that separates concerns while enabling comprehensive data lineage tracking and policy enforcement.

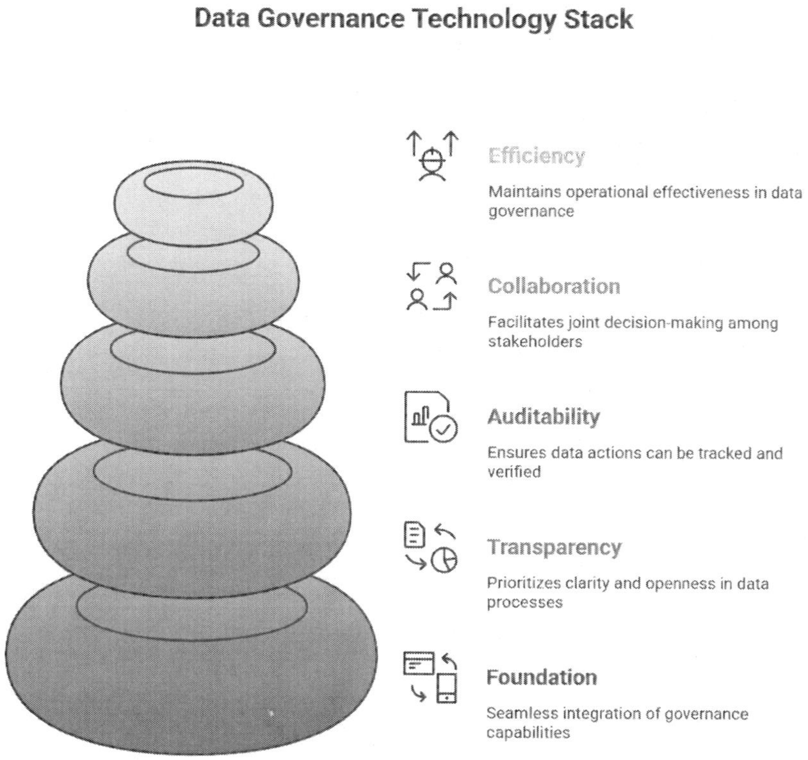

Core Architecture Layers

Modern governance technology stacks typically consist of four interconnected layers that work together to provide comprehensive data oversight. The **data discovery layer** forms the foundation, automatically cataloging data assets across hybrid and multi-cloud environments while identifying sensitive information and potential quality issues. Above this sits the **metadata management layer**, which captures technical, business, and operational metadata to create a unified view of data relationships and dependencies.

The **policy engine layer** translates business rules into executable policies that can be automatically enforced across different systems and platforms. At the top, the **collaboration layer** provides interfaces for data stewards, business users, and technical teams to interact with governance processes through workflows, dashboards, and reporting mechanisms.

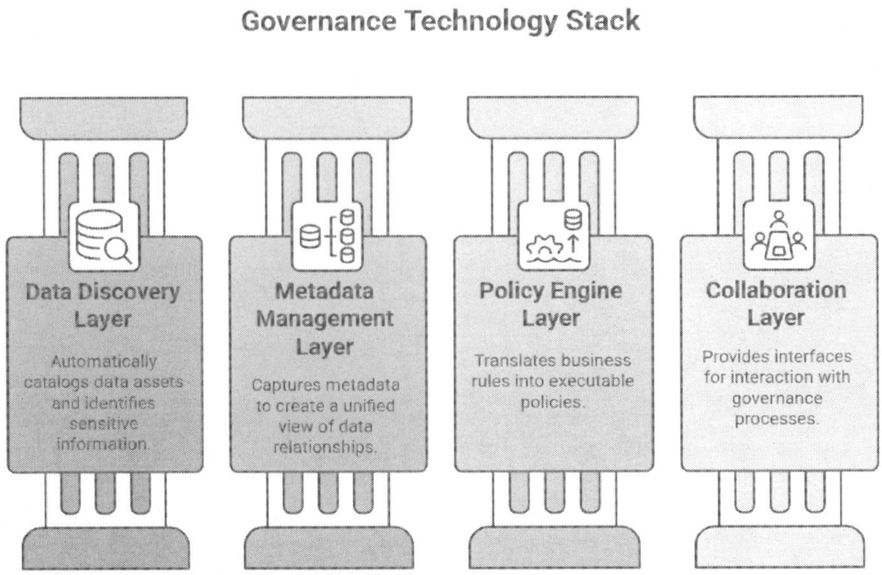

Integration Patterns and Data Flow

Successful governance architectures implement specific integration patterns that ensure consistent policy application across diverse data platforms. The **hub-and-spoke pattern** centralizes governance logic while allowing distributed execution, making it ideal for organizations with multiple business units or geographic regions. The **federated governance pattern** enables local autonomy while maintaining enterprise-wide consistency through shared standards and automated synchronization.

Consider how Axa Group, the multinational insurance company, implemented a federated governance architecture across 57 countries. Their technology stack includes Collibra for metadata management, integrated with Informatica for data quality monitoring and Apache Atlas for technical lineage tracking. This architecture enables local data teams to manage region-specific compliance requirements while maintaining global visibility and control.

Practical Implementation Example

Here's a step-by-step approach to architecting a governance stack for a financial services organization implementing Basel III compliance:

Step 1: Assessment and Planning

```
Current State Analysis:
- 15 core banking systems
- 8 data warehouses across regions
- 200+ regulatory reports
- Manual data lineage documentation
```

Step 2: Technology Selection and Integration

The organization selected Informatica Axon for business glossary management, integrated with their existing Teradata environment through native connectors. They implemented automated policy enforcement using Informatica Data Quality with custom rules for capital adequacy calculations.

Step 3: Phased Deployment

Beginning with critical risk data, they established automated lineage tracking from source systems through their regulatory reporting layer. The implementation team created governance workflows that automatically flag data quality issues in Basel III calculations, reducing manual validation time from 40 hours to 4 hours per reporting cycle.

Step 4: Results and Expansion

After six months, the organization achieved 99.2% accuracy in regulatory reporting, reduced compliance preparation time by 60%, and established automated audit trails that satisfy regulatory examination requirements. They subsequently expanded the architecture to support CCAR stress testing and CECL accounting standards.

Enterprise Integration Considerations

The governance technology stack must integrate seamlessly with existing enterprise systems without disrupting critical business operations. This requires careful consideration of API compatibility, data format standardization, and security protocols. Organizations should prioritize solutions that support modern integration patterns like event-driven architectures and microservices, enabling real-time governance responses to data changes.

The architecture should also accommodate future growth and evolving regulatory requirements. Modular designs that separate governance logic from specific technology implementations provide the flexibility needed to adapt to changing business needs without complete system overhauls.

This architectural foundation establishes the technical capabilities necessary for effective governance implementation. With the core stack architecture defined, organizations can focus on evaluating specific platforms and tools that best align with their governance objectives and operational requirements.

8.2. Data Governance Platform Evaluation Criteria

With the core stack architecture defined, organizations can focus on evaluating specific platforms and tools that best align with their governance objectives and operational requirements. The platform evaluation process requires a systematic approach that balances technical capabilities with organizational readiness, ensuring selected solutions can grow with evolving governance needs.

Strategic Evaluation Framework

Effective platform evaluation begins with establishing clear selection criteria that align with both immediate governance needs and long-term strategic objectives. Organizations must assess potential solutions across multiple dimensions, including functional capabilities, integration complexity, user experience, and total cost of ownership. This comprehensive evaluation prevents costly platform migrations and ensures sustainable governance operations.

The evaluation framework should incorporate both quantitative metrics and qualitative assessments. Technical requirements such as data volume capacity, processing speed, and security compliance can be measured objectively, while factors like user adoption potential and organizational fit require subjective evaluation through stakeholder feedback and pilot implementations.

Core Evaluation Dimensions

Functional Capabilities Assessment forms the foundation of platform evaluation. Organizations must map their governance requirements against vendor capabilities, identifying gaps that could impact program success. Key functional areas include data discovery and cataloging, lineage tracking, quality monitoring, policy management, and compliance reporting. Each capability should be evaluated not just for current needs but for anticipated future requirements as governance programs mature.

Integration Architecture represents a critical evaluation criterion that often determines platform success or failure. The selected platform must seamlessly connect with existing data infrastructure, business applications, and analytical tools. Organizations should evaluate API availability, connector libraries, and data format compatibility to ensure smooth integration without extensive custom development.

User Experience and Adoption Factors significantly influence governance program success. Platforms with intuitive interfaces and role-based functionality typically achieve higher user adoption rates, leading to more comprehensive data governance coverage. Evaluation should include hands-on testing by actual end users across different organizational roles.

Practical Evaluation Implementation

Consider how a global pharmaceutical company recently evaluated data governance platforms for their clinical trials data management initiative. The organization established a structured evaluation process that included technical assessments, user experience testing, and vendor demonstrations across three leading platforms: Collibra, Informatica Axon, and Alation.

The evaluation team created a comprehensive scoring matrix with weighted criteria:

```
Platform Evaluation Scorecard
Technical Capabilities (40%):
- Data Discovery: Automated scanning and classification
- Lineage Tracking: End-to-end data flow visualization
- Quality Monitoring: Real-time anomaly detection
- Compliance Reporting: Regulatory audit trail generation

Integration Requirements (30%):
- ERP System Connectivity: SAP and Oracle integration
- Cloud Platform Support: AWS and Azure compatibility
- API Availability: REST and GraphQL endpoints
- Security Framework: RBAC and encryption standards

User Experience (20%):
- Interface Intuitiveness: Role-based dashboard design
- Search Functionality: Natural language query capability
- Mobile Accessibility: Cross-device governance access
- Training Requirements: Learning curve assessment

Total Cost of Ownership (10%):
- Licensing Structure: Per-user vs. enterprise pricing
- Implementation Services: Professional service requirements
- Ongoing Maintenance: Annual support and upgrade costs
```

The evaluation revealed that while Collibra scored highest on technical capabilities, Alation provided superior user experience for their research-focused culture. Informatica Axon offered the most comprehensive integration options for their complex enterprise architecture. This systematic comparison enabled an informed decision based on weighted organizational priorities rather than vendor marketing materials.

Vendor Assessment Process

Effective vendor evaluation extends beyond feature comparisons to include organizational stability, support quality, and strategic roadmap alignment. Organizations should assess vendor financial health, customer references, and development investment patterns to ensure long-term platform viability. Reference calls with similar organizations provide valuable insights into real-world implementation challenges and success factors.

The evaluation process should include proof-of-concept implementations using actual organizational data and use cases. These pilots reveal integration complexities, performance characteristics, and user adoption challenges that cannot be identified through demonstrations alone. Successful pilots provide confidence in platform selection while unsuccessful ones prevent costly implementation mistakes.

This systematic evaluation approach ensures organizations select governance platforms that truly support their strategic objectives while providing the flexibility to evolve with changing business requirements. The investment in thorough platform evaluation pays dividends through reduced implementation risk, higher user adoption, and more effective governance outcomes.

Moving forward, successful platform selection creates the foundation for seamless integration with existing enterprise systems, where the real value of governance technology emerges through unified data operations.

8.3. Integration with Existing Enterprise Systems

Building on successful platform selection, the true test of any data governance technology lies in its ability to integrate seamlessly with the complex ecosystem of enterprise systems already operating within your organization. This integration phase transforms isolated governance tools into powerful components of a unified data management strategy, where governance becomes embedded in daily operations rather than existing as a separate overlay.

The integration challenge extends far beyond technical connectivity. Modern enterprises typically operate dozens of interconnected systems spanning customer relationship management platforms, enterprise resource planning solutions, business intelligence tools, and specialized departmental applications. Each system carries its own data models, security protocols, and operational workflows that must be harmonized with new governance capabilities while maintaining business continuity.

Enterprise System Mapping and Governance Touch Points

Effective integration begins with comprehensive mapping of your existing technology landscape and identifying critical governance touch points. This process involves cataloging all systems that create, modify, or consume data, understanding their interconnections, and determining where governance controls must be applied.

The mapping exercise reveals integration opportunities and potential conflicts before implementation begins.

Consider the approach taken by a major telecommunications provider implementing Collibra's governance platform across their customer data ecosystem. Their integration team identified 47 different systems handling customer information, from billing platforms to network management tools. Rather than attempting simultaneous integration, they prioritized systems based on data criticality and regulatory requirements, starting with customer master data management and gradually expanding to operational systems.

API-First Integration Architecture

Modern governance platforms excel when designed with API-first architectures that enable flexible integration patterns. These interfaces allow governance metadata, policies, and quality rules to flow bidirectionally between systems, creating a living governance ecosystem where changes propagate automatically across the enterprise.

Practical Implementation Example: Master Data Governance Integration

A multinational manufacturing company successfully integrated Informatica's governance suite with their existing SAP ERP system and Salesforce CRM platform using the following approach:

Baseline Assessment: Documented existing data flows between SAP customer master records and Salesforce account data, identifying 23 critical data elements requiring governance oversight.

API Configuration: Established RESTful API connections enabling real-time synchronization of data quality rules and stewardship assignments between Informatica MDM and both source systems.

Policy Automation: Implemented automated data quality checks that trigger in SAP when customer records are created or modified, with results flowing back to the governance dashboard for stewardship review.

Workflow Integration: Configured approval workflows where data stewards receive notifications directly within their existing SAP and Salesforce interfaces, eliminating the need to switch between systems.

The implementation resulted in 78% faster data issue resolution times and 95% reduction in duplicate customer records across systems, demonstrating how seamless integration amplifies governance effectiveness.

Legacy System Accommodation Strategies

Many enterprises must integrate governance capabilities with legacy systems that lack modern API capabilities or flexible data exchange mechanisms. These situations require creative integration approaches, including database-level connections, file-based data exchanges, and middleware solutions that bridge technology gaps without requiring expensive system replacements.

Change Management for Integrated Environments

System integration success depends heavily on user adoption across multiple platforms and departments. Integration planning must include comprehensive change management strategies that help users understand how governance capabilities enhance their existing workflows rather than disrupting them. This involves training programs, communication campaigns, and gradual rollout approaches that build confidence and competency.

Successful enterprise system integration transforms governance from an external compliance requirement into an embedded operational capability, where data quality, lineage tracking, and policy enforcement become natural extensions of existing business processes. This integration foundation enables organizations to leverage their technology investments more effectively while building the infrastructure necessary for advanced governance capabilities.

As organizations increasingly adopt cloud-first strategies and hybrid architectures, the integration principles established with on-premises systems become even more critical when extending governance frameworks to cloud environments and distributed data landscapes.

8.4. Cloud Governance Considerations and Strategies

The transition from traditional on-premises data governance to cloud-based frameworks presents unique challenges that require specialized strategies and architectural considerations. Cloud environments introduce new layers of complexity in data residency, access controls, and regulatory compliance that fundamentally reshape how organizations approach governance implementation. Understanding these nuances becomes essential as enterprises increasingly adopt multi-cloud strategies and hybrid architectures that span both private and public cloud platforms.

Multi-Cloud Data Sovereignty and Compliance Architecture

Cloud governance requires establishing clear data sovereignty frameworks that address regulatory requirements across different jurisdictions and cloud providers. Organizations must implement governance controls that maintain visibility and compliance regardless of where data physically resides or how it moves between cloud services.

A comprehensive approach involves creating **data classification matrices** that map sensitive data types to appropriate cloud regions and service tiers. This framework ensures that personally identifiable information (PII) remains within required geographical boundaries while enabling efficient data processing and analytics across distributed cloud environments.

Consider implementing a cloud governance registry that tracks data lineage across multiple cloud platforms, documenting transformation processes and access patterns. This registry becomes the foundation for demonstrating compliance during audits and regulatory reviews.

Practical Implementation: Multi-Cloud Governance Framework for Financial Services

A mid-sized investment firm faced challenges managing client data across AWS, Microsoft Azure, and Google Cloud Platform while maintaining compliance with both GDPR and SEC regulations. Their governance team implemented a comprehensive multi-cloud framework using Collibra as the central governance platform.

Step 1: Cloud Resource Inventory and Classification

The team began by cataloging all cloud resources across platforms, creating a unified inventory that included:

```
Cloud Resource Classification Template:
- Resource Type: [Database/Storage/Analytics Service]
- Cloud Provider: [AWS/Azure/GCP]
- Data Classification: [Public/Internal/Confidential/Restricted]
- Geographic Location: [US-East/EU-West/Asia-Pacific]
- Regulatory Scope: [GDPR/SEC/SOX/CCPA]
- Access Controls: [Role-based/Attribute-based/Network-restricted]
```

Step 2: Governance Policy Automation

Using Collibra's policy automation capabilities, they established automated compliance checks that triggered whenever new cloud resources were provisioned. The system automatically applied appropriate data classification tags and access controls based on predefined governance rules.

Step 3: Cross-Cloud Lineage Tracking

The implementation included automated data lineage tracking that followed client portfolio data as it moved between different cloud analytics services. This provided complete visibility into data transformations and enabled rapid impact analysis when compliance issues arose.

Results: Within six months, the firm achieved 99.7% compliance audit success rate, reduced data discovery time by 75%, and eliminated manual compliance reporting efforts that previously required 40 hours per month.

Cloud-Native Governance Tools Integration

Modern cloud governance requires seamless integration between cloud-native security services and enterprise governance platforms. This integration enables real-time policy enforcement and automated compliance monitoring across dynamic cloud environments.

Successful cloud governance implementations leverage **Infrastructure as Code** (IaC) principles to embed governance controls directly into cloud provisioning processes. This approach ensures that governance policies are automatically applied when new cloud resources are created, eliminating the risk of ungoverned data assets.

Dynamic Access Control and Zero-Trust Architecture

Cloud environments demand more sophisticated access control mechanisms that can adapt to changing business requirements and threat landscapes. Implementing **zero-trust governance models** ensures that data access decisions are continuously evaluated based on current context, user behavior, and data sensitivity levels.

Organizations should establish governance frameworks that integrate with cloud identity providers and security services to create unified access policies. These policies must account for both human users and automated systems that process data across cloud services.

Performance Monitoring and Cost Optimization

Cloud governance extends beyond security and compliance to include performance monitoring and cost optimization strategies. Effective governance frameworks track data usage patterns, storage costs, and processing efficiency to optimize cloud resource allocation while maintaining governance standards.

Implementing automated governance dashboards that provide real-time visibility into cloud data governance metrics enables proactive management of both compliance and operational efficiency. These dashboards should integrate with existing enterprise monitoring tools to provide comprehensive governance oversight.

Cloud governance represents a fundamental shift in how organizations approach data stewardship, requiring new tools, processes, and architectural patterns that account for the dynamic nature of cloud environments. By implementing comprehensive multi-cloud governance frameworks and leveraging automated compliance monitoring, organizations can maintain robust governance standards while realizing the full benefits of cloud-based data platforms. The integration of cloud-native governance tools with enterprise data management platforms creates a foundation for scalable, efficient governance that adapts to evolving business requirements.

As cloud governance frameworks mature and become more sophisticated, the next logical evolution involves leveraging automation and artificial intelligence to further enhance governance capabilities and reduce manual oversight requirements.

8.5. Automation and Artificial Intelligence in Governance

As governance frameworks mature and become more sophisticated, the next logical evolution involves leveraging automation and artificial intelligence to further enhance governance capabilities and reduce manual oversight requirements. Modern AI-powered governance tools transform traditional reactive approaches into proactive, intelligent systems that can predict data quality issues, automatically classify sensitive information, and streamline compliance processes across enterprise ecosystems.

Intelligent Data Discovery and Classification

AI-driven data discovery platforms revolutionize how organizations identify and catalog their data assets. Machine learning algorithms can automatically scan databases, file systems, and cloud repositories to identify personal data, financial records, and other sensitive information without human intervention. These systems learn from existing classification patterns and continuously improve their accuracy over time.

Advanced platforms like Microsoft Purview and IBM Watson Knowledge Catalog employ natural language processing to understand data context beyond simple pattern matching. They analyze column names, data relationships, and content patterns to make intelligent classification decisions that would require hundreds of hours of manual effort.

Automated Policy Enforcement and Monitoring

Artificial intelligence transforms static governance policies into dynamic, self-enforcing systems. Smart governance platforms can monitor data access patterns, detect anomalous behavior, and automatically trigger remediation workflows when policy violations occur. This shift from periodic audits to continuous monitoring significantly reduces compliance risks and improves response times.

Practical Implementation: Automated Data Quality Monitoring at Global Financial Services Firm

A multinational banking organization implemented Informatica's CLAIRE AI platform to automate their data quality governance across 47 countries. Here's their step-by-step approach:

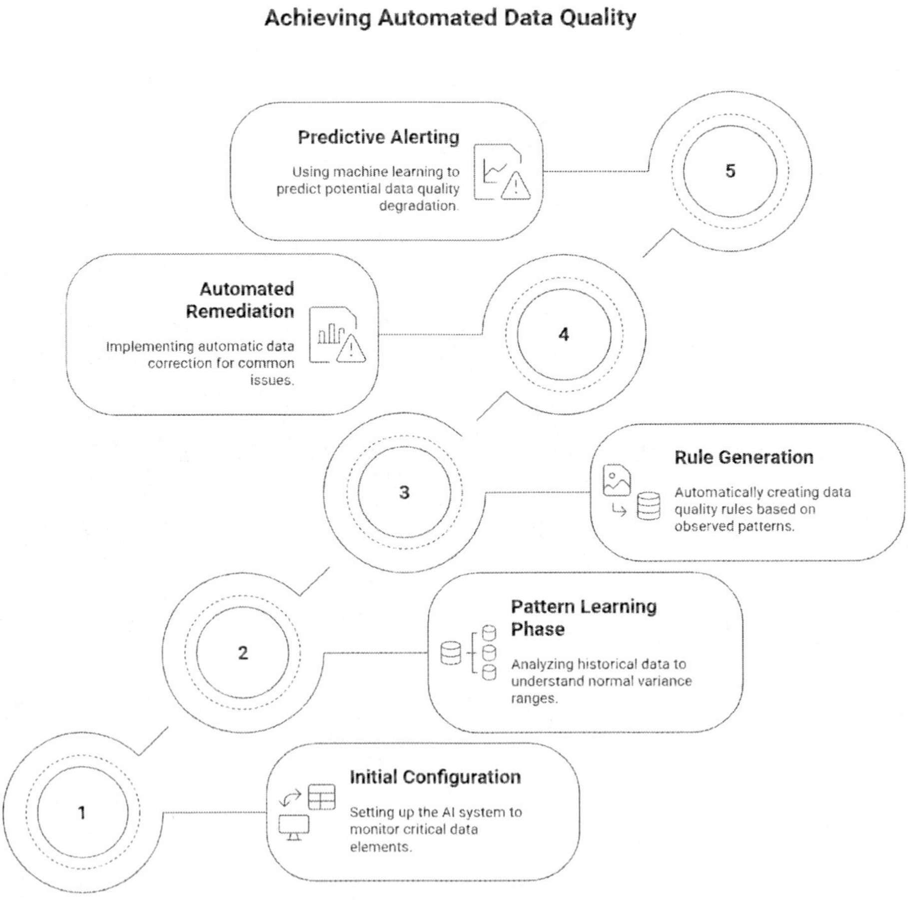

Initial Configuration: The team configured CLAIRE to monitor critical customer data elements including account numbers, transaction amounts, and regulatory identifiers across their core banking systems.

Pattern Learning Phase: Over 30 days, the AI system analyzed historical data patterns, learning normal variance ranges for key metrics like daily transaction volumes and customer demographic distributions.

Rule Generation: CLAIRE automatically generated 340 data quality rules based on observed patterns, replacing 180 manually-created rules that required constant maintenance.

Automated Remediation: The system implemented automatic data correction for common issues like standardizing address formats and phone number patterns, reducing manual intervention by 73%.

Predictive Alerting: Machine learning models began predicting potential data quality degradation 48 hours before issues manifested, enabling proactive remediation.

```
# Sample AI-Generated Data Quality Rule
Rule: Customer_Phone_Validation
Trigger: New customer record creation
Logic: IF phone_format NOT IN (learned_patterns)
       AND confidence_score < 0.85
Action: Flag for manual review + Auto-suggest corrections
Success Metric: 89% reduction in invalid phone numbers
```

Results: The implementation reduced data quality incidents by 67%, decreased manual validation effort by 8,200 hours annually, and improved regulatory reporting accuracy from 94% to 99.2%.

Predictive Governance Analytics

AI enables governance teams to move beyond reactive problem-solving to predictive risk management. Machine learning models analyze usage patterns, access behaviors, and data lineage to identify potential governance risks before they impact business operations. These systems can predict which datasets are likely to experience quality degradation, which users might inadvertently access restricted information, and where compliance gaps may emerge.

Predictive analytics also optimize resource allocation by identifying high-impact governance activities and automating routine tasks. This intelligence allows governance teams to focus their expertise on strategic initiatives while AI handles operational monitoring and basic remediation tasks.

Conversational Governance Interfaces

Natural language processing transforms how stakeholders interact with governance systems. Modern platforms offer chatbot interfaces that allow business users to ask questions about data policies, request access permissions, or report potential issues using everyday language. These conversational interfaces democratize governance participation by removing technical barriers that previously limited engagement to IT specialists.

The integration of automation and artificial intelligence into governance frameworks represents a fundamental shift toward intelligent, self-managing data ecosystems. Organizations implementing AI-powered governance solutions experience significant reductions in manual effort while achieving higher compliance rates and improved data quality outcomes. These technologies enable governance teams to scale their impact across growing data volumes and increasingly complex regulatory environments.

Building on these automated governance capabilities, organizations must ensure their AI-enhanced frameworks integrate seamlessly with master data management systems to maintain consistency and accuracy across all enterprise data assets.

8.6. Master Data Management Integration

Building upon AI-enhanced governance capabilities, the integration of Master Data Management (MDM) systems represents a critical convergence point where data governance policies meet operational reality. This integration transforms abstract governance principles into concrete data stewardship practices that ensure consistency, accuracy, and reliability across all enterprise data touchpoints.

Strategic MDM-Governance Alignment Framework

The intersection of MDM and data governance creates a powerful synergy that addresses fundamental challenges in enterprise data management. While governance establishes the "what" and "why" of data management, MDM provides the "how" through systematic approaches to managing critical business entities such as customers, products, suppliers, and locations. This alignment requires careful orchestration of governance policies with MDM technical capabilities to create a unified data ecosystem.

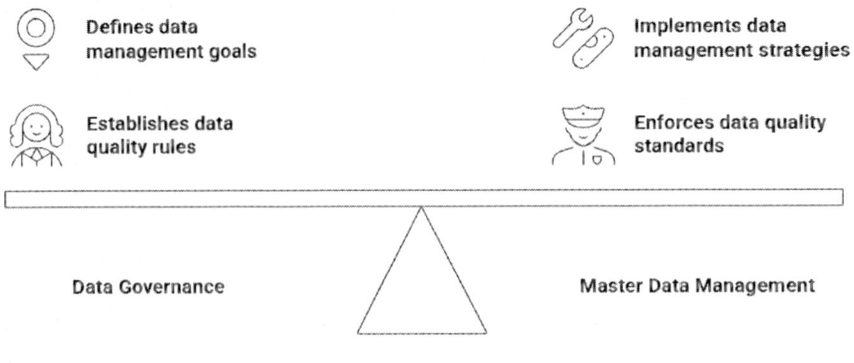

Modern MDM platforms serve as enforcement engines for governance policies, automatically applying data quality rules, standardization procedures, and validation protocols defined by governance committees. The integration ensures that master data entities maintain consistent definitions, structures, and quality standards across all consuming systems and business processes.

Practical Implementation: Global Retail Chain MDM Integration

Consider the implementation at **Carrefour**, the French multinational retail corporation, which successfully integrated its data governance framework with Informatica MDM to manage product master data across 30+ countries. Their approach demonstrates how governance policies can be embedded directly into MDM workflows.

Step 1: Governance Policy Translation

```
Product Data Governance Rule: "All product descriptions must include
mandatory attributes: SKU, category, brand, price currency, and local
compliance certifications"

MDM Technical Implementation:
- Configure Informatica MDM validation rules
- Set mandatory field constraints
- Establish country-specific validation logic
- Create automated compliance checking workflows
```

Step 2: Stewardship Workflow Integration

The governance team established data stewardship roles within the MDM platform, assigning regional product managers as domain stewards with specific responsibilities for maintaining product master data quality in their markets.

Step 3: Policy Enforcement Automation

```
# Example validation rule configuration in MDM platformdef
validate_product_completeness(product_record):
    mandatory_fields = [ 'sku', 'category', 'brand', 'price_currency',
'compliance_cert']
    missing_fields = [field for field in mandatory_fields if not
product_record.get(field)]

    if missing_fields:
        trigger_steward_notification(product_record.id, missing_fields)
        set_record_status( 'INCOMPLETE')
        return False
    return True
```

Results Achieved:

95% reduction in product data inconsistencies across regional systems
40% faster time-to-market for new product launches

Automated compliance reporting for regulatory requirements
Standardized product information across 12,000+ stores globally

Governance Committee Integration Patterns

Successful MDM-governance integration requires establishing clear decision-making structures that bridge governance oversight with operational MDM management. The governance committee must maintain strategic oversight while empowering MDM administrators to execute tactical decisions within established policy boundaries.

Policy Development Sessions involve collaborative workshops where business stakeholders, data stewards, and MDM technical teams jointly develop data standards that are both governmentally sound and technically feasible. These sessions typically follow a structured approach: business requirement identification, technical feasibility assessment, policy drafting, and implementation planning.

Decision Escalation Frameworks ensure that complex data conflicts or policy exceptions flow efficiently from MDM operational teams to appropriate governance authorities. This includes establishing clear criteria for automatic resolution versus human intervention, defining escalation timelines, and maintaining audit trails for all governance decisions.

Risk Mitigation and Performance Measurement

MDM-governance integration introduces specific risk categories that require proactive management. **Data synchronization risks** emerge when governance policy changes must be propagated across multiple MDM instances or federated systems. **Stewardship accountability gaps** can occur when MDM workflows bypass established governance review processes.

Performance measurement focuses on governance effectiveness metrics embedded within MDM operations, including policy compliance rates, data quality improvement trends, and stewardship response times. These metrics provide tangible evidence of governance value while identifying areas requiring additional attention or policy refinement.

The integration of master data management with governance frameworks creates a self-reinforcing system where policy compliance becomes embedded in operational workflows, ensuring that governance principles translate into measurable data quality improvements. This foundation enables organizations to establish comprehensive data lineage tracking and impact analysis capabilities that provide complete visibility into how master data flows and transforms throughout the enterprise ecosystem.

8.7. Data Lineage and Impact Analysis Tools

Building upon comprehensive data lineage tracking capabilities, organizations must implement sophisticated impact analysis tools that enable proactive risk assessment and change management across their data ecosystem. These advanced analytical platforms transform static lineage documentation into dynamic intelligence that predicts downstream effects of proposed changes, identifies critical data dependencies, and supports informed decision-making throughout the enterprise.

Automated Impact Assessment Capabilities

Modern impact analysis tools leverage machine learning algorithms to automatically evaluate the ripple effects of proposed data changes across interconnected systems. **Impact radius mapping** creates visual representations of how modifications to source systems, data models, or business rules will propagate through downstream applications, reports, and analytics platforms. These tools analyze historical change patterns to predict the scope and complexity of implementation efforts, enabling governance teams to prioritize changes based on business criticality and technical complexity.

Advanced platforms integrate with existing development workflows to provide real-time impact assessments during the planning phase of data projects. When data architects propose schema modifications or business analysts request new data transformations, the impact analysis engine automatically identifies affected systems, dependent processes, and stakeholder groups that require notification or approval.

Dependency Visualization and Risk Scoring

Sophisticated visualization engines create interactive dependency maps that reveal hidden connections between seemingly unrelated data assets. These tools employ graph database technologies to model complex relationships and generate **dependency heat maps** that highlight areas of highest risk and interconnectedness. Risk scoring algorithms evaluate factors such as data volume, update frequency, business criticality, and downstream consumer count to prioritize governance attention on the most vulnerable data pathways.

Practical Implementation: Global Telecommunications Impact Analysis Framework

A major European telecommunications provider implemented Informatica's Enterprise Data Catalog with custom impact analysis extensions to manage their complex customer data ecosystem spanning 12 countries and 45 million subscribers. The organization faced challenges tracking how changes to customer billing systems affected downstream analytics, regulatory reporting, and partner data feeds.

Implementation Steps:

Baseline Mapping Phase: The governance team cataloged 1,200 data sources across billing, CRM, and network management systems, establishing automated lineage discovery for critical customer data elements including subscriber identifiers, usage metrics, and billing cycles.

Impact Analysis Configuration: Custom algorithms were developed to weight dependencies based on regulatory requirements, with GDPR-related data flows receiving 3x risk multipliers and financial reporting pathways assigned critical status flags.

Stakeholder Integration: The platform integrated with ServiceNow to automatically create change tickets and notify affected business units when proposed modifications exceeded predetermined risk thresholds.

Results Achieved: The implementation reduced unplanned system outages by 67% and decreased average change implementation time from 14 days to 6 days. Most significantly, the organization achieved 100% compliance with regulatory notification requirements during the 18-month implementation period, avoiding potential penalties exceeding €2.3 million.

Change Simulation and Testing Protocols

Leading impact analysis platforms provide **sandbox simulation capabilities** that allow governance teams to model proposed changes in isolated environments before production implementation. These testing protocols validate impact predictions against actual system behavior, continuously improving the accuracy of predictive algorithms through feedback loops and machine learning refinement.

```
Impact Analysis Workflow Template:

1. Change Request Submission
   - Automated lineage scan initiation
   - Risk score calculation
   - Stakeholder notification triggers

2. Impact Assessment Phase
   - Dependency mapping generation
   - Business impact evaluation
   - Technical complexity scoring

3. Approval Workflow Routing
   - Risk-based approval paths
   - Stakeholder review assignments
   - Timeline estimation updates

4. Implementation Monitoring
   - Real-time impact tracking
```

```
  - Exception handling protocols
  - Post-change validation
```

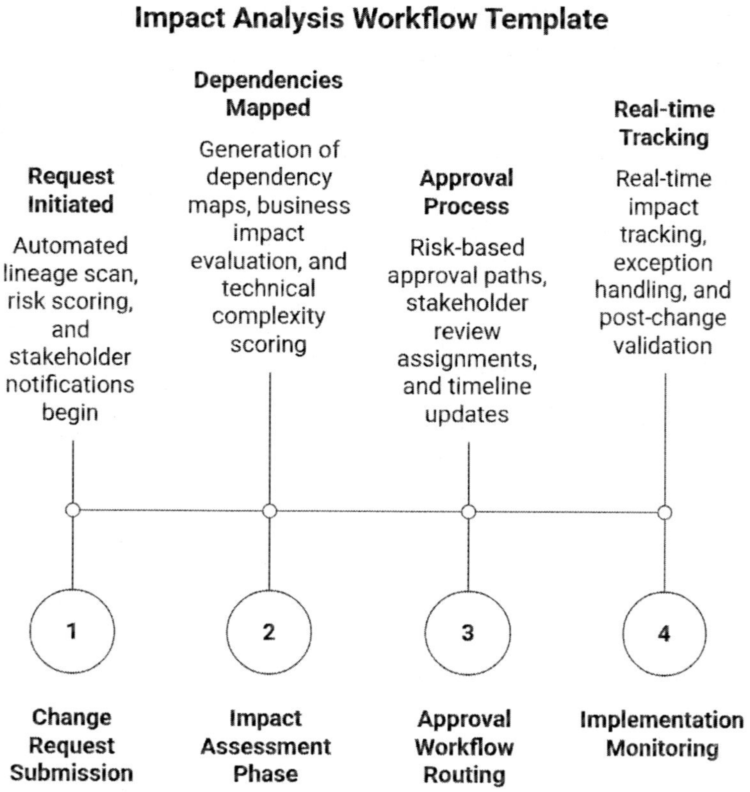

The strategic implementation of impact analysis tools transforms reactive data management into proactive governance, enabling organizations to anticipate challenges, minimize disruptions, and maintain data ecosystem stability while supporting business agility and innovation requirements.

With robust impact analysis capabilities established, organizations must next develop comprehensive technology implementation roadmaps that coordinate tool deployment, integration sequencing, and organizational change management to ensure successful governance platform adoption across the enterprise.

8.8. Technology Implementation Roadmaps

Building upon the foundation of coordinated tool deployment and integration sequencing, successful technology implementation requires a structured roadmap that transforms governance platforms from isolated systems into integrated enterprise assets. The most effective implementation strategies follow a phased approach that balances technical complexity with organizational readiness, ensuring each deployment milestone builds sustainable adoption momentum while delivering measurable business value.

Implementation Phase Architecture

Effective technology roadmaps organize deployment activities into four distinct phases: **foundation establishment**, **core platform deployment**, **integration expansion**, and **optimization maturity**. Each phase contains specific deliverables, success criteria, and stakeholder engagement requirements that prevent common implementation failures such as scope creep, user resistance, and technical debt accumulation.

The foundation establishment phase focuses on infrastructure readiness, data source identification, and initial metadata capture. Organizations typically spend 6-8 weeks preparing their technical environment, establishing connectivity protocols, and conducting preliminary data profiling activities. This phase concludes with a comprehensive technical baseline that informs subsequent deployment decisions.

Core platform deployment introduces primary governance capabilities including data cataloging, lineage tracking, and basic quality monitoring. Implementation teams configure user interfaces, establish initial workflows, and conduct pilot testing with select business units. This phase typically requires 10-12 weeks and produces a functioning governance platform ready for broader organizational adoption.

Practical Implementation Demonstration

Consider the implementation roadmap executed by **Axa Group**, the French multinational insurance company, during their enterprise-wide Collibra deployment across 57 countries. Their roadmap addressed the complexity of managing data governance across diverse regulatory environments while maintaining operational efficiency.

Phase 1: Foundation (Weeks 1-6)

```
Technical Preparation:
- Established secure cloud infrastructure in 4 regional data centers
- Configured LDAP integration for 45,000 employee accounts
- Mapped 127 critical data sources across insurance product lines
- Created initial data domain structure for 8 business units

Governance Framework Setup:
- Defined 12 data steward roles with specific responsibilities
- Established escalation procedures for data quality issues
- Created approval workflows for sensitive data access requests
```

Phase 2: Core Deployment (Weeks 7-18)

The implementation team deployed Collibra's core modules systematically, beginning with the European headquarters before expanding to regional offices. They established automated data lineage tracking for their customer data platform, which processes over 50 million policy records monthly.

Key deployment activities included configuring business glossaries in six languages, establishing data quality rules for regulatory compliance (Solvency II requirements), and creating role-based access controls that accommodate different national privacy regulations. The platform successfully cataloged 2,400 data assets within the first deployment cycle.

Phase 3: Integration Expansion (Weeks 19-30)

Integration activities connected Collibra with existing enterprise systems including Salesforce, SAP, and legacy mainframe applications. The team implemented automated metadata harvesting that reduced manual cataloging efforts by 75% while improving data lineage accuracy across complex insurance product workflows.

Phase 4: Optimization (Weeks 31-40)

Advanced features including machine learning-powered data classification and automated compliance reporting were activated. The platform now automatically identifies personally identifiable information across 15 different data types and generates regulatory reports for multiple jurisdictions simultaneously.

Change Management Integration

Technology roadmaps must incorporate comprehensive change management strategies that address user adoption challenges throughout each implementation phase. Successful deployments establish **governance champions** within each business unit who receive advanced training and serve as local implementation advocates. These champions facilitate user onboarding, collect feedback for platform improvements, and identify opportunities for expanding governance capabilities.

Training programs should align with roadmap phases, delivering just-in-time education that coincides with feature availability. Organizations achieve higher adoption rates when training focuses on specific business scenarios rather than generic platform capabilities. For example, training sessions might demonstrate how marketing teams can use the governance platform to ensure campaign data quality rather than explaining abstract data lineage concepts.

Risk Mitigation Protocols

Implementation roadmaps must include specific risk mitigation strategies that address common deployment challenges. Technical risks include data integration failures, performance bottlenecks, and security vulnerabilities. Organizational risks encompass user resistance, insufficient executive support, and competing technology priorities.

Effective roadmaps establish **implementation checkpoints** every two weeks where technical progress, user feedback, and business value metrics are reviewed. These checkpoints enable early identification of potential issues and allow implementation teams to adjust deployment strategies before problems become critical.

The systematic approach to technology implementation roadmaps transforms governance platforms from technical projects into strategic business enablers. By following structured phases, incorporating practical change management strategies, and maintaining focus on measurable outcomes, organizations can achieve successful governance technology deployments that deliver sustainable value across the enterprise. This foundational implementation capability sets the stage for examining how these technical investments integrate with broader organizational governance maturity and long-term sustainability strategies.

9. IMPLEMENTATION STRATEGY AND DEPLOYMENT

The most meticulously designed data governance framework remains nothing more than an elaborate blueprint until it encounters the complex reality of organizational implementation. While previous chapters have equipped you with the strategic foundations and architectural elements of effective governance, this chapter bridges the critical gap between theoretical design and operational reality, transforming your governance vision into a living, breathing organizational capability.

Implementation represents the moment where governance frameworks meet human nature, organizational culture, and existing business processes. Success hinges not merely on technical excellence, but on your ability to navigate change resistance, secure stakeholder buy-in, and orchestrate a transformation that touches every corner of your enterprise. The difference between governance initiatives that flourish and those that falter often lies in the sophistication of the deployment strategy rather than the elegance of the framework itself.

This chapter provides a comprehensive roadmap for translating governance designs into sustainable organizational practices. You'll discover proven methodologies for phased implementation that minimize disruption while maximizing adoption rates, alongside sophisticated pilot program strategies that validate your approach before enterprise-wide rollout. We'll explore the nuanced art of change management in data-centric environments, revealing how to cultivate champions, overcome skepticism, and build the cultural momentum necessary for long-term success.

Through detailed examination of training frameworks, communication strategies, and progress measurement techniques, you'll gain the tactical expertise needed to execute flawless deployments. Real-world insights from Latin American banking sector transformations illuminate common pitfalls and breakthrough solutions, providing you with battle-tested strategies for navigating the most challenging aspects of governance implementation. By chapter's end, you'll possess the complete toolkit for turning governance aspirations into measurable business results.

9.1. Phased Implementation Approaches

Successful data governance implementation rarely succeeds through enterprise-wide rollouts that attempt to transform entire organizational data landscapes overnight. Instead, leading organizations adopt structured, phased approaches that build momentum through demonstrable wins while minimizing operational disruption. This methodology allows governance teams to refine processes, address unforeseen challenges, and secure stakeholder buy-in before expanding to broader organizational scope.

The phased implementation strategy recognizes that data governance transformation involves complex organizational change management, technical integration challenges, and cultural shifts that require careful orchestration. By breaking down comprehensive governance frameworks into manageable phases, organizations can allocate resources effectively, measure progress incrementally, and adapt their approach based on real-world feedback and emerging requirements.

Foundation Phase: Establishing Governance Infrastructure

The initial phase focuses on creating the essential infrastructure that supports all subsequent governance activities. This includes establishing governance councils, defining roles and responsibilities, and implementing core governance technologies. Organizations typically begin with a **governance charter** that outlines authority structures, decision-making processes, and escalation procedures.

During this phase, governance teams select and configure foundational tools that will support ongoing data stewardship activities. The infrastructure must accommodate future expansion while providing immediate value to early adopters. Key deliverables include governance operating procedures, stakeholder communication frameworks, and baseline data quality metrics that establish the starting point for improvement initiatives.

Domain-Specific Implementation: Targeted Governance Deployment

The second phase involves selecting specific data domains or business areas for focused governance implementation. This targeted approach allows organizations to demonstrate value within defined boundaries while building expertise and refining processes. Common starting points include customer data, financial reporting information, or regulatory compliance datasets where governance impact can be clearly measured and communicated.

Practical Implementation Example: Financial Services Data Lineage Initiative

A mid-sized investment firm implemented domain-specific governance using Informatica Axon to address regulatory reporting challenges. The implementation followed this structured approach:

Implementing Domain-Specific Governance

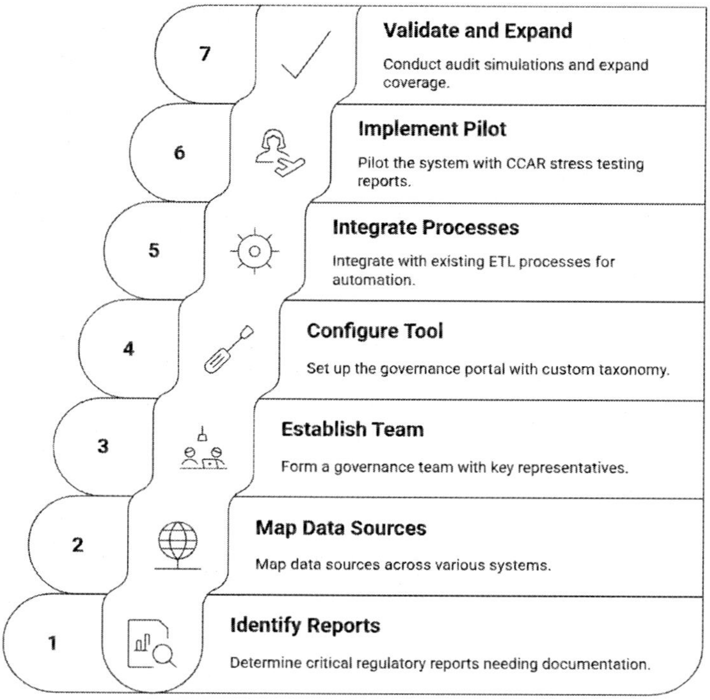


```
Phase 1: Scope Definition (Weeks 1-2)
- Identified 15 critical regulatory reports requiring data lineage
documentation
- Mapped data sources across trading systems, risk management
platforms, and client databases
- Established governance team with representatives from Risk,
Compliance, and IT

Phase 2: Tool Configuration (Weeks 3-6)
- Configured Informatica Axon governance portal with custom regulatory
taxonomy
- Integrated with existing Informatica PowerCenter ETL processes for
automated lineage capture
- Created role-based access controls aligned with regulatory
requirements

Phase 3: Pilot Implementation (Weeks 7-12)
- Selected CCAR stress testing reports as initial use case
- Documented complete data lineage from source systems to regulatory
submissions
- Implemented data quality rules and automated monitoring for critical
data elements

Phase 4: Validation and Expansion (Weeks 13-16)
- Conducted regulatory audit simulation demonstrating complete data
```

```
traceability
- Achieved 40% reduction in audit preparation time
- Expanded coverage to include Dodd-Frank reporting requirements
```

The results demonstrated clear regulatory value: audit preparation time decreased from 120 hours to 72 hours per reporting cycle, while data lineage documentation completeness increased from 35% to 98% for covered reports. This success provided the business case for expanding governance to additional regulatory domains.

Cross-Functional Integration: Scaling Governance Capabilities

The third phase extends governance processes across organizational boundaries, integrating data stewardship activities with existing business workflows. This phase requires careful attention to change management, as governance processes become embedded in daily operational activities. Success depends on demonstrating how governance enhances rather than impedes business productivity.

Integration activities include connecting governance tools with enterprise data platforms, implementing automated policy enforcement mechanisms, and establishing feedback loops that continuously improve governance effectiveness. Organizations must balance standardization with flexibility, ensuring governance processes adapt to diverse business requirements while maintaining consistency in core principles.

Enterprise Optimization: Comprehensive Governance Maturity

The final implementation phase focuses on optimizing governance processes for enterprise-wide efficiency and effectiveness. This includes implementing advanced analytics for governance performance monitoring, establishing predictive data quality capabilities, and creating self-service governance tools that empower business users while maintaining appropriate oversight.

Phased implementation approaches provide organizations with structured pathways for governance transformation that minimize risk while maximizing stakeholder engagement. By building governance capabilities incrementally, organizations can adapt their approach based on real-world experience, ensuring their final governance framework aligns with actual business needs rather than theoretical ideals. This methodology has proven particularly effective in complex enterprise environments where comprehensive governance requirements must be balanced with operational continuity and resource constraints.

With foundational implementation phases established, the next critical step involves designing and executing focused pilot programs that validate governance approaches within controlled environments before broader organizational deployment.

9.2. Pilot Program Design and Execution

Building on the established governance foundation, pilot programs serve as the critical bridge between theoretical frameworks and enterprise-wide implementation. These controlled environments allow organizations to test governance approaches, refine processes, and demonstrate value before committing to full-scale deployment. Successful pilot execution requires strategic selection of scope, participants, and success metrics that align with broader organizational objectives.

Strategic Pilot Scope Selection

Effective pilot programs begin with careful scope definition that balances ambition with achievability. The ideal pilot encompasses a specific business domain or data asset that represents broader organizational challenges while remaining manageable within resource constraints. Financial services organizations often select customer onboarding data as pilot scope, given its cross-functional nature and regulatory importance, while healthcare systems frequently choose patient master data management due to its critical operational impact.

The scope selection process involves evaluating data complexity, stakeholder engagement levels, and potential business impact. High-visibility data domains with engaged business sponsors typically yield better pilot outcomes than technically complex but organizationally isolated datasets. Manufacturing companies have found success piloting governance around product master data, where quality improvements directly translate to operational efficiency gains.

Stakeholder Engagement and Team Formation

Pilot success hinges on assembling cross-functional teams that represent both technical expertise and business domain knowledge. The core pilot team should include data stewards from the target business area, IT representatives familiar with underlying systems, and governance coordinators who can facilitate process adherence. Additionally, executive sponsors must remain actively engaged to address resource constraints and organizational resistance.

Practical Implementation Example: Healthcare Data Quality Pilot

A regional healthcare network implemented a comprehensive data quality pilot using Informatica Data Quality platform to address patient matching challenges across three hospital systems. The six-month pilot followed this structured approach:

```
Phase 1: Baseline Assessment (Weeks 1-2)
- Current patient matching accuracy: 78%
```

```
- Duplicate patient records: 12,000 across systems
- Data quality rules: 15 basic validations

Phase 2: Governance Framework Implementation (Weeks 3-6)
- Deployed Informatica DQ with 45 custom matching rules
- Established weekly data steward review sessions
- Created patient data quality scorecard dashboard

Phase 3: Process Integration (Weeks 7-16)
- Integrated DQ checks into patient registration workflow
- Trained 25 registration staff on new validation procedures
- Implemented real-time duplicate prevention alerts

Phase 4: Results Measurement (Weeks 17-24)
- Patient matching accuracy improved to 94%
- Duplicate records reduced to 3,200 (73% improvement)
- Registration errors decreased by 41%
```

The pilot team included patient access directors, health information management staff, IT database administrators, and clinical data analysts. Weekly governance committee meetings reviewed progress against established KPIs, while monthly executive briefings maintained leadership visibility and support.

Pilot Execution Framework and Monitoring

Successful pilot execution requires structured project management combined with flexible adaptation to emerging challenges. The execution framework should establish clear phases with defined deliverables, decision points, and escalation procedures. Regular checkpoint reviews allow teams to assess progress, adjust approaches, and maintain stakeholder alignment throughout the pilot duration.

Monitoring mechanisms must capture both quantitative metrics and qualitative feedback from pilot participants. Data quality improvements, process efficiency gains, and user satisfaction scores provide measurable evidence of governance value. Equally important are stakeholder interviews and feedback sessions that reveal implementation challenges, training needs, and change management requirements for broader deployment.

Risk mitigation during pilot execution involves establishing clear escalation paths for technical issues, resource constraints, and stakeholder conflicts. Pilot teams should maintain regular communication with executive sponsors and prepare contingency plans for common implementation challenges such as system integration difficulties or user resistance to new processes.

Success Measurement and Documentation

Comprehensive pilot documentation serves multiple purposes: validating governance approaches, supporting broader deployment planning, and creating organizational learning assets. Documentation should capture baseline measurements, implementation decisions, lessons learned, and recommendations for scaling governance practices. This knowledge base becomes invaluable during subsequent deployment phases and helps avoid repeating pilot-phase mistakes.

The pilot program validates governance frameworks through controlled testing while building organizational confidence in data stewardship practices. Teams gain hands-on experience with governance tools, develop cross-functional collaboration skills, and create documented proof points for executive decision-making. These foundational experiences become crucial assets as organizations prepare for comprehensive change management initiatives and user adoption strategies across the enterprise.

9.3. Change Management and User Adoption Strategies

The transition from governance design to organizational reality hinges on one critical factor: people. While technical frameworks and policy documents provide the blueprint, sustainable data governance success depends entirely on how effectively organizations navigate the human dimension of change. Data professionals who have witnessed failed governance initiatives understand that resistance to change, not technical complexity, represents the primary barrier to implementation success.

Understanding Resistance Patterns in Data Governance

Resistance to data governance manifests differently across organizational levels and functional areas. Financial analysts may resist standardized reporting formats that alter familiar workflows, while IT teams might push back against new data access controls that complicate existing integrations. Marketing departments often express concern about data sharing requirements that could slow campaign execution, and operations teams may question quality standards that introduce additional validation steps.

These resistance patterns follow predictable psychological stages. Initial denial occurs when stakeholders dismiss governance requirements as temporary initiatives. Anger emerges when mandatory compliance disrupts established routines. Bargaining follows as teams attempt to negotiate exceptions or modified requirements. Depression sets in when the scope of required changes becomes apparent. Finally, acceptance develops as benefits become visible and new processes stabilize.

Stakeholder Engagement Framework

Effective change management begins with comprehensive stakeholder mapping that identifies influence networks, communication preferences, and motivational drivers. **Data stewards** require different messaging than **business analysts**, who need distinct approaches from **executive sponsors**. This mapping process reveals informal leaders whose endorsement can accelerate adoption across peer networks.

The engagement framework operates through structured touchpoints that maintain momentum while addressing concerns proactively. Weekly governance champions meetings provide forums for addressing implementation challenges, while monthly executive briefings demonstrate progress against business objectives. Quarterly success showcases highlight teams achieving measurable improvements through governance adoption.

Practical Implementation: Global Manufacturing Company Change Strategy

A multinational manufacturing company faced significant resistance when implementing Collibra's data governance platform across 15 countries and 40,000 employees. The organization's previous attempts at standardization had failed due to cultural differences and competing regional priorities.

Phase 1: Pilot Program Design

The implementation team selected three diverse pilot groups: German engineering (highly structured), Brazilian sales (relationship-focused), and Indian IT operations (process-oriented). Each pilot received customized change management approaches:

```
German Engineering Pilot:
- Technical documentation emphasizing precision and compliance
- Detailed process maps showing quality improvements
- Integration with existing SAP workflows
- Metrics focused on defect reduction and efficiency gains

Brazilian Sales Pilot:
- Relationship-building sessions with local governance champions
- Success stories from similar Latin American organizations
- Flexible implementation timeline accommodating cultural holidays
- Benefits messaging centered on customer satisfaction improvements

Indian IT Operations Pilot:
- Process improvement workshops led by respected technical leaders
- Integration demonstrations with existing ITSM tools
- Career development opportunities through governance certification
- Recognition programs celebrating early adopters
```

Phase 2: Resistance Mitigation

Each pilot group exhibited distinct resistance patterns. German engineers questioned data quality rules that conflicted with legacy standards. Brazilian sales teams worried about reduced autonomy in customer data management. Indian operations staff expressed concerns about increased workload without additional resources.

The change management team addressed these concerns through targeted interventions. For German engineers, they organized technical deep-dives demonstrating how Collibra's lineage capabilities would enhance their existing quality processes. Brazilian sales received assurance that governance would improve, not restrict, customer insights through better data integration. Indian operations teams gained additional staffing and automation tools to offset governance overhead.

Phase 3: Success Amplification

Within six months, all three pilots demonstrated measurable improvements. German engineering reduced data-related defects by 34%, Brazilian sales increased cross-selling effectiveness by 28%, and Indian operations decreased incident resolution time by 41%. These success stories became powerful change catalysts for broader organizational rollout.

The company leveraged these results through a comprehensive communication strategy. Video testimonials from pilot participants shared authentic experiences with governance benefits. Executive sponsors referenced specific improvements in quarterly business reviews. Regional governance champions facilitated peer-to-peer knowledge transfer sessions.

Measuring Adoption Success

Effective change management requires continuous measurement of adoption progress through both quantitative and qualitative indicators. **User engagement metrics** track system utilization, policy compliance rates, and training completion percentages. **Behavioral indicators** assess collaboration patterns, data sharing frequency, and proactive issue reporting.

Advanced organizations implement sentiment analysis tools that monitor internal communications for governance-related discussions. These insights reveal emerging concerns before they become widespread resistance, enabling proactive intervention strategies.

The manufacturing company's measurement approach included monthly pulse surveys, quarterly focus groups, and annual governance maturity assessments. This multi-layered feedback system identified adoption barriers early and guided continuous improvement efforts.

Change management transforms data governance from imposed compliance burden into embraced organizational capability. Success requires understanding human psychology, addressing resistance proactively, and celebrating achievements that reinforce positive behaviors. Organizations that master these change dynamics create sustainable governance cultures that adapt and evolve with business needs.

With foundational change management principles established, organizations must now focus on building the specific capabilities and competencies that enable long-term governance success through comprehensive training and development programs.

9.4. Training and Capability Development

Building upon the foundational principles established, organizations must now focus on building the specific capabilities and competencies that enable long-term governance success through comprehensive training and development programs. Effective data governance requires more than policies and procedures; it demands a workforce equipped with the knowledge, skills, and mindset necessary to operationalize governance principles across all organizational levels.

Competency Framework Development

Successful training initiatives begin with a structured competency framework that defines role-specific knowledge requirements and skill progressions. Organizations should establish tiered competency levels that align with governance responsibilities, from basic data literacy for general employees to advanced stewardship capabilities for data custodians. This framework serves as the foundation for targeted learning paths that address specific organizational needs while building enterprise-wide data culture.

The competency framework should encompass technical skills such as data quality assessment, metadata management, and privacy compliance procedures, alongside soft skills including stakeholder communication, change management, and cross-functional collaboration. By mapping these competencies to specific roles within the governance structure, organizations create clear development pathways that support career advancement while strengthening overall governance capabilities.

Role-Based Training Programs

Different stakeholder groups require tailored training approaches that reflect their unique responsibilities and interaction patterns with governance processes. **Data stewards** need deep technical training on data lineage tools, quality monitoring platforms, and issue resolution workflows. **Business users** require focused sessions on data classification standards, access request procedures, and compliance obligations relevant to their functional areas.

Executive stakeholders benefit from strategic governance briefings that emphasize business value realization, risk mitigation outcomes, and competitive advantage opportunities. These sessions should focus on governance metrics, ROI demonstrations, and strategic alignment rather than operational details, enabling leadership to make informed decisions about program investments and priority setting.

Practical Implementation: Collibra-Based Training Program

Consider a financial services organization implementing a comprehensive training program using Collibra's data governance platform. The training initiative begins with **baseline assessment workshops** where participants complete skills evaluations covering data cataloging, stewardship workflows, and policy interpretation capabilities.

The program structure includes:

Foundation modules covering Collibra navigation, asset discovery, and basic metadata management

Intermediate workshops focusing on data quality rule configuration, workflow customization, and stakeholder collaboration features

Advanced certification tracks for power users covering API integrations, custom dashboard development, and governance reporting

```
Training Progression Example:
Week 1-2: Platform orientation and basic navigation
- Asset browsing and search functionality
- User profile configuration and team assignments
- Basic metadata viewing and commenting

Week 3-4: Stewardship workflow participation
- Issue identification and escalation procedures
- Quality assessment and remediation tracking
- Stakeholder notification and collaboration tools

Week 5-6: Advanced governance operations
- Custom attribute configuration and management
- Automated workflow design and testing
- Performance metrics configuration and monitoring
```

Participants demonstrate competency through hands-on exercises including cataloging critical business assets, configuring data quality monitoring rules, and facilitating cross-departmental data lineage mapping sessions. The program measures success through platform usage analytics, governance process participation rates, and stakeholder satisfaction surveys.

Continuous Learning and Development

Sustainable capability development requires ongoing learning opportunities that adapt to evolving governance requirements and technological advances. Organizations should establish **communities of practice** where data stewards share experiences, discuss challenges, and collaborate on innovative solutions. These communities foster knowledge transfer while building internal expertise networks that support governance sustainability.

Regular capability assessments ensure training programs remain aligned with organizational needs and emerging governance challenges. By tracking skill development progress and identifying capability gaps, organizations can adjust training priorities and resource allocation to maintain governance effectiveness as business requirements evolve.

The investment in comprehensive training and capability development transforms governance from a compliance obligation into an organizational competency that drives innovation and competitive advantage. Through structured learning programs and continuous development initiatives, organizations build the human capital necessary to sustain long-term governance success while adapting to changing business environments and regulatory requirements.

This capability-building foundation sets the stage for establishing robust communication frameworks that ensure governance insights and updates reach all stakeholders effectively, maintaining program visibility and engagement across the enterprise ecosystem.

9.5. Communication Plans and Stakeholder Updates

Building on established communication frameworks, the tactical execution of stakeholder updates requires structured methodologies that transform governance progress into actionable intelligence for diverse organizational audiences. Effective communication plans must balance transparency with strategic messaging, ensuring that technical achievements translate into business value propositions that resonate across departmental boundaries.

Multi-Tiered Communication Architecture

Successful governance communication operates through layered channels that deliver targeted information to specific stakeholder segments. Executive briefings focus on strategic outcomes and risk mitigation metrics, while operational teams receive detailed process updates and implementation timelines. This architectural approach prevents information overload while ensuring each audience receives relevant, actionable insights.

The most effective governance programs establish **communication cadences** that align with organizational rhythms. Monthly steering committee updates provide strategic oversight, while weekly tactical briefings keep implementation teams synchronized. Quarterly business reviews demonstrate governance value through concrete performance metrics and compliance achievements.

Stakeholder Engagement Through Governance Dashboards

Modern governance communication leverages visual analytics platforms to create real-time stakeholder visibility. Organizations implementing Alation's governance suite have discovered that interactive dashboards reduce meeting overhead by 40% while increasing stakeholder engagement. These platforms transform static reports into dynamic communication tools that stakeholders can explore based on their specific interests and responsibilities.

Practical Implementation: Financial Services Compliance Dashboard

A multinational bank implemented a comprehensive stakeholder communication system using Informatica's governance platform to manage GDPR compliance updates across 15 countries. The implementation followed this structured approach:

```
Phase 1: Stakeholder Mapping and Communication Requirements
- Legal Teams: Weekly compliance status reports with risk indicators
- Business Units: Monthly data usage impact assessments
- IT Operations: Daily technical implementation progress updates
- Executive Committee: Quarterly strategic governance scorecards

Phase 2: Dashboard Configuration and Content Strategy
- Risk Heat Maps: Visual representation of compliance gaps by geography
- Progress Tracking: Implementation milestones with completion
percentages
- Impact Analysis: Business process changes and resource requirements
- Escalation Alerts: Automated notifications for critical issues
```

The bank configured Informatica's governance dashboard to automatically generate stakeholder-specific reports. Legal teams received detailed compliance matrices showing data processing activities across jurisdictions, while business units accessed simplified impact assessments highlighting operational changes. Executive dashboards focused on strategic metrics including compliance readiness scores and budget utilization.

Results demonstrated the power of targeted communication: compliance preparation time decreased by 35%, stakeholder satisfaction scores increased from 6.2 to 8.4 out of 10, and governance program visibility across the organization improved significantly. The automated reporting reduced manual communication overhead by 60% while increasing update frequency and accuracy.

Interactive Governance Forums and Feedback Mechanisms

Beyond traditional reporting, leading organizations establish **governance communities of practice** that facilitate two-way communication between stakeholders and governance teams. These forums create opportunities for stakeholders to influence governance evolution while maintaining program momentum through collaborative problem-solving.

Pharmaceutical giant Novartis implemented monthly governance forums where business stakeholders present data challenges and governance teams demonstrate solution approaches. These sessions generate valuable feedback that shapes governance priorities while building stakeholder investment in program success. The interactive format transforms governance from a top-down mandate into a collaborative business capability.

Crisis Communication and Issue Escalation

Effective governance communication plans include robust escalation procedures that ensure critical issues receive appropriate attention without overwhelming stakeholders with routine updates. **Escalation matrices** define trigger conditions, notification pathways, and response timeframes for different issue severities.

Organizations must establish clear protocols for communicating governance failures, data breaches, or compliance violations. These protocols balance transparency requirements with reputation management, ensuring stakeholders receive timely, accurate information while maintaining organizational credibility.

Structured communication plans and stakeholder engagement strategies form the operational backbone of successful governance implementation, creating transparency and accountability that sustain program momentum. These frameworks ensure that governance achievements translate into recognized business value while building the stakeholder support necessary for long-term success.

With communication frameworks established and stakeholder engagement patterns optimized, the next critical phase involves developing comprehensive measurement systems that quantify governance progress and demonstrate tangible business impact.

9.6. Measuring Implementation Progress and Success

With governance structures established and stakeholder engagement patterns optimized, the next critical phase involves developing comprehensive measurement systems that quantify governance progress and demonstrate tangible business impact. Effective measurement transforms abstract governance concepts into concrete organizational value, providing the evidence needed to sustain executive support and justify continued investment in data stewardship initiatives.

Establishing Baseline Metrics and Progress Indicators

Successful governance measurement begins with establishing clear baseline metrics that capture the organization's current data maturity state. These foundational measurements serve as reference points for tracking improvement over time and identifying areas requiring focused attention.

Implementation maturity indicators provide quantitative assessment of governance deployment across organizational dimensions. Key metrics include policy adoption rates measured by department, data steward certification completion percentages, and governance tool utilization statistics. Organizations typically track these metrics monthly during initial deployment phases, transitioning to quarterly assessments once frameworks stabilize.

Operational effectiveness measurements capture how governance processes impact daily business operations. These include data request fulfillment times, data quality issue resolution speeds, and compliance audit preparation timelines. Leading organizations establish service level agreements for governance processes, measuring performance against these commitments to demonstrate operational value.

Practical Implementation: Building a Governance Scorecard System

A global pharmaceutical company successfully implemented a comprehensive governance measurement system using Informatica's Data Quality suite combined with custom PowerBI dashboards. Their approach demonstrates practical scorecard development for complex organizational environments.

The implementation began with establishing **four measurement tiers**:

Strategic metrics tracked quarterly, focusing on business impact indicators like regulatory compliance scores and data-driven decision velocity

Tactical metrics measured monthly, including data steward activity levels and policy exception rates

Operational metrics monitored weekly, covering data quality scores and issue resolution times

Technical metrics captured daily, encompassing system availability and data lineage completeness

```
Governance Scorecard Framework:
Strategic Tier:
- Regulatory Compliance Score: 94% (Target: 95%)
- Data-Driven Decision Rate: 67% (Target: 70%)
- Business Process Efficiency Gain: 23% YoY

Tactical Tier:
- Policy Adherence Rate: 89% (Target: 85%)
- Data Steward Engagement Score: 4.2/5.0
- Exception Request Resolution: 3.2 days avg

Operational Tier:
- Data Quality Score: 91.5% (Target: 90%)
- Issue Resolution Time: 1.8 days avg
- Stakeholder Satisfaction: 4.1/5.0
```

The pharmaceutical company's **before-and-after comparison** revealed significant improvements within twelve months. Initially, regulatory compliance preparation required 180 days with 15% of submissions requiring corrections. Post-implementation, preparation time decreased to 45 days with only 3% requiring corrections, directly attributable to improved data governance processes.

Value Demonstration Through Business Impact Correlation

Effective measurement systems establish clear connections between governance activities and business outcomes. Organizations achieve this by implementing **correlation tracking mechanisms** that link governance metrics to operational performance indicators.

Revenue impact analysis connects data quality improvements to customer satisfaction scores and sales performance. One telecommunications provider documented how improving customer data accuracy from 76% to 94% correlated with a 12% increase in cross-selling success rates and 8% reduction in customer churn.

Risk mitigation quantification measures how governance processes reduce organizational exposure to data-related risks. This includes tracking privacy breach incidents, regulatory fine reductions, and audit finding decreases. Financial services organizations particularly benefit from demonstrating how governance investments reduce regulatory risk exposure.

Advanced Measurement Techniques and Continuous Improvement

Sophisticated measurement approaches incorporate **predictive analytics** to forecast governance program success and identify potential implementation challenges before they impact operations. Organizations use machine learning algorithms to analyze historical governance metrics and predict future performance trends.

Stakeholder sentiment analysis provides qualitative measurement complementing quantitative metrics. Regular surveys, focus groups, and governance committee feedback sessions generate insights into program effectiveness from user perspectives. These measurements often reveal implementation gaps not captured by technical metrics alone.

Measuring implementation progress requires systematic approaches that balance quantitative performance indicators with qualitative stakeholder feedback. Organizations that invest in comprehensive measurement systems create sustainable governance programs that continuously demonstrate business value while identifying opportunities for enhancement. The pharmaceutical company's scorecard approach illustrates how structured measurement transforms governance from overhead expense into strategic business enabler.

Having established robust measurement frameworks, organizations must prepare to address the inevitable implementation obstacles that arise during governance deployment, requiring proactive strategies for overcoming resistance and technical challenges.

9.7. Overcoming Common Implementation Challenges

Even the most meticulously planned data governance initiatives encounter predictable obstacles that can derail progress and erode stakeholder confidence. Understanding these challenges before they emerge allows organizations to develop targeted mitigation strategies that transform potential roadblocks into opportunities for strengthening governance foundations.

Resistance to Cultural Change

The transition from ad-hoc data practices to formalized governance structures often triggers organizational resistance that manifests in subtle but damaging ways. Data professionals frequently encounter pushback when established workflows face scrutiny, particularly in departments that have operated with minimal oversight for years.

A practical approach involves implementing **change champions** within each business unit who can translate governance requirements into department-specific benefits. These champions should possess both technical credibility and interpersonal skills necessary for facilitating difficult conversations about data ownership and accountability.

Consider establishing regular feedback sessions where concerns can be voiced openly, addressing misconceptions before they solidify into opposition. Document these interactions to identify patterns that may indicate broader implementation issues requiring strategic intervention.

Technology Integration Complexities

Legacy systems present unique challenges when implementing governance frameworks, particularly when data lineage tracking and quality monitoring capabilities must be retrofitted into existing architectures. Organizations often underestimate the technical complexity of integrating governance tools with established data pipelines.

Practical Implementation Example: Master Data Governance in Manufacturing

A global automotive manufacturer faced significant challenges implementing Informatica Data Quality across their supply chain systems. Their existing SAP environment contained inconsistent vendor records across 15 regional implementations, creating compliance risks and operational inefficiencies.

The implementation team adopted a phased approach:

Assessment Phase: Cataloged all vendor data sources using Informatica's profiling capabilities, revealing 47% duplicate records and inconsistent naming conventions across regions

Pilot Implementation: Selected three high-impact regions for initial deployment, focusing on critical vendor categories affecting production schedules

Rule Development: Created standardized data quality rules within Informatica, including vendor name standardization, address validation, and duplicate detection algorithms

Integration Testing: Established automated workflows connecting SAP master data management with Informatica's governance platform

Validation Process: Implemented business user review cycles for flagged records, creating accountability loops between regional procurement teams and central data stewards

The results demonstrated measurable improvements: vendor record accuracy increased from 53% to 94% within six months, procurement cycle times decreased by 23%, and compliance audit findings dropped by 78%. Most importantly, regional teams began proactively requesting governance tool access for other data domains, indicating successful cultural adoption.

Stakeholder Alignment Challenges

Competing priorities across business units often create governance implementation friction, particularly when resource allocation decisions must be made. Finance departments may prioritize cost reduction while marketing focuses on customer insights, creating tension around data investment priorities.

Successful organizations address this challenge through **governance value mapping** exercises that demonstrate how data quality improvements support multiple departmental objectives simultaneously. Create concrete examples showing how master data standardization reduces both operational costs and improves customer experience metrics.

Measurement and Communication Gaps

Many governance initiatives struggle with demonstrating tangible business value, particularly during early implementation phases when benefits may not be immediately visible. This challenge becomes acute when executive sponsors require quarterly progress reports that show measurable returns on governance investments.

Establish baseline metrics before implementation begins, focusing on both technical indicators like data quality scores and business outcomes such as decision-making speed or regulatory compliance costs. Regular communication should emphasize incremental improvements while building toward larger transformation goals.

The key to overcoming implementation challenges lies in anticipating resistance patterns, preparing targeted responses, and maintaining consistent communication about governance value. Organizations that proactively address these obstacles create stronger foundations for long-term governance success, building momentum that carries initiatives through inevitable periods of uncertainty and change.

These implementation strategies provide the groundwork for examining how specific regional contexts and industry requirements shape governance deployment approaches, particularly in sectors where regulatory compliance drives governance adoption.

9.8. Deployment Lessons from Latin American Banking Sector

The Latin American banking sector's unique regulatory landscape and economic volatility have created a distinctive environment for data governance implementation, offering valuable lessons for organizations worldwide. Between 2018 and 2022, major banks across Brazil, Mexico, and Colombia faced simultaneous pressures from evolving privacy regulations like Brazil's LGPD, anti-money laundering requirements, and digital transformation initiatives that fundamentally reshaped their governance deployment strategies.

Regional Regulatory Complexity and Governance Adaptation

Latin American banks operate within a complex web of cross-border regulations that demand sophisticated governance frameworks. **Banco Santander Brasil's** approach to LGPD compliance illustrates how regulatory requirements can drive comprehensive governance transformation. The bank implemented a **federated governance model** that accommodated both local Brazilian privacy requirements and global Santander standards, creating governance structures that could adapt to varying regulatory intensities across different jurisdictions.

The deployment strategy centered on **regulatory impact mapping**, where governance teams identified specific data flows that crossed regulatory boundaries and established automated compliance monitoring. This approach proved essential when Mexico's data protection law evolved in 2020, allowing the bank to rapidly adapt existing governance frameworks rather than rebuilding from scratch.

Technology Infrastructure and Economic Constraints

Resource constraints common in emerging markets forced Latin American banks to develop innovative deployment approaches that maximize governance impact while minimizing technology investments. **BBVA Mexico's** data governance implementation demonstrates how economic realities shaped deployment priorities and technology selection strategies.

Rather than implementing comprehensive governance platforms immediately, BBVA adopted a **progressive enhancement approach** that built governance capabilities incrementally. The bank began with basic data cataloging using open-source tools, then gradually integrated commercial governance platforms as business value became demonstrable and budgets expanded.

Practical Implementation: Cross-Border Data Governance Framework

A comprehensive example emerges from **Itaú Unibanco's** regional expansion strategy, which required deploying governance frameworks across multiple Latin American countries with varying regulatory requirements.

Phase 1: Regulatory Baseline Assessment

The implementation began with a systematic analysis of regulatory requirements across target markets. Governance teams created a **regulatory compliance matrix** that mapped data protection, banking supervision, and anti-money laundering requirements for each jurisdiction.

```
Regulatory Compliance Matrix Template:
Country: [Brazil/Colombia/Chile]
Data Protection: [LGPD/Local Requirements]
Banking Supervision: [Central Bank Requirements]
AML/KYC: [Local Financial Intelligence Unit Standards]
Cross-Border Restrictions: [Data Residency/Transfer Limitations]
Governance Impact: [High/Medium/Low Priority Areas]
```

Phase 2: Federated Governance Architecture

Itaú developed a **hub-and-spoke governance model** where central governance standards provided consistency while local governance committees addressed jurisdiction-specific requirements. Each country operation maintained governance autonomy for local compliance while adhering to group-wide data quality and risk management standards.

The deployment utilized **Collibra's** multi-tenant architecture to create separate governance domains for each country while maintaining centralized reporting and policy management. Local governance teams could customize data classification schemes and privacy controls without disrupting regional consistency.

Phase 3: Stakeholder Engagement and Cultural Adaptation

Recognizing cultural differences in hierarchical decision-making across Latin American markets, Itaú adapted traditional governance committee structures. In Brazil, governance decisions followed consensus-building approaches, while Colombian operations required more directive leadership styles. The bank created **culturally-**

adapted governance playbooks that maintained consistent outcomes while respecting local business practices.

Change Management in Hierarchical Organizations

Latin American banking culture often emphasizes hierarchical decision-making and personal relationships, requiring governance deployment strategies that work within these organizational dynamics. Successful implementations leveraged **executive sponsorship cascading**, where senior leaders personally endorsed governance initiatives within their direct reports, creating accountability chains that aligned with cultural expectations.

Bancolombia's governance deployment illustrates this approach through **governance ambassador programs** where respected senior managers championed governance adoption within their business units. These ambassadors received intensive governance training and became local experts who could address resistance and provide peer-level guidance during implementation.

The Latin American banking sector's governance deployment experiences demonstrate how regional context, regulatory complexity, and cultural factors fundamentally shape implementation strategies. Organizations worldwide can apply these lessons by adapting governance frameworks to local constraints, leveraging progressive implementation approaches when resources are limited, and recognizing that successful governance deployment requires deep understanding of organizational culture and stakeholder dynamics. These insights prove particularly valuable for multinational organizations seeking to deploy consistent governance standards across diverse regional markets.

With these regional deployment lessons established, we can now synthesize the comprehensive implementation strategies and practical frameworks that form the foundation for sustainable data governance success across any organizational context.

10. OPERATIONALIZING DATA GOVERNANCE

The most sophisticated data governance framework in the world is worthless if it cannot function effectively in the daily operations of your organization. While previous chapters have equipped you with the strategic vision, structural foundations, and deployment methodologies necessary for governance success, this chapter addresses the critical challenge of making governance work seamlessly within your organization's operational rhythm.

Operationalizing data governance represents the bridge between ambitious strategic plans and tangible business outcomes. It transforms governance from a conceptual framework into a living, breathing system that enhances decision-making, ensures compliance, and drives continuous improvement across every data touchpoint in your enterprise. This operational transformation requires careful orchestration of people, processes, and technology to create sustainable governance practices that become second nature to your organization.

The journey from governance design to operational excellence demands a fundamental shift in perspective. Rather than viewing governance as an overlay on existing operations, successful organizations integrate governance principles into the fabric of their daily workflows. This integration creates a multiplier effect where governance activities strengthen business processes while business operations reinforce governance objectives.

In this chapter, you will discover how to architect day-to-day governance operations that scale with your organization's growth and complexity. We will explore the design of automated workflows that reduce manual overhead while maintaining governance rigor, examine the critical roles and responsibilities that make data stewardship effective, and establish escalation procedures that resolve issues before they impact business outcomes. Through real-world examples, including a comprehensive case study from the telecommunications industry, you will gain practical insights into building governance operations that deliver measurable value while positioning your organization for sustained competitive advantage.

10.1. Day-to-Day Governance Operations

The transition from strategic governance frameworks to operational reality requires a fundamental shift in focus from high-level policies to granular, executable processes that integrate seamlessly into existing organizational workflows. While governance frameworks provide the architectural blueprint, day-to-day operations represent the living, breathing implementation that determines whether your governance initiative thrives or withers. This operational layer encompasses the routine activities, decision-making processes, and continuous monitoring that transform governance from a compliance checkbox into a value-generating business capability.

Successful operational governance hinges on establishing clear, repeatable processes that data stewards, business users, and IT professionals can execute consistently without requiring specialized governance expertise. These operations must be embedded within existing business processes rather than creating parallel governance workflows that compete for attention and resources.

Establishing Operational Governance Rhythms

The cornerstone of effective day-to-day governance lies in creating predictable operational rhythms that align with business cycles and data consumption patterns. These rhythms encompass regular data quality assessments, stewardship reviews, policy compliance checks, and stakeholder communications that maintain governance momentum without overwhelming operational teams.

Consider the implementation at **Royal Bank of Canada**, where governance operations follow a structured quarterly rhythm. Their approach includes monthly data stewardship councils, bi-weekly quality scorecards, and weekly issue escalation reviews. This predictable cadence allows business units to plan governance activities alongside operational priorities, ensuring consistent participation and reducing governance fatigue.

Governance Committee Operations

Effective governance committees operate as decision-making engines rather than discussion forums. Operational excellence requires structured agendas, pre-meeting preparation protocols, and clear decision authority matrices. The **Commonwealth Bank of Australia** exemplifies this approach through their tiered committee structure: tactical data stewardship councils handle routine decisions weekly, while strategic governance boards address policy changes monthly.

Their operational model includes standardized decision templates, voting protocols, and escalation criteria that prevent bottlenecks while maintaining accountability. Committee members receive pre-meeting data packages with quality metrics, policy compliance reports, and recommended actions, enabling efficient decision-making during scheduled sessions.

Integration with Enterprise Workflows

Operational governance succeeds when it becomes invisible to end users while remaining highly visible to governance practitioners. This requires deep integration with existing enterprise systems, business processes, and user interfaces. Rather than forcing users to access separate governance tools, successful implementations embed governance controls within familiar applications and workflows.

Modern governance platforms like **Collibra** achieve this through API-driven integrations that surface data lineage, quality scores, and usage policies directly within business intelligence tools, data preparation platforms, and analytical applications. Users receive governance-relevant information contextually without disrupting their primary workflows.

Day-to-day governance operations transform abstract policies into concrete business value through consistent execution of well-designed processes, clear stakeholder accountability, and seamless integration with existing enterprise workflows. The practical skills gained from mastering operational governance include workflow automation, stakeholder engagement protocols, and performance measurement techniques that ensure governance remains both effective and sustainable.

Building upon these operational foundations, the next critical element involves designing automated workflows that reduce manual governance overhead while maintaining rigorous oversight and control mechanisms.

10.2. Workflow Design and Process Automation

Building upon these operational foundations, the next critical element involves designing automated workflows that reduce manual governance overhead while maintaining rigorous oversight and control mechanisms. Effective workflow design transforms governance from a reactive, manual process into a proactive, systematic operation that scales with organizational growth while ensuring consistent policy enforcement across all data touchpoints.

Automated Governance Workflow Architecture

Modern data governance workflows require sophisticated orchestration that balances automation efficiency with human oversight. The architecture begins with **trigger-based initiation**, where specific data events automatically launch governance processes. These triggers include new data source onboarding, policy violations detected through monitoring systems, or scheduled compliance reviews that require stakeholder participation.

Successful workflow design incorporates multiple decision points where automated systems can handle routine approvals while escalating complex scenarios to appropriate governance committees. The workflow engine must maintain detailed audit trails, capturing every decision point, approval timestamp, and stakeholder interaction for regulatory compliance and process improvement analysis.

Practical Implementation: Automated Data Classification Workflow

A multinational pharmaceutical company implemented an automated data classification workflow using Informatica Axon to handle their complex regulatory requirements across different global jurisdictions. The workflow addresses the challenge of classifying thousands of research datasets while ensuring compliance with both FDA regulations and European Medicines Agency standards.

Step 1: Automated Discovery and Initial Classification

```
Workflow Trigger: New dataset uploaded to research repository
Initial Processing:
- Automated metadata extraction
- Content scanning for regulatory keywords
- Preliminary sensitivity scoring (1-5 scale)
- Assignment to appropriate classification queue
```

Step 2: Rule-Based Decision Matrix

The system applies predefined classification rules based on content analysis, source system, and regulatory jurisdiction. Datasets containing patient information automatically receive "Highly Sensitive" classification, while anonymized research summaries route through standard classification workflows.

Step 3: Stakeholder Engagement Automation

For datasets requiring human review, the workflow automatically assigns tasks to relevant data stewards based on therapeutic area expertise and geographic responsibility. The system sends notifications with embedded decision forms, allowing stewards to review classification recommendations and provide approvals directly within their email systems.

Step 4: Quality Assurance and Escalation

Before final classification, the workflow routes decisions through a secondary review process. Chief Data Officers receive automated reports on classification volumes, processing times, and any escalated decisions requiring executive attention.

Results and Outcomes:

The automated workflow reduced classification processing time from 14 days to 3 days while improving consistency scores by 85%. The pharmaceutical company reported significant improvements in regulatory audit preparation, with classification documentation automatically generated and maintained throughout the workflow process.

Integration with Enterprise Data Workflows

Governance workflows must seamlessly integrate with existing enterprise data processes rather than creating parallel systems that increase complexity. This integration occurs through **API-based connectivity** with data pipelines, business intelligence platforms, and operational systems. The governance workflow engine monitors data movement, automatically applying policy controls and generating compliance reports without disrupting business operations.

Effective integration also requires **bidirectional communication** between governance workflows and business processes. When governance policies change, the workflow system must automatically update downstream processes and notify affected stakeholders of new requirements or procedures.

Risk Mitigation and Escalation Procedures

Robust workflow design includes sophisticated escalation mechanisms that ensure critical governance issues receive appropriate attention. The escalation matrix considers factors such as data sensitivity levels, potential regulatory impact, and business criticality when determining routing paths and response timeframes.

Performance Measurement and Continuous Improvement

Governance workflows generate extensive operational metrics that enable continuous optimization. Key performance indicators include processing times, approval rates, escalation frequencies, and stakeholder satisfaction scores. These metrics inform workflow refinements and identify bottlenecks that may require process redesign or additional automation capabilities.

This systematic approach to workflow automation transforms data governance from a manual, inconsistent process into a reliable, scalable operation that adapts to changing business requirements while maintaining rigorous oversight standards. Organizations implementing these automated workflows report significant improvements in governance efficiency and stakeholder engagement levels.

The foundation established through automated workflows creates the operational framework necessary for defining clear data stewardship roles and responsibilities that ensure consistent governance execution across all organizational levels.

10.3. Data Stewardship Activities and Responsibilities

Building on the operational framework established through automated workflows, the successful implementation of data stewardship requires clearly defined activities and responsibilities that translate governance policies into daily operational excellence. Data stewards serve as the critical bridge between strategic governance objectives and tactical data management execution, requiring structured role definitions, accountability frameworks, and measurable performance standards.

Core Stewardship Role Definitions

Effective data stewardship operates through a three-tiered responsibility model that ensures comprehensive coverage across all data domains. **Business data stewards** focus on domain-specific data quality, business rule validation, and stakeholder communication within their functional areas. **Technical data stewards** handle system-level data integrity, metadata management, and integration quality controls. **Executive data stewards** provide strategic oversight, resolve cross-functional conflicts, and ensure alignment with organizational objectives.

Each stewardship tier requires distinct competencies and decision-making authority. Business stewards must possess deep domain expertise combined with analytical skills to identify data anomalies and business impact. Technical stewards need systems knowledge and data architecture understanding to implement quality controls and monitor data lineage. Executive stewards require strategic thinking capabilities and organizational influence to drive governance adoption and resource allocation.

Daily Operational Activities Framework

Data stewardship effectiveness depends on consistent execution of routine activities that maintain data quality and governance compliance. Daily monitoring activities include data quality dashboard reviews, exception report analysis, and stakeholder communication regarding data issues. Weekly responsibilities encompass trend analysis, quality metric reporting, and coordination with IT teams on system improvements. Monthly activities involve governance committee participation, policy review sessions, and stakeholder feedback collection.

Practical Implementation: Financial Services Data Quality Stewardship

Consider how JPMorgan Chase implemented comprehensive data stewardship for their regulatory reporting systems following post-financial crisis compliance requirements. The implementation demonstrates practical stewardship activity structuring:

```
Daily Stewardship Workflow:
1. Morning quality dashboard review (30 minutes)
   - Critical data element validation
   - Overnight batch job exception analysis
   - Regulatory deadline impact assessment

2. Issue triage and prioritization (45 minutes)
   - Business impact scoring
   - Technical feasibility evaluation
   - Stakeholder notification protocols

3. Afternoon stakeholder communication (60 minutes)
   - Business unit consultation
   - IT coordination meetings
   - Executive status updates
```

The bank's stewards utilized Informatica Data Quality platform integrated with their existing risk management systems. Business stewards focused on regulatory data definitions and business rule validation, while technical stewards managed data lineage tracking and system integration quality. Executive stewards coordinated with compliance teams and provided board-level reporting on data governance effectiveness.

Results demonstrated measurable improvement: regulatory reporting accuracy increased from 87% to 99.2% within six months, compliance audit findings decreased by 75%, and data-related operational risk incidents dropped by 60%. The structured stewardship approach enabled the bank to transform regulatory compliance from a reactive burden into a proactive competitive advantage.

Performance Measurement and Accountability

Successful stewardship requires quantifiable performance metrics that align individual contributions with organizational governance objectives. Key performance indicators include data quality improvement rates, issue resolution timeframes, stakeholder satisfaction scores, and policy compliance percentages. These metrics provide objective assessment criteria while supporting continuous improvement initiatives.

Accountability frameworks must balance individual responsibility with collaborative governance success. Regular performance reviews should evaluate both technical competency and stakeholder engagement effectiveness. Recognition programs that celebrate stewardship achievements help maintain motivation and demonstrate organizational commitment to data governance excellence.

Data stewardship activities and responsibilities form the operational backbone of effective governance, transforming strategic policies into measurable business outcomes through structured role definitions, consistent daily practices, and performance-driven accountability. This operational foundation enables organizations to maintain governance momentum while adapting to evolving business requirements and regulatory landscapes.

The structured stewardship framework naturally leads to establishing robust issue escalation and resolution procedures that ensure data governance challenges receive appropriate attention and timely resolution across all organizational levels.

10.4. Issue Escalation and Resolution Procedures

The foundation of operational data governance rests on the ability to address challenges swiftly and systematically. While establishing data stewardship roles and accountability frameworks creates the structure for governance success, the true test of any governance program lies in how effectively it handles issues when they arise. Data quality problems, policy violations, and access disputes are inevitable in complex enterprise environments, making robust escalation procedures not just beneficial but essential for maintaining stakeholder confidence and organizational momentum.

Building Multi-Tiered Escalation Frameworks

Effective issue escalation requires a structured approach that matches problem severity with appropriate response levels. The most successful organizations implement a three-tiered escalation model that begins with frontline data stewards and progresses through domain experts to executive leadership when necessary.

The first tier addresses routine operational issues through automated workflows and designated data stewards. These might include standard data quality exceptions, routine access requests, or minor policy clarifications that can be resolved within established service level agreements. The second tier engages domain experts and governance coordinators for complex technical issues, cross-functional disputes, or situations requiring specialized knowledge. The third tier involves executive sponsors and governance committees for strategic decisions, major policy changes, or issues with significant business impact.

Practical Implementation: Healthcare Data Breach Response

Consider the implementation at Kaiser Permanente, where a comprehensive escalation framework proved critical during a data quality incident affecting patient medication records. The organization's three-tier approach demonstrates how structured escalation procedures transform potential crises into manageable governance processes.

Step 1: Initial Detection and Classification

When automated data quality monitoring detected inconsistencies in medication dosage fields across multiple systems, the alert triggered the first-tier response protocol:

```
Issue Classification: Tier 1 - Data Quality Exception
Severity: High (Patient Safety Impact)
Auto-Assignment: Regional Data Steward
SLA: 2-hour response, 8-hour resolution target
```

Step 2: Steward Assessment and Escalation Decision

The regional data steward quickly determined that the issue affected multiple clinical domains and required specialized pharmaceutical data expertise, triggering automatic escalation to Tier 2:

```
Escalation Trigger: Cross-domain impact detected
New Assignment: Clinical Data Governance Team
Additional Resources: Pharmacy informatics specialist
Updated SLA: 4-hour comprehensive assessment
```

Step 3: Executive Notification and Resolution

When the assessment revealed potential regulatory compliance implications, the framework automatically engaged Tier 3 protocols, notifying the Chief Medical Officer and triggering the organization's patient safety governance committee.

The structured approach enabled Kaiser Permanente to resolve the medication data inconsistencies within 12 hours, implement corrective measures across all affected systems, and document lessons learned for future prevention. Most importantly, the clear escalation path ensured that patient safety concerns received immediate executive attention while routine technical remediation proceeded efficiently.

Technology-Enabled Resolution Workflows

Modern data governance platforms provide sophisticated workflow capabilities that automate much of the escalation process. Tools like Collibra's workflow engine enable organizations to create dynamic escalation rules based on issue type, data sensitivity, and business impact. These platforms can automatically assign issues to appropriate stewards, track resolution progress, and escalate based on predefined criteria such as time elapsed or stakeholder involvement.

The integration of artificial intelligence into these workflows increasingly enables predictive escalation, where systems anticipate potential issues before they become critical problems. This proactive approach transforms governance from reactive problem-solving to preventive risk management.

Documentation and Continuous Improvement

Every escalated issue represents a learning opportunity that strengthens the overall governance framework. Successful organizations maintain detailed issue registries that capture not only the technical resolution but also the effectiveness of the escalation process itself. This documentation enables continuous refinement of escalation criteria, response procedures, and stakeholder communication protocols.

Well-designed escalation and resolution procedures create organizational confidence in the data governance program while ensuring that critical issues receive appropriate attention and resources. By establishing clear pathways for addressing challenges, organizations transform potential governance obstacles into opportunities for framework improvement and stakeholder engagement. The next critical element involves structuring governance committees and meetings that provide the organizational foundation for these escalation procedures to function effectively within broader decision-making processes.

10.5. Governance Committee Structures and Meetings

The foundation of any successful data governance program rests on establishing well-structured committees and conducting purposeful meetings that drive consistent decision-making across the enterprise. While escalation procedures provide the framework for addressing issues, the governance committee structure serves as the operational backbone where strategic decisions are made, policies are refined, and accountability is enforced. Effective committee structures transform governance from theoretical concepts into actionable organizational practices that deliver measurable business value.

Strategic Committee Hierarchy Design

A properly designed governance committee structure operates through multiple tiers, each with distinct responsibilities and decision-making authority. The **executive steering committee** provides strategic direction and resource allocation, typically meeting quarterly to review governance performance metrics and approve major policy changes. Below this level, the **data governance council** handles tactical decisions and cross-functional coordination, meeting monthly to address operational challenges and policy interpretations. **Domain-specific working groups** focus on specialized areas such as customer data, financial reporting, or regulatory compliance, convening bi-weekly to address technical implementation details and domain-specific standards.

The most effective structures avoid the common pitfall of creating too many overlapping committees that dilute accountability and slow decision-making. Instead, successful organizations establish clear **committee charters** that define scope, authority levels, and escalation pathways. Each committee should have explicit decision rights, measurable deliverables, and defined success criteria that align with broader governance objectives.

Practical Implementation: Financial Services Committee Structure

A leading European investment bank recently restructured its data governance committees to address regulatory compliance challenges following new data protection requirements. The organization implemented a three-tier structure that significantly improved decision-making speed and accountability.

Step 1: Executive Data Council Formation

The bank established a monthly executive council comprising the Chief Data Officer, Chief Risk Officer, and business line heads. This council focused exclusively on strategic decisions affecting multiple business units, with each meeting limited to 90 minutes and structured around three key agenda items: regulatory compliance status, cross-business data initiatives, and resource allocation decisions.

Step 2: Functional Working Groups Creation

Four specialized working groups were created: Risk Data Management, Customer Data Privacy, Regulatory Reporting, and Data Quality Standards. Each group met bi-weekly with rotating leadership from relevant business units. The Risk Data Management group, for example, included representatives from trading, compliance, and IT infrastructure teams.

Step 3: Decision Authority Matrix Implementation

The bank developed a comprehensive decision matrix using Informatica's governance platform to track committee decisions and approvals. The matrix clearly defined which committee could approve data classification changes, policy exceptions, and system access requests. For instance, customer data privacy policy changes required approval from both the Customer Data Privacy working group and the Executive Data Council.

```
Committee Decision Matrix Example:
- Policy Changes (Major): Executive Data Council approval required
- Data Classification Updates: Functional Working Group + CDO sign-off
- System Access Exceptions: Domain Working Group authority
- Cross-Business Data Sharing: Executive Data Council + Legal review
```

Results and Outcomes

Within six months, the bank reduced governance decision cycle times by 40% and achieved 95% compliance with new regulatory reporting requirements. The structured committee approach eliminated the previous ad-hoc decision-making process that had created inconsistent data policies across business units.

Meeting Effectiveness and Engagement Strategies

Successful governance meetings require careful orchestration to maintain engagement and drive actionable outcomes. **Meeting cadence optimization** ensures committees convene frequently enough to maintain momentum without overwhelming participants with excessive time commitments. High-performing organizations typically establish standard meeting templates that include governance scorecards, issue escalation reviews, and forward-looking action planning.

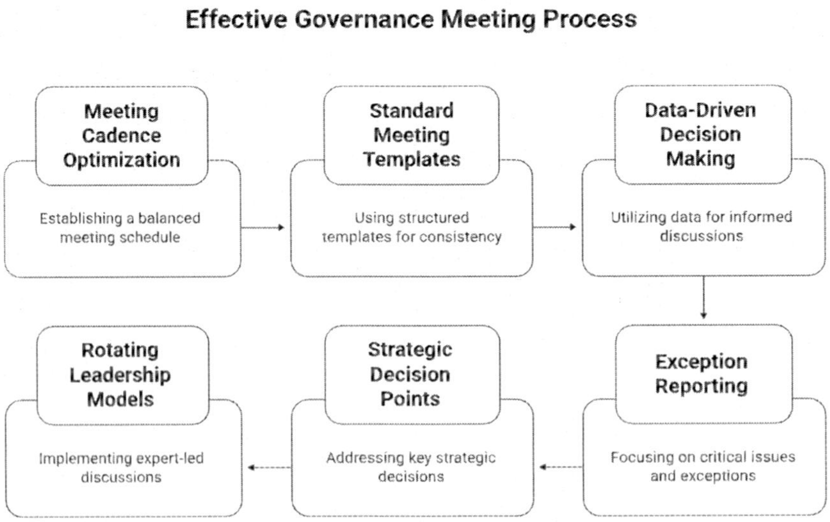

The most effective meetings incorporate **data-driven decision making** by presenting governance metrics, compliance dashboards, and performance trends that enable informed discussions. Rather than lengthy status updates, successful committees focus on exception reporting and strategic decision points that require collective input and approval.

Effective governance committees also implement **rotating leadership models** where subject matter experts lead discussions within their domains of expertise. This approach ensures that technical decisions receive appropriate attention while maintaining executive oversight of strategic direction and resource allocation.

The establishment of robust committee structures and purposeful meeting cadences creates the organizational foundation necessary for sustained governance success. These structures provide the decision-making framework that transforms governance policies into operational reality while ensuring accountability and continuous improvement. As governance programs mature, the focus naturally shifts toward measuring the effectiveness of these committee structures through comprehensive performance monitoring and reporting systems that demonstrate tangible business value and guide ongoing optimization efforts.

10.6. Performance Monitoring and Reporting

Building on the foundation of effective committee structures, IT leaders must establish sophisticated monitoring and reporting systems that transform governance activities into measurable business outcomes. Performance monitoring in data governance extends far beyond simple compliance checklists, requiring comprehensive frameworks that capture both operational effectiveness and strategic value creation. The challenge lies in designing reporting mechanisms that speak to diverse stakeholder groups while maintaining the technical rigor necessary for continuous improvement.

Establishing Multi-Dimensional Performance Frameworks

Effective governance monitoring requires a balanced approach that measures both leading and lagging indicators across multiple dimensions. **Operational metrics** track day-to-day governance activities, including data stewardship response times, policy compliance rates, and issue resolution cycles. **Strategic metrics** focus on business impact, measuring improvements in decision-making speed, risk reduction, and revenue generation from enhanced data quality. **Cultural metrics** assess governance maturity through stakeholder engagement levels, training completion rates, and cross-functional collaboration effectiveness.

The most successful implementations establish cascading metrics that align with organizational hierarchies. Executive dashboards emphasize business value and risk mitigation, while operational teams focus on process efficiency and quality improvements. This multi-layered approach ensures that governance reporting resonates with each audience while maintaining consistency in underlying measurement methodologies.

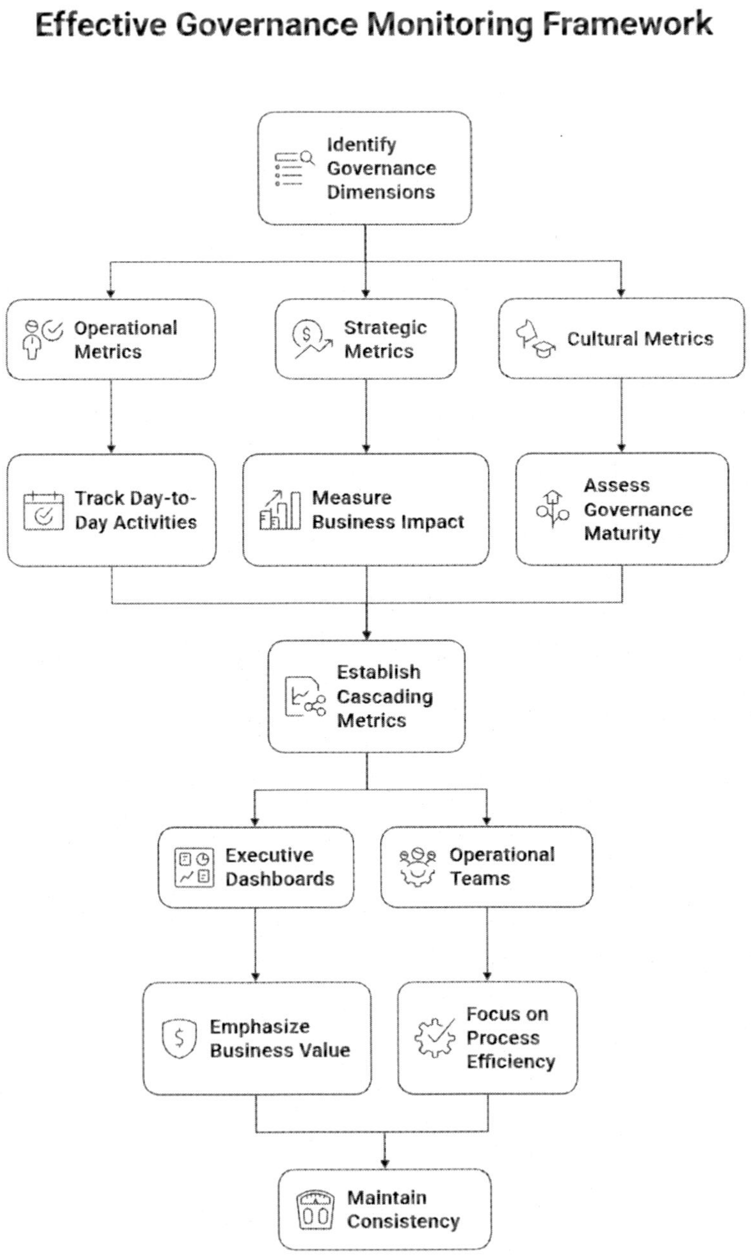

Real-Time Monitoring Implementation

Modern governance programs leverage automated monitoring capabilities to provide continuous visibility into data ecosystem health. **Data lineage tracking** enables real-time assessment of downstream impact when quality issues occur, while **automated policy enforcement** systems generate immediate alerts when governance violations are detected. These capabilities transform governance from a reactive discipline into a proactive strategic function.

Practical Implementation: Automated Governance Monitoring with Informatica

A global pharmaceutical company implemented comprehensive performance monitoring using Informatica's Axon Data Governance platform to track their clinical trial data governance program. The implementation followed these specific steps:

```
1. Baseline Establishment Phase
   - Configured 47 key performance indicators across data quality,
     compliance, and operational efficiency dimensions
   - Established automated data profiling for 12 critical clinical
     datasets with daily refresh cycles
   - Created role-based dashboards for regulatory affairs, clinical
     operations, and IT governance teams

2. Automated Alert Configuration
   - Set up real-time notifications for data quality thresholds
     below 95% completeness for patient enrollment data
   - Configured escalation workflows triggering within 2 hours
     for critical compliance violations
   - Implemented automated lineage impact analysis for regulatory
     submission datasets

3. Performance Dashboard Deployment
   - Executive dashboard: Monthly governance ROI metrics showing
     $2.3M in avoided regulatory penalties
   - Operational dashboard: Daily stewardship activity tracking
     with average 4-hour issue resolution time
   - Compliance dashboard: Real-time regulatory readiness scoring
     across 15 global markets
```

The results demonstrated measurable governance value: regulatory submission preparation time decreased by 40%, data quality incidents dropped by 65%, and compliance audit preparation time reduced from weeks to days. The automated monitoring system now processes over 500,000 daily data quality checks across their clinical trial portfolio.

Stakeholder Communication Strategies

Effective governance reporting requires tailored communication approaches that translate technical metrics into business language. **Executive reporting** focuses on strategic outcomes, risk mitigation, and competitive advantages gained through improved data management. **Operational reporting** emphasizes process efficiency, resource utilization, and tactical improvements that enhance daily workflows. **Technical reporting** provides detailed analytics on system performance, data quality trends, and infrastructure optimization opportunities.

The most impactful reporting programs establish regular communication rhythms that align with business planning cycles. Monthly governance scorecards track tactical performance, quarterly business reviews assess strategic progress, and annual governance assessments evaluate overall program maturity and future roadmap priorities.

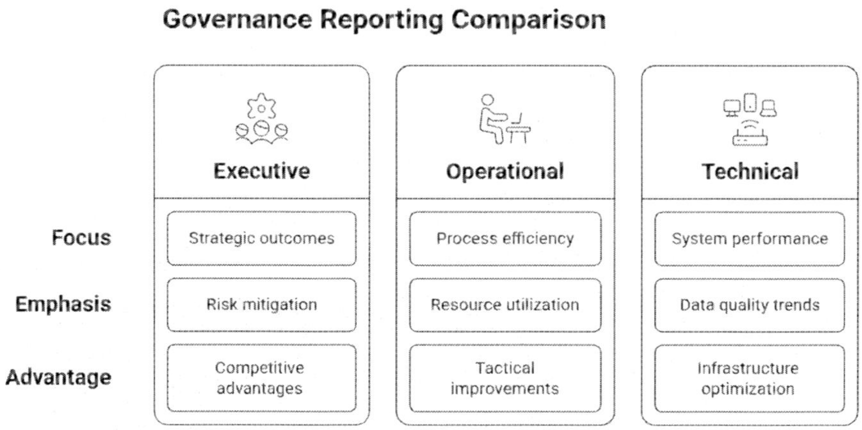

Advanced Analytics for Governance Optimization

Leading organizations leverage predictive analytics to anticipate governance challenges before they impact business operations. **Trend analysis** identifies emerging data quality patterns that may indicate systemic issues, while **predictive modeling** forecasts resource requirements for upcoming governance initiatives. These advanced capabilities enable proactive governance management that prevents issues rather than simply responding to them.

Performance monitoring and reporting systems serve as the nervous system of effective data governance, providing the feedback loops necessary for continuous optimization and stakeholder confidence. Organizations that invest in sophisticated monitoring capabilities gain the visibility needed to demonstrate governance value while identifying opportunities for ongoing improvement.

This comprehensive monitoring foundation enables organizations to implement systematic continuous process improvement methodologies that drive governance maturity to increasingly sophisticated levels.

10.7. Continuous Process Improvement

Building on the systematic monitoring foundation established in previous sections, organizations must now transform governance insights into actionable improvements that continuously elevate data stewardship maturity. The transition from reactive problem-solving to proactive enhancement represents a critical evolution in operational governance, requiring structured methodologies that embed improvement cycles directly into daily data operations.

Effective continuous improvement in data governance operates on multiple organizational levels simultaneously. At the tactical level, data stewards identify process bottlenecks and quality gaps through daily operations. At the strategic level, governance committees evaluate framework effectiveness against evolving business requirements. This multi-layered approach ensures that improvements address both immediate operational needs and long-term strategic alignment.

Implementing Kaizen Methodology for Data Governance

The Japanese concept of **Kaizen**, meaning continuous improvement, provides a powerful framework for systematically enhancing data governance processes. Unlike traditional project-based improvements, Kaizen embeds enhancement activities into regular workflows, creating sustainable momentum for governance evolution.

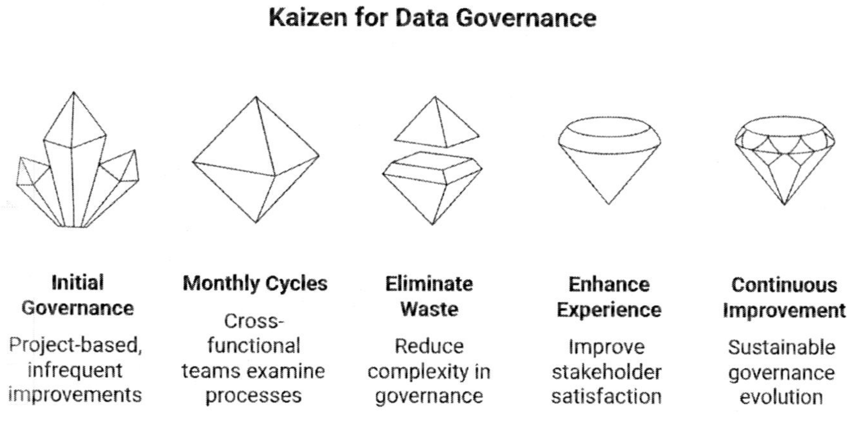

Organizations implementing Kaizen for data governance typically establish monthly improvement cycles where cross-functional teams examine specific governance processes. These sessions focus on eliminating waste, reducing complexity, and enhancing stakeholder experience. The methodology's emphasis on small, incremental changes makes it particularly suitable for governance environments where major disruptions can impact critical business operations.

Practical Implementation: Data Quality Improvement Cycle at Global Manufacturing

A multinational automotive manufacturer implemented a comprehensive continuous improvement program using Informatica Data Quality platform integrated with their existing governance framework. Their approach demonstrates how systematic enhancement cycles can transform governance effectiveness.

Phase 1: Baseline Assessment and Tool Configuration

The organization began by configuring Informatica's data profiling capabilities to automatically assess quality metrics across their master data domains. They established baseline measurements for completeness, accuracy, consistency, and timeliness across product catalogs, supplier databases, and manufacturing specifications.

```
Data Quality Assessment Framework:
- Completeness Score: 76% (Baseline)
- Accuracy Rate: 82% (Baseline)
- Consistency Index: 69% (Baseline)
- Timeliness Metric: 84% (Baseline)
- Business Rule Compliance: 71% (Baseline)
```

Phase 2: Automated Improvement Workflow Implementation

Using Informatica's workflow capabilities, they created automated improvement cycles that triggered monthly quality assessments, identified degradation patterns, and generated improvement recommendations. The system integrated with their existing governance committee structure, automatically routing quality issues to appropriate data stewards based on domain ownership.

Phase 3: Stakeholder Engagement and Process Refinement

Monthly improvement sessions brought together manufacturing engineers, procurement specialists, and IT data architects to review quality trends and implement targeted enhancements. Each session followed a structured agenda examining root causes, evaluating solution options, and implementing changes within defined governance approval processes.

Results After 18 Months:

Completeness Score increased to 94%
Accuracy Rate improved to 96%
Consistency Index reached 91%
Manufacturing defect rates decreased by 23%
Supplier onboarding time reduced by 40%

Governance Maturity Assessment Integration

Continuous improvement effectiveness depends heavily on accurate maturity assessment capabilities that provide objective evaluation of governance evolution. Leading organizations implement quarterly maturity assessments using frameworks like DAMA-DMBOK or custom evaluation models tailored to their specific industry requirements.

These assessments examine governance capabilities across multiple dimensions including policy adherence, stakeholder engagement, technology utilization, and business value delivery. The assessment results drive targeted improvement initiatives that address specific maturity gaps while building on existing strengths.

Performance Feedback Loop Architecture

Sustainable improvement requires robust feedback mechanisms that capture stakeholder experiences and translate them into actionable enhancement opportunities. Organizations achieve this through structured feedback collection processes that gather input from data consumers, stewards, and business users regarding governance process effectiveness.

Advanced feedback systems utilize sentiment analysis and automated survey tools to continuously monitor stakeholder satisfaction with governance services. This real-time feedback enables rapid identification of emerging issues and proactive resolution before they impact broader organizational operations.

Continuous process improvement transforms data governance from static policy enforcement into dynamic capability enhancement that adapts to evolving business needs. Organizations that master these improvement methodologies create self-reinforcing governance systems that become more effective over time while maintaining stakeholder engagement and business alignment.

This systematic approach to governance enhancement establishes the operational foundation necessary for achieving excellence in specific industry contexts, where specialized requirements demand tailored governance strategies that balance standardization with sector-specific optimization needs.

10.8. Operational Excellence in Telecommunications Industry

The telecommunications industry presents unique operational challenges that require sophisticated data governance frameworks capable of managing massive data volumes, real-time processing demands, and stringent regulatory compliance across multiple jurisdictions. Unlike other sectors, telecom operators must balance customer privacy requirements with network optimization needs while maintaining service quality standards that directly impact millions of users simultaneously.

Network Performance Data Governance Framework

Telecommunications companies generate petabytes of network performance data daily, requiring governance frameworks that can handle high-velocity data streams while ensuring data quality for critical operational decisions. The governance structure must accommodate real-time analytics for network optimization, fraud detection, and customer experience management while maintaining historical data integrity for regulatory reporting and capacity planning.

A comprehensive telecom data governance framework typically incorporates **network data classification schemes** that distinguish between operational metrics, customer interaction data, and regulatory compliance information. Each category requires different retention policies, access controls, and quality standards aligned with both business objectives and regulatory mandates such as GDPR, CCPA, and telecommunications-specific regulations.

Practical Implementation: Real-Time Network Quality Management

Consider Vodafone's implementation of a comprehensive data governance framework for their European operations, which demonstrates operational excellence in telecommunications data management. Their approach integrates multiple governance tools to manage network performance data across 17 countries while maintaining consistent quality standards and regulatory compliance.

Step 1: Data Classification and Inventory

```
Network Performance Data Categories:
- Real-time KPIs: Call drop rates, data throughput, latency metrics
- Customer Experience Data: Service quality scores, complaint patterns
- Infrastructure Data: Cell tower performance, equipment health status
- Regulatory Data: Coverage maps, emergency services compliance
```

Step 2: Governance Tool Integration

Vodafone utilized Collibra for data catalog management combined with Informatica for data quality monitoring across their network operations centers. The implementation process involved:

Policy Development Sessions: Cross-functional teams including network engineers, data scientists, and compliance officers collaborated to define data quality rules specific to telecommunications operations

Automated Quality Monitoring: Real-time data quality checks were implemented for critical network metrics, with automatic escalation procedures for quality threshold breaches

Regulatory Compliance Integration: Governance workflows were designed to automatically generate regulatory reports while maintaining data lineage for audit purposes

Before Implementation: Network quality data existed in siloed systems with inconsistent definitions, leading to conflicting performance reports and delayed issue resolution.

After Implementation: Unified data definitions across all European operations, 40% reduction in network issue resolution time, and automated regulatory compliance reporting with 99.7% accuracy.

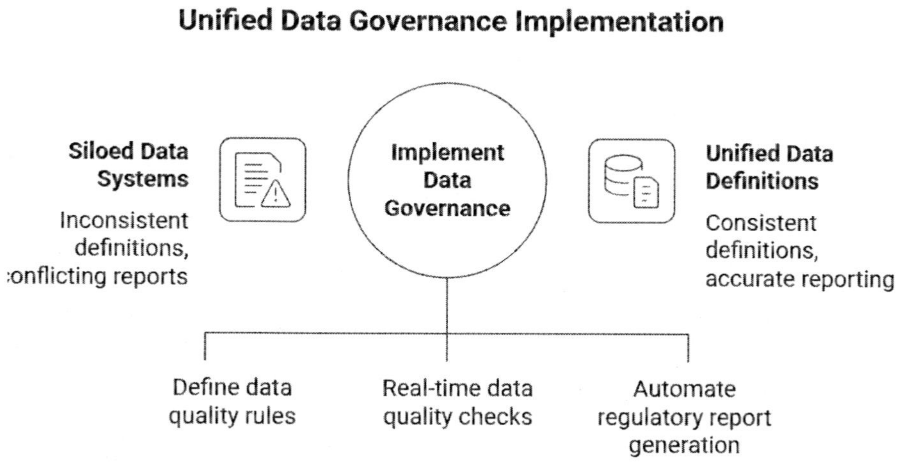

Governance Committee Structure for Telecom Operations

Telecommunications data governance requires specialized committee structures that can make rapid decisions while maintaining oversight of critical network operations. The **Network Data Governance Council** typically includes representatives from network operations, customer experience, regulatory affairs, and data protection teams, with escalation procedures for service-impacting data quality issues.

The governance framework must also address **cross-border data management** challenges unique to telecommunications, where customer data may traverse multiple jurisdictions during normal network operations. This requires sophisticated data classification schemes and automated policy enforcement mechanisms that can adapt to different regulatory environments in real-time.

Risk Mitigation and Performance Measurement

Telecom-specific governance frameworks incorporate specialized risk assessment methodologies that evaluate both operational and compliance risks. Performance metrics focus on data availability, quality accuracy for network optimization algorithms, and compliance audit readiness. Success measurements typically include network performance improvement percentages, regulatory compliance scores, and customer experience enhancement metrics directly attributable to improved data governance practices.

The telecommunications industry's operational excellence in data governance demonstrates how sector-specific requirements can be successfully addressed through tailored governance frameworks that maintain standardization while accommodating unique industry demands. This approach provides a foundation for sustainable governance practices that evolve with technological advancement and regulatory changes.

This telecommunications governance model establishes the groundwork for understanding how operational excellence principles can be adapted across various industry contexts, setting the stage for comprehensive governance framework evaluation and continuous improvement strategies.

11. MEASURING SUCCESS AND DEMONSTRATING VALUE

The most sophisticated data governance framework in the world means nothing if you cannot demonstrate its value to the organization. After months of careful planning, stakeholder engagement, and systematic implementation, governance leaders inevitably face the critical question from executive sponsors: "What tangible benefits are we seeing from this investment?" This pivotal moment separates successful governance programs that secure ongoing support and resources from those that quietly fade into bureaucratic obscurity.

Measuring success in data governance extends far beyond simple compliance checkboxes or technical quality scores. Today's business leaders demand clear evidence that governance initiatives drive measurable improvements in decision-making speed, risk reduction, operational efficiency, and revenue generation. The challenge lies in translating abstract concepts like "data stewardship" and "information quality" into concrete metrics that resonate with financial stakeholders and operational managers alike.

This chapter provides the essential framework for quantifying governance impact through comprehensive measurement strategies that span technical performance, business outcomes, and organizational maturity. You will discover how to construct meaningful key performance indicators that align with strategic objectives, calculate return on investment using proven methodologies, and present compelling evidence of value creation through executive dashboards and governance scorecards.

The measurement approaches explored here transform governance from a perceived cost center into a recognized profit driver. Through practical examples from retail and e-commerce organizations, you will learn how leading companies demonstrate millions in value through improved data quality, accelerated compliance, and enhanced customer insights. These proven techniques enable governance teams to secure executive buy-in, justify budget allocations, and build sustainable programs that continuously evolve with organizational needs.

11.1. Key Performance Indicators for Data Governance

Effective data governance measurement requires a strategic approach that transforms technical activities into quantifiable business outcomes. While many organizations focus on implementation metrics, true governance success depends on establishing **Key Performance Indicators (KPIs)** that align with enterprise objectives and demonstrate clear value to stakeholders across all organizational levels.

The foundation of governance measurement lies in creating a balanced scorecard approach that captures both operational excellence and strategic impact. Organizations must move beyond simple compliance checklists to develop comprehensive metrics that reflect data quality improvements, risk reduction, and business enablement capabilities. This multidimensional approach ensures that governance initiatives remain relevant and continue to receive executive support throughout their lifecycle.

Operational Excellence Metrics

Operational KPIs form the backbone of governance measurement by tracking the fundamental health of data management processes. These metrics include **data quality scores**, which measure accuracy, completeness, consistency, and timeliness across critical data domains. Organizations typically establish baseline measurements and track improvement trajectories over time, with targets aligned to business requirements rather than arbitrary benchmarks.

Data stewardship effectiveness represents another crucial operational metric, measuring how well designated stewards fulfill their responsibilities. This includes tracking issue resolution times, data definition completeness, and stakeholder satisfaction ratings. Leading organizations also monitor **policy compliance rates**, measuring adherence to established data handling procedures and governance standards across different business units.

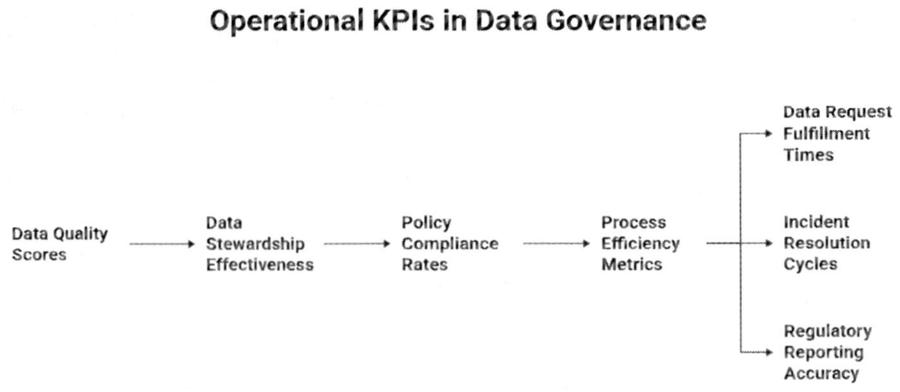

Process efficiency metrics capture the operational impact of governance investments. These include **data request fulfillment times**, **incident resolution cycles**, and **regulatory reporting accuracy**. By tracking these metrics consistently, organizations can identify bottlenecks and optimize their governance processes for maximum effectiveness.

Strategic Impact Indicators

Strategic KPIs demonstrate governance value at the enterprise level by connecting data management activities to business outcomes. **Data-driven decision velocity** measures how quickly organizations can access reliable data for critical business decisions. This metric often correlates directly with competitive advantage and market responsiveness capabilities.

Risk mitigation effectiveness quantifies governance impact on enterprise risk management. Organizations track metrics such as **data breach prevention rates**, **regulatory violation reductions**, and **audit finding improvements**. These indicators provide compelling evidence of governance value to risk-conscious executives and board members.

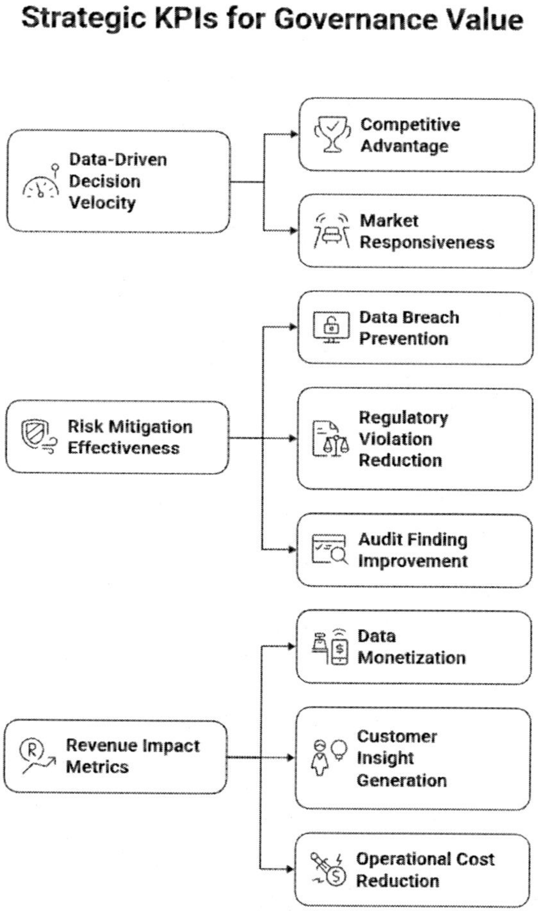

Revenue impact metrics connect governance activities to financial performance. Organizations measure **data monetization opportunities**, **customer insight generation rates**, and **operational cost reductions** achieved through improved data management. These metrics transform governance from a cost center into a recognized profit contributor.

Practical Implementation Example

Consider the implementation at **ING Bank**, which developed a comprehensive governance KPI framework during their digital transformation initiative. The organization established a three-tier measurement system using **Collibra** as their governance platform, tracking over 150 data quality rules across critical banking domains.

Step 1: Baseline Establishment

ING began by measuring data quality across core banking processes, discovering that customer data accuracy was only 73% across their European operations. They established automated quality monitoring using Collibra's data quality engine, creating real-time dashboards for business stakeholders.

Step 2: Governance Process Metrics

The bank implemented stewardship effectiveness tracking, measuring how quickly data stewards resolved quality issues. Initial resolution times averaged 14 days, with significant variation across business units. They established targets of 5-day resolution for critical issues and 10 days for standard problems.

Step 3: Business Impact Correlation

ING connected governance metrics to business outcomes by tracking **customer onboarding cycle times** and **regulatory reporting accuracy**. They discovered that improved data quality reduced customer onboarding from 8 days to 3 days, directly impacting customer satisfaction and competitive positioning.

Results Achieved:

Within 18 months, ING demonstrated measurable improvements: customer data accuracy increased to 94%, regulatory reporting accuracy reached 99.2%, and data-related customer complaints decreased by 67%. The governance program generated an estimated €15 million in operational savings while significantly reducing compliance risks.

This comprehensive KPI framework provides organizations with the measurement foundation necessary to demonstrate governance value and maintain stakeholder support. By balancing operational metrics with strategic indicators, governance programs can prove their worth while identifying continuous improvement opportunities.

Moving forward, organizations must translate these performance indicators into compelling business value narratives that resonate with different stakeholder groups, particularly when calculating return on investment and justifying continued governance investments.

11.2. Business Value Metrics and ROI Calculation

Quantifying the business impact of data governance requires a sophisticated approach that goes beyond simple cost-benefit calculations. Organizations must develop comprehensive measurement frameworks that capture both tangible financial returns and intangible strategic benefits. The key lies in establishing baseline metrics before governance implementation, then tracking improvements across multiple dimensions of business performance.

Financial Impact Measurement Framework

Successful ROI calculations for data governance initiatives require a multi-layered approach that captures direct cost savings, revenue enhancements, and risk mitigation benefits. Organizations should establish measurement categories that include operational efficiency gains, compliance cost reductions, and strategic decision-making improvements.

Direct cost savings typically emerge from reduced data preparation time, decreased system maintenance overhead, and eliminated redundant data storage. Revenue enhancements manifest through improved customer segmentation accuracy, faster time-to-market for new products, and enhanced cross-selling effectiveness. Risk mitigation benefits include avoided regulatory penalties, reduced security breach costs, and prevented reputational damage.

Practical ROI Calculation Methodology

Consider a multinational pharmaceutical company implementing comprehensive data governance using Informatica's Data Quality and Master Data Management platform. The organization faced challenges with clinical trial data inconsistencies, regulatory reporting delays, and duplicated patient records across research sites.

Step 1: Baseline Assessment

The company documented pre-governance metrics including 40 hours weekly spent on data reconciliation, $2.3 million annual compliance penalties, and 15% error rates in clinical trial reporting. Data preparation consumed 60% of analyst time, while regulatory submissions averaged 8.2 weeks completion time.

Step 2: Implementation Tracking

After deploying Informatica's governance platform with automated data quality rules and master data workflows, the organization tracked improvements monthly. Data reconciliation time dropped to 8 hours weekly, compliance penalties decreased to $200,000 annually, and reporting error rates fell to 3%.

Step 3: Value Calculation

```
Annual Labor Savings = (32 hours weekly × 52 weeks × $75/hour) =
$124,800
Compliance Cost Reduction = ($2.3M - $200K) = $2,100,000
Efficiency Gains = (Reduced submission time × project volume ×
opportunity cost)
Total Annual Benefits = $3,847,000
Implementation Costs = $1,200,000 (platform + services)
Three-Year ROI = 320%
```

Strategic Value Indicators

Beyond financial metrics, organizations must measure strategic value indicators that demonstrate governance's contribution to competitive advantage. These include data-driven decision velocity, cross-functional collaboration effectiveness, and innovation pipeline acceleration. Companies should track metrics such as time-to-insight reduction, self-service analytics adoption rates, and business user data confidence scores.

Leading organizations establish governance scorecards that balance quantitative financial metrics with qualitative strategic indicators. These scorecards typically include customer satisfaction improvements attributable to better data quality, employee productivity gains from streamlined data access, and market responsiveness enhancements from faster analytics capabilities.

Stakeholder-Specific Value Narratives

Different stakeholder groups require tailored value propositions that resonate with their specific priorities and concerns. CFOs focus on cost reduction and risk mitigation metrics, while CMOs emphasize customer insight quality and campaign effectiveness improvements. IT leaders appreciate infrastructure optimization and operational efficiency gains, whereas business unit heads value decision-making speed and competitive intelligence enhancements.

Successful governance leaders develop stakeholder-specific dashboards that highlight relevant metrics and translate technical improvements into business language. These narratives should connect governance investments to strategic business outcomes, demonstrating clear causation between governance activities and performance improvements.

This comprehensive approach to measuring governance value creates a foundation for sustained executive support and continued investment. By establishing robust measurement frameworks and calculating meaningful ROI, organizations transform data governance from a cost center into a recognized profit driver. The next critical step involves assessing organizational governance maturity to identify improvement opportunities and benchmark progress against industry standards.

11.3. Governance Maturity Assessment Models

Building on the foundation of demonstrating governance value as a profit driver, organizations must now systematically evaluate their current governance capabilities to identify improvement opportunities and benchmark progress against industry standards. Governance maturity assessment models provide structured frameworks for measuring organizational readiness, capability gaps, and progress toward data governance excellence.

Understanding Governance Maturity Dimensions

Effective governance maturity models evaluate organizations across multiple dimensions that collectively determine overall governance effectiveness. The **Data Management Body of Knowledge (DAMA-DMBOK)** framework identifies key assessment areas including data architecture, data modeling, data storage, data integration, data quality, and data governance itself. Each dimension requires specific evaluation criteria that reflect both technical capabilities and organizational readiness.

Leading organizations like Axa Group have successfully implemented comprehensive maturity assessments using the **Capability Maturity Model Integration (CMMI)** adapted for data governance. Their assessment framework evaluates five maturity levels: Initial (ad-hoc processes), Managed (documented procedures), Defined (standardized processes), Quantitatively Managed (measured performance), and Optimizing (continuous improvement). This structured approach enables systematic capability building and progress tracking across global operations.

Implementing Structured Assessment Frameworks

The most effective maturity assessments combine quantitative metrics with qualitative evaluations to provide comprehensive organizational insights. Assessment frameworks should evaluate governance processes, technology infrastructure, organizational culture, and stakeholder engagement levels. Regular assessments create baseline measurements and track improvement trajectories over time.

Practical Implementation Example: Healthcare System Maturity Assessment

A major healthcare network implemented a comprehensive governance maturity assessment using Informatica's Data Governance and Catalog platform combined with custom evaluation frameworks. The assessment process included:

Initial Capability Mapping: The organization evaluated 127 data domains across five hospitals, identifying current governance practices, data quality levels, and compliance readiness. Using Informatica's assessment tools, they discovered that only 23% of critical data elements had defined ownership, and data quality scores averaged 2.1 out of 5.0.

Stakeholder Readiness Evaluation: Leadership conducted structured interviews with 45 department heads using standardized maturity questionnaires. The assessment revealed significant variation in governance understanding, with clinical departments scoring 1.8/5.0 on governance awareness while IT departments scored 4.2/5.0.

Technology Infrastructure Assessment: Technical teams evaluated existing systems against governance requirements, discovering that legacy systems lacked data lineage tracking and audit capabilities. The assessment identified 12 critical integration points requiring governance framework updates.

Maturity Scoring and Gap Analysis: The comprehensive assessment resulted in an overall maturity score of 2.3/5.0 (Managed level), with specific gaps in data stewardship (1.9/5.0), policy enforcement (2.1/5.0), and cross-departmental collaboration (1.7/5.0).

Improvement Roadmap Development: Based on assessment results, the organization developed a 24-month improvement plan targeting specific maturity gaps, with quarterly reassessments to track progress and adjust strategies.

Assessment Tool Selection and Customization

Organizations must carefully select assessment tools that align with their governance objectives and organizational structure. Commercial platforms like Collibra's Data Intelligence Cloud provide built-in maturity assessment capabilities, while custom frameworks offer greater flexibility for unique organizational requirements. The key lies in balancing standardization with organizational specificity.

Successful assessment implementations require clear scoring methodologies, consistent evaluation criteria, and regular calibration across assessment teams. Organizations should establish assessment governance committees to ensure consistent application of evaluation standards and maintain assessment integrity over time.

This systematic approach to governance maturity assessment provides organizations with clear visibility into current capabilities, specific improvement targets, and measurable progress indicators. Regular maturity assessments transform governance from subjective evaluation into objective capability management, enabling data-driven decisions about governance investments and priorities.

Moving beyond internal capability assessment, organizations must also measure how effectively their governance initiatives satisfy stakeholder needs and drive meaningful engagement across the enterprise.

11.4. Stakeholder Satisfaction and Engagement Metrics

Building beyond internal capability assessment, organizations must also measure how effectively their governance initiatives satisfy stakeholder needs and drive meaningful engagement across the enterprise. Stakeholder satisfaction serves as a critical barometer for governance success, revealing whether your framework creates genuine value or merely adds bureaucratic overhead. Unlike traditional IT metrics that focus on system performance, stakeholder satisfaction metrics capture the human dimension of governance, measuring trust, usability, and perceived value across diverse organizational constituencies.

Effective stakeholder satisfaction measurement requires a multi-dimensional approach that captures both quantitative engagement levels and qualitative feedback on governance utility. The most successful organizations establish baseline satisfaction scores before governance implementation, then track improvement trends across different stakeholder groups including business users, data stewards, compliance officers, and executive sponsors.

Designing Comprehensive Stakeholder Feedback Systems

Modern stakeholder feedback systems extend far beyond simple surveys, incorporating continuous listening mechanisms that capture real-time sentiment and engagement patterns. Leading organizations implement **pulse surveys** every quarter, supplemented by in-depth annual assessments that explore governance impact on daily workflows and decision-making processes. These systems must account for varying stakeholder perspectives, as business analysts may prioritize data accessibility while compliance teams focus on control mechanisms.

The most effective feedback systems segment stakeholders by role, department, and governance touchpoints, enabling targeted measurement strategies. For instance, data consumers require different satisfaction metrics than data producers, and executive stakeholders need different engagement measures than operational users. This segmentation allows governance teams to identify specific pain points and tailor improvement initiatives accordingly.

Practical Implementation: Stakeholder Engagement Dashboard in Collibra

Consider how a multinational pharmaceutical company implemented comprehensive stakeholder satisfaction tracking using Collibra's governance platform. The organization faced declining engagement in their clinical trial data governance program, with researchers increasingly bypassing established data request processes.

Step 1: Baseline Assessment Configuration

The governance team configured Collibra's workflow engine to automatically capture stakeholder interaction data, including data request fulfillment times, approval bottlenecks, and user feedback scores. They established baseline metrics across four stakeholder categories: clinical researchers, regulatory affairs specialists, data stewards, and IT administrators.

```
Stakeholder Satisfaction Framework:
- Researchers: Data accessibility (target: 85% same-day fulfillment)
- Regulatory: Compliance confidence (target: 95% audit readiness)
- Stewards: Process efficiency (target: 70% reduction in manual tasks)
- IT Teams: System reliability (target: 99.5% platform uptime)
```

Step 2: Continuous Feedback Integration

The team embedded satisfaction surveys directly into Collibra's user interface, triggering brief assessments after major governance interactions. They implemented a **Net Promoter Score (NPS)** system specifically for governance processes, asking stakeholders how likely they were to recommend the data governance framework to colleagues.

Step 3: Engagement Analytics Implementation

Using Collibra's analytics capabilities, they tracked engagement patterns including login frequency, data catalog usage, policy acknowledgment rates, and collaboration metrics. The platform automatically generated monthly stakeholder engagement reports showing trends across different user segments.

Results and Transformation

Within six months, the pharmaceutical company observed remarkable improvements. Researcher satisfaction scores increased from 3.2 to 4.1 (out of 5), while data request fulfillment improved by 40%. Most significantly, voluntary compliance with governance processes increased from 65% to 89%, demonstrating genuine stakeholder buy-in rather than reluctant adherence.

Measuring Cross-Functional Collaboration Effectiveness

Successful governance creates synergies between previously siloed departments, making cross-functional collaboration metrics essential for comprehensive stakeholder assessment. These metrics capture how governance facilitates knowledge sharing, reduces departmental friction, and enables coordinated decision-making across organizational boundaries.

Effective collaboration metrics include **joint project success rates**, where governance-enabled teams demonstrate higher completion rates and better outcomes compared to non-governed initiatives. Organizations also measure **knowledge transfer efficiency**, tracking how quickly insights and best practices spread across departments through governance-enabled communication channels.

The most sophisticated organizations implement **collaboration network analysis**, mapping how governance initiatives strengthen or weaken relationships between different stakeholder groups. This analysis reveals whether governance creates productive partnerships or inadvertently introduces new organizational barriers.

Sustaining Long-term Stakeholder Engagement

Long-term stakeholder satisfaction requires continuous evolution of governance practices based on changing business needs and stakeholder expectations. Organizations must establish **feedback loop mechanisms** that translate stakeholder insights into concrete governance improvements, demonstrating responsive leadership and maintaining engagement momentum.

Regular stakeholder forums, governance advisory committees, and user experience sessions create ongoing dialogue opportunities that prevent satisfaction erosion. These mechanisms also serve as early warning systems, identifying potential dissatisfaction before it impacts governance effectiveness or organizational compliance.

This comprehensive approach to stakeholder satisfaction measurement transforms governance from a top-down mandate into a collaborative framework that genuinely serves organizational needs. By systematically tracking and responding to stakeholder feedback, organizations build sustainable governance practices that strengthen over time rather than decay through neglect.

Key Takeaways: Stakeholder satisfaction metrics provide essential insights into governance effectiveness beyond technical performance measures. Implementing comprehensive feedback systems, leveraging platform capabilities for continuous monitoring, and maintaining responsive improvement processes ensures governance frameworks deliver genuine value while sustaining long-term organizational engagement.

Understanding stakeholder satisfaction patterns naturally leads to examining how governance initiatives drive measurable improvements in data quality and operational outcomes, forming the foundation for comprehensive quality improvement tracking and reporting systems.

11.5. Quality Improvement Tracking and Reporting

Building on the foundation of measurable governance improvements, organizations must establish systematic approaches to track and report quality enhancements across their data ecosystem. Effective quality improvement tracking transforms scattered data quality efforts into cohesive, measurable programs that demonstrate tangible business value while providing actionable insights for continuous enhancement.

Establishing Quality Baseline Measurements

The cornerstone of effective quality improvement tracking lies in establishing comprehensive baseline measurements that capture the current state of data quality across all critical business domains. Organizations must implement systematic data profiling initiatives that assess accuracy, completeness, consistency, timeliness, and validity metrics across their enterprise data landscape.

Modern data quality platforms like Informatica Data Quality and Talend Data Quality provide automated profiling capabilities that generate detailed quality scorecards. These tools examine data patterns, identify anomalies, and establish statistical baselines that serve as reference points for measuring improvement over time. The key is implementing consistent measurement methodologies that remain stable throughout the governance lifecycle, ensuring reliable trend analysis and meaningful progress tracking.

Implementing Continuous Quality Monitoring Frameworks

Sustainable quality improvement requires continuous monitoring frameworks that provide real-time visibility into data quality trends and immediate alerts when quality thresholds are breached. Organizations should deploy automated quality monitoring solutions that integrate seamlessly with existing data pipelines, providing continuous assessment without disrupting operational workflows.

Practical Implementation Example: Global Manufacturing Quality Tracking System

A multinational manufacturing corporation implemented a comprehensive quality tracking system using Collibra's Data Quality framework to monitor product specification data across 47 manufacturing facilities worldwide.

Step 1: Quality Dimension Definition

The organization defined five critical quality dimensions for product specification data:

- Completeness: All required specification fields populated
- Accuracy: Specifications match engineering standards
- Consistency: Uniform specification formats across facilities
- Timeliness: Specifications updated within 24 hours of changes
- Validity: Specifications conform to regulatory requirements

Step 2: Automated Monitoring Implementation

```
Quality Rule Configuration:
- Completeness Rule: CHECK (specification_weight IS NOT NULL AND
specification_dimensions IS NOT NULL)
- Accuracy Rule: VALIDATE specification_tolerance BETWEEN
engineering_min AND engineering_max
- Consistency Rule: MATCH specification_format AGAINST
global_standard_template
- Timeliness Rule: VERIFY last_updated_timestamp < CURRENT_TIMESTAMP -
INTERVAL 24 HOURS
- Validity Rule: CONFIRM regulatory_compliance_flag = 'APPROVED'
```

Step 3: Dashboard Development

The implementation included executive dashboards showing:

- Real-time quality scores by facility and product line
- Trend analysis comparing monthly quality improvements
- Exception reporting highlighting facilities requiring intervention
- Cost impact analysis linking quality improvements to operational savings

Results Achieved:

- 34% improvement in specification completeness within six months
- 28% reduction in production delays caused by specification errors
- $2.3 million annual savings from reduced rework and quality incidents
- 89% reduction in time spent on manual quality audits

Developing Comprehensive Quality Reporting Systems

Effective quality reporting systems must cater to diverse stakeholder needs, providing technical details for data stewards while delivering executive summaries that highlight business impact. Organizations should implement tiered reporting structures that automatically generate appropriate content for different audience levels, ensuring consistent messaging while maintaining relevant detail depth.

Quality reports should incorporate trend analysis, root cause identification, and predictive indicators that help stakeholders understand not just current quality levels but also projected improvement trajectories. Integration with business intelligence platforms enables dynamic reporting that allows stakeholders to drill down into specific quality issues while maintaining overall program visibility.

Measuring Quality Impact on Business Outcomes

The most compelling quality improvement tracking connects data quality enhancements directly to business performance improvements. Organizations must establish clear linkages between quality metrics and key performance indicators such as customer satisfaction, operational efficiency, revenue recognition accuracy, and regulatory compliance scores.

Advanced analytics platforms can correlate quality improvements with business outcomes, demonstrating how enhanced data accuracy translates into reduced customer service calls, improved inventory management, or accelerated financial reporting cycles. This correlation analysis provides the business case justification necessary for continued governance investment and expansion.

Quality improvement tracking and reporting transforms data governance from a technical initiative into a strategic business capability. By establishing baseline measurements, implementing continuous monitoring, developing comprehensive reporting systems, and measuring business impact, organizations create sustainable frameworks that demonstrate ongoing value while driving continuous enhancement. This systematic approach to quality tracking provides the foundation for expanding measurement capabilities into compliance and risk reduction assessments, where governance value extends beyond operational efficiency into regulatory protection and strategic risk mitigation.

11.6. Compliance and Risk Reduction Measurements

Building on operational measurement frameworks, compliance and risk reduction metrics represent the most tangible demonstration of governance value to executive leadership and regulatory bodies. These measurements transform abstract governance activities into quantifiable risk mitigation outcomes that directly protect organizational assets and reputation. Unlike operational metrics that focus on efficiency gains, compliance measurements demonstrate how governance frameworks prevent costly violations, reduce audit findings, and strengthen regulatory relationships.

Regulatory Compliance Tracking Framework

Effective compliance measurement requires establishing baseline risk exposure before implementing governance controls, then tracking improvement trajectories across multiple regulatory domains. Organizations must develop comprehensive tracking systems that monitor compliance posture in real-time while providing historical trend analysis for regulatory reporting and strategic planning.

The most sophisticated compliance measurement frameworks integrate automated monitoring with manual assessment processes, creating multi-layered validation systems that catch potential violations before they escalate into regulatory incidents. These systems track compliance across data protection regulations, industry-specific requirements, and internal policy adherence, providing granular visibility into governance effectiveness.

Practical Implementation: GDPR Compliance Measurement System

A multinational financial services organization implemented a comprehensive GDPR compliance measurement framework using Collibra's governance platform integrated with their existing risk management systems. The implementation demonstrates how systematic compliance measurement transforms reactive compliance management into proactive risk prevention.

Step 1: Baseline Risk Assessment Configuration

The organization configured Collibra's Data Privacy module to automatically scan their enterprise data landscape, identifying personal data processing activities across 47 business systems. They established baseline measurements including:

```
Compliance Baseline Metrics:
- Personal data processing activities: 1,247 identified
- Data subject rights request response time: 18.3 days average
- Privacy policy alignment score: 67%
- Data retention policy compliance: 43%
- Cross-border transfer documentation: 28% complete
```

Step 2: Automated Monitoring Implementation

The team configured automated compliance monitoring workflows that track key GDPR requirements in real-time. The system monitors data processing activities, consent management, and retention policy adherence through integration with enterprise data systems.

Step 3: Risk Reduction Measurement Protocol

After six months of governance implementation, the organization documented significant compliance improvements:

```
Post-Implementation Results:
- Data subject rights response time: 6.2 days average (66% improvement)
- Privacy policy alignment score: 94% (40% improvement)
- Data retention compliance: 89% (107% improvement)
- Cross-border transfer documentation: 96% complete (243% improvement)
- Regulatory inquiry response time: 2.1 days (previously 14 days)
```

The measurement system automatically generates compliance scorecards that demonstrate governance value through risk reduction metrics, showing potential avoided fines of €2.3 million based on industry benchmarks for similar violations.

Risk Mitigation Performance Indicators

Beyond regulatory compliance, governance frameworks must demonstrate broader risk reduction capabilities across operational, reputational, and strategic dimensions. Organizations develop sophisticated risk measurement systems that quantify governance impact on business continuity, data security incidents, and operational resilience.

These measurements include tracking data breach prevention effectiveness, measuring improvement in audit findings, and quantifying reduction in compliance-related operational disruptions. Advanced organizations correlate governance maturity scores with risk incident frequencies, creating predictive models that demonstrate governance investment returns through prevented losses.

Integration with Enterprise Risk Management

The most effective compliance measurement systems integrate seamlessly with existing enterprise risk management frameworks, ensuring governance metrics contribute to comprehensive organizational risk reporting. This integration enables executives to understand governance value within broader business context while supporting strategic decision-making about governance investment priorities.

Modern governance platforms facilitate this integration through automated risk scoring algorithms that translate governance metrics into enterprise risk language, enabling seamless communication between governance professionals and executive leadership about compliance posture and risk reduction achievements.

This comprehensive approach to compliance and risk measurement establishes governance as a strategic risk management function rather than operational overhead, demonstrating clear value through prevented incidents and strengthened regulatory relationships. These measurement capabilities provide the foundation for executive reporting systems that communicate governance value in business terms that resonate with organizational leadership and support continued governance investment.

The systematic measurement of compliance and risk reduction creates compelling evidence of governance value that extends beyond operational efficiency into strategic risk protection, setting the stage for sophisticated executive dashboard systems that translate these detailed metrics into strategic insights for organizational leadership.

11.7. Executive Dashboards and Governance Scorecards

Building on comprehensive risk management frameworks, successful data governance programs require sophisticated visualization and communication strategies that translate complex technical metrics into executive-level insights. Executive dashboards and governance scorecards represent the culmination of measurement efforts, transforming detailed operational data into strategic intelligence that drives organizational decision-making and demonstrates governance value to senior leadership.

The evolution from operational metrics to executive reporting requires a fundamental shift in perspective, moving beyond technical accuracy to focus on business impact and strategic alignment. While data stewards monitor detailed quality scores and compliance rates, executives need consolidated views that highlight trends, identify risks, and showcase governance contributions to organizational objectives.

Strategic Dashboard Architecture for Governance Communication

Executive dashboards must balance comprehensiveness with clarity, presenting governance performance through multiple lenses that resonate with different leadership priorities. The most effective governance dashboards organize information hierarchically, starting with high-level business impact indicators before drilling down into supporting operational metrics.

The architecture begins with **governance health indicators** that provide immediate visual confirmation of program status. These typically include overall governance maturity scores, critical risk alerts, and trend indicators showing improvement or degradation over time. Supporting layers present detailed performance across governance domains, including data quality achievements, compliance status, and stakeholder engagement levels.

Modern dashboard implementations leverage real-time data integration capabilities, connecting directly to governance platforms like Collibra or Informatica to ensure executives receive current information without manual reporting delays. This integration eliminates the traditional lag between governance activities and executive visibility, enabling more responsive decision-making.

Practical Implementation: Financial Services Governance Dashboard

A major European bank implemented a comprehensive governance dashboard using Collibra's reporting capabilities combined with Tableau for executive visualization. The implementation addressed regulatory compliance requirements while demonstrating business value through improved decision-making speed and risk reduction.

Step 1: Metric Selection and Prioritization

The governance team identified five primary executive metrics: regulatory compliance percentage, data quality index, governance cost per managed asset, risk mitigation value, and business process improvement impact. Each metric connected directly to board-level objectives and regulatory requirements.

Step 2: Dashboard Design and Stakeholder Validation

Working with executive assistants and department heads, the team designed dashboard layouts that matched leadership review patterns. The main view presented monthly governance performance against targets, with drill-down capabilities for detailed analysis.

```
Executive Governance Scorecard - Monthly View
============================================
Overall Governance Health: 87% (Target: 85%)
Critical Alerts: 2 (Previous: 5)
Compliance Status: 94% (Regulatory Requirement: 90%)
Quality Improvement: +12% (Quarter-over-Quarter)
Cost Avoidance: €2.3M (Risk Mitigation Value)
```

Step 3: Integration and Automation

The dashboard connected to Collibra's API for real-time governance metrics, regulatory compliance databases for current status updates, and financial systems for cost tracking. Automated alerts triggered when metrics fell below thresholds, ensuring immediate executive awareness of critical issues.

Results and Organizational Impact

Within six months, executive engagement with governance initiatives increased significantly, with board-level discussions shifting from compliance concerns to strategic data utilization opportunities. The bank achieved a 15% reduction in regulatory inquiry response time and documented €2.3 million in avoided costs through proactive risk identification.

Governance Scorecard Development and Stakeholder Alignment

Effective governance scorecards extend beyond dashboard visualization to provide structured performance evaluation frameworks that align with organizational maturity models and strategic objectives. These scorecards establish clear expectations, track progress against governance goals, and facilitate meaningful discussions about program effectiveness.

The scorecard development process begins with stakeholder interviews to understand decision-making priorities and information preferences. Executive sponsors typically focus on risk reduction and business enablement, while operational leaders emphasize efficiency improvements and process optimization. Successful scorecards accommodate these different perspectives through layered reporting that serves multiple audiences.

Scorecard metrics should demonstrate clear connections between governance activities and business outcomes, avoiding purely technical measurements that lack executive relevance. The most impactful scorecards include forward-looking indicators that predict future performance trends, enabling proactive management rather than reactive responses to governance challenges.

This sophisticated approach to governance measurement and communication transforms data stewardship from a technical necessity into a strategic capability that drives organizational success. Executive dashboards and governance scorecards provide the visibility and accountability mechanisms necessary to sustain long-term governance commitment and investment.

By implementing comprehensive dashboard and scorecard systems, organizations create feedback loops that continuously improve governance effectiveness while demonstrating tangible value to leadership stakeholders. These measurement frameworks become essential tools for navigating organizational change and maintaining governance momentum across evolving business priorities.

The transition from measurement frameworks to value demonstration requires understanding how different industries apply these governance insights to achieve specific business objectives and competitive advantages.

11.8. Value Realization in Retail and E-commerce Organizations

The retail and e-commerce sector presents unique opportunities for demonstrating data governance value through customer-centric metrics and operational efficiency gains. Unlike traditional industries where governance benefits may take quarters to materialize, retail organizations can often showcase immediate returns through improved customer experiences, inventory optimization, and personalized marketing effectiveness.

Customer Data Integration and Personalization Impact

Retail organizations typically struggle with fragmented customer data across multiple touchpoints including online platforms, mobile applications, in-store systems, and loyalty programs. Effective data governance creates unified customer profiles that directly translate to measurable business outcomes. When Target implemented comprehensive customer data governance in 2019, they achieved a 23% increase in personalized marketing campaign effectiveness within six months by establishing consistent customer identifiers and data quality standards across all channels.

The governance framework enabled real-time customer behavior tracking, allowing for dynamic pricing adjustments and inventory recommendations that increased average order values by 18%. This demonstrates how governance investments in data integration and quality management generate immediate revenue impact rather than merely reducing compliance risks.

Inventory Optimization Through Master Data Management

E-commerce giants like Zalando have leveraged master data management governance to transform their inventory operations. By implementing standardized product categorization, attribute management, and supplier data governance, they reduced inventory holding costs by 15% while improving product discovery rates by 31%. The governance framework established consistent product hierarchies across 17 European markets, enabling automated demand forecasting and dynamic pricing strategies.

Practical Implementation Example: Multi-Channel Product Data Governance

Consider implementing a comprehensive product master data governance program using Informatica MDM for a mid-sized fashion retailer:

Assessment Phase: Audit existing product data across e-commerce platform, ERP system, and POS terminals

Standardization Framework: Develop unified product taxonomy with mandatory attributes (SKU, color codes, size standards, material composition)

Data Steward Assignment: Designate category managers as product data stewards with specific accountability metrics

Quality Rules Implementation: Configure automated validation rules in Informatica MDM:

```
Product_Validation_Rules:
  - SKU_Format: "^[A-Z]{3}[0-9]{6}$"
  - Color_Standardization: Reference_Table_Lookup
  - Price_Consistency: Cross_Channel_Validation
  - Image_Requirements: Minimum_Resolution_Standards
```

Governance Workflow: Establish approval processes where new products require data steward validation before publication

Performance Monitoring: Track data quality scores, time-to-market for new products, and cross-channel consistency rates

Expected Outcomes: Organizations typically observe 25-40% reduction in product listing errors, 20% faster product launch cycles, and improved customer satisfaction scores due to accurate product information.

Customer Privacy and Trust Monetization

GDPR and CCPA compliance in retail extends beyond regulatory adherence to competitive advantage. Companies like Sephora transformed privacy governance into a trust-building mechanism, achieving 34% higher customer lifetime value among users who actively manage their privacy preferences through transparent data governance interfaces. Their privacy-by-design approach includes customer-controlled data sharing preferences that enable more effective personalization while building stronger customer relationships.

Cross-Border Governance Complexity

International retailers face unique governance challenges managing data across different regulatory environments. H&M's global data governance framework accommodates varying privacy regulations across 74 markets while maintaining consistent customer experience standards. Their federated governance model allows local compliance adaptations while preserving global data integration capabilities, resulting in 28% improved operational efficiency in new market entries.

This retail-focused approach to value demonstration provides concrete evidence that data governance delivers measurable business outcomes beyond traditional compliance metrics. The combination of customer experience improvements, operational efficiencies, and competitive advantages creates compelling business cases for continued governance investment and expansion.

As we conclude this comprehensive examination of measuring success and demonstrating value across various organizational contexts, the foundation is set for synthesizing these measurement approaches into actionable governance strategies that sustain long-term organizational transformation.

12. SUSTAINING MOMENTUM & CONTINUOUS IMPROVEMENT

The journey from implementing data governance to achieving lasting organizational transformation represents one of the most challenging yet rewarding phases of any governance initiative. While many organizations successfully launch their governance frameworks with enthusiasm and executive support, the true test lies in maintaining that momentum over months and years as initial excitement wanes, priorities shift, and new challenges emerge.

This critical juncture separates governance programs that fade into bureaucratic obscurity from those that evolve into indispensable business capabilities. The difference isn't merely about maintaining compliance or keeping processes running; it's about cultivating a governance ecosystem that grows stronger, smarter, and more valuable over time. Organizations that master this art of sustained governance don't just protect their data assets—they transform them into engines of continuous innovation and competitive advantage.

The path to governance sustainability requires a fundamental shift from implementation thinking to evolutionary thinking. Rather than viewing governance as a project with a defined endpoint, successful organizations embrace it as a living capability that must adapt, scale, and mature alongside their business. This transformation demands sophisticated strategies for maintaining stakeholder engagement when the novelty wears off, evolving governance practices as new technologies emerge, and building institutional knowledge that survives leadership changes.

Throughout this chapter, we'll explore the essential elements that distinguish enduring governance programs from those that stagnate or collapse under their own weight. From developing robust succession planning frameworks to integrating emerging technologies seamlessly into existing governance structures, these strategies will equip you with the tools needed to transform your governance initiative from a temporary program into a permanent competitive advantage that continues delivering value long after the initial implementation celebrations have ended.

12.1. Long-term Sustainability Strategies

Data governance programs that achieve initial success often face their greatest challenge not in the launch phase, but in maintaining momentum over the long term. The transition from implementation to operational excellence requires a fundamental shift in mindset, moving from project-based thinking to sustainable, embedded practices that evolve with organizational needs. Successful IT leaders recognize that sustainability isn't about maintaining the status quo, but about creating adaptive frameworks that strengthen over time.

The foundation of long-term sustainability lies in establishing governance as a core business capability rather than an IT initiative. This transformation requires deliberate strategies that embed data stewardship into daily operations, create self-reinforcing feedback loops, and build organizational muscle memory around data quality and compliance practices.

Building Self-Sustaining Governance Ecosystems

Creating governance systems that operate independently of constant oversight requires careful design of automated processes and embedded accountability structures. Organizations that achieve sustainability focus on building governance into existing workflows rather than creating parallel processes that compete for attention and resources.

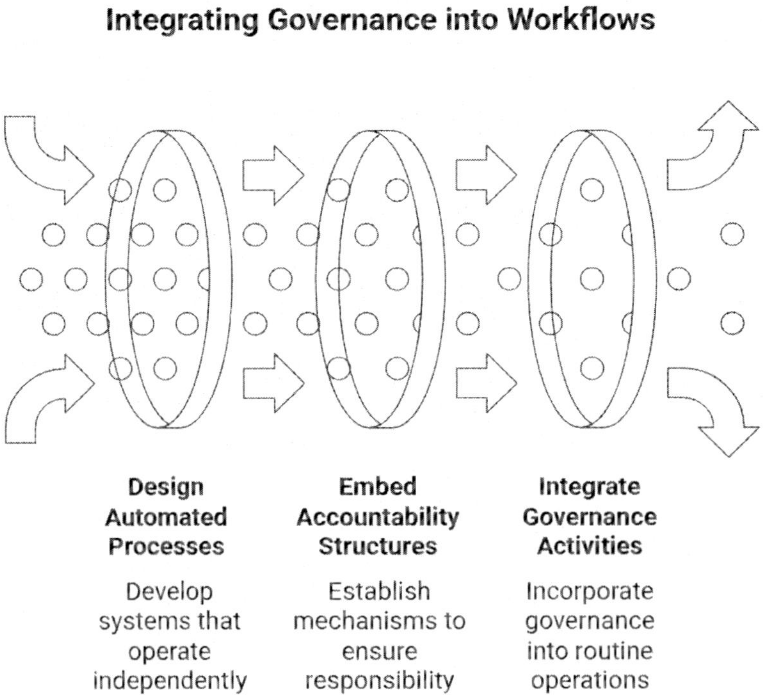

The most effective approach involves identifying natural integration points where governance activities align with existing business processes. For instance, integrating data quality checks into monthly financial close procedures ensures that governance activities become part of routine operations rather than additional overhead.

Practical Implementation: Automated Governance Monitoring at Royal Bank of Scotland

Royal Bank of Scotland's approach to sustainable data governance demonstrates how automation can maintain program momentum without constant manual intervention. Their implementation showcases a comprehensive monitoring framework that evolved from manual oversight to intelligent automation.

Initial State Assessment:

The bank's data governance team spent 40% of their time manually checking data quality reports and chasing down stewards for compliance updates. This reactive approach created bottlenecks and inconsistent enforcement across business units.

Implementation Steps:

Automated Dashboard Creation in Collibra:

```
Data Quality Rule: Customer_SSN_Validation
Trigger: Daily batch processing completion
Action: Generate quality score for customer master data
Escalation: Auto-assign to regional data steward if score < 95%
SLA: 24-hour resolution requirement
```

Self-Service Monitoring Portal:

Business users gained access to real-time data lineage and quality metrics through Collibra's governance portal, reducing dependency on central teams by 60%.

Intelligent Alert System:

The system learned from historical patterns to distinguish between critical issues requiring immediate attention and routine fluctuations that could be addressed during regular maintenance windows.

Measurable Outcomes:

- Data steward response time improved from 3.2 days to 8 hours
- Governance team capacity increased by 35% for strategic initiatives
- Data quality incidents decreased by 45% within six months
- Business user self-sufficiency in data discovery increased by 70%

This transformation enabled the governance program to scale across 12 additional business units without proportional increases in central team resources.

Embedding Governance in Organizational DNA

Sustainable governance requires integration with performance management systems, career development paths, and organizational reward structures. Leading organizations establish data stewardship competencies as formal job requirements and create advancement opportunities that recognize governance contributions.

The most successful programs develop governance champions at multiple organizational levels, creating a distributed leadership model that doesn't rely on single points of failure. These champions serve as local experts who can adapt central policies to specific business contexts while maintaining overall consistency.

Continuous Capability Development

Long-term sustainability demands ongoing investment in people development and process refinement. Organizations must balance standardization with flexibility, ensuring that governance frameworks can adapt to new technologies, regulatory requirements, and business models without losing their foundational strength.

Regular capability assessments help identify skill gaps and emerging needs before they become critical issues. This proactive approach to talent development ensures that governance teams remain effective as organizational complexity increases and data landscapes evolve.

As governance frameworks mature and demonstrate consistent value, organizations must prepare for the next challenge: adapting these established systems to meet evolving business requirements and emerging technological opportunities.

12.2. Governance Evolution and Adaptation

Successful data governance frameworks must evolve continuously to remain relevant and effective in dynamic business environments. Organizations that treat governance as a static implementation inevitably find their frameworks becoming obsolete, losing stakeholder support, and failing to address emerging data challenges. The most resilient governance programs build adaptation mechanisms directly into their operational DNA, creating systematic approaches for framework evolution that respond to changing business priorities, regulatory requirements, and technological advances.

Building Adaptive Framework Architecture

The foundation of governance evolution lies in designing frameworks with inherent flexibility. Rather than creating rigid policies and procedures, leading organizations implement modular governance architectures that can be reconfigured as needs change. This approach involves establishing core governance principles that remain constant while allowing supporting processes, tools, and organizational structures to adapt based on business evolution.

Effective adaptive frameworks incorporate feedback loops at multiple levels, from operational data steward input to executive steering committee oversight. These mechanisms ensure that governance evolution is driven by real-world experience rather than theoretical assumptions. Organizations achieve this by implementing quarterly governance health checks, annual framework assessments, and continuous monitoring of governance effectiveness metrics.

Practical Implementation: Governance Framework Modernization at Global Manufacturing Corporation

A multinational manufacturing company faced significant challenges when their five-year-old data governance framework could no longer support new IoT sensor data streams, cloud analytics platforms, and emerging AI initiatives. Their original framework, built around traditional data warehouse structures, required comprehensive evolution to address modern data architectures.

Step 1: Current State Assessment

The governance team conducted a comprehensive framework audit using the DAMA-DMBOK assessment methodology, evaluating eleven knowledge areas against current business requirements. They discovered that their data quality rules were incompatible with real-time sensor data, their metadata management couldn't handle unstructured IoT telemetry, and their approval workflows created bottlenecks for machine learning model development.

Step 2: Framework Redesign Using Collibra

Working within their existing Collibra Data Governance Center, the team implemented a hybrid governance model that maintained strict controls for regulatory compliance data while introducing agile governance tracks for innovation initiatives. They created new data domains specifically for IoT sensor data, established automated quality monitoring for streaming data sources, and developed expedited approval processes for AI model training datasets.

```
# Adaptive Governance Policy Template
Policy Type: Dynamic Data Classification
Scope: IoT and Real-time Data Streams
Review Cycle: Quarterly

Classification Rules:
- Critical Safety Data: Traditional approval workflow (72 hours)
- Operational Telemetry: Automated classification with human oversight
- Experimental Sensor Data: Self-service with post-processing
validation

Escalation Triggers:
- Data volume exceeds 10TB daily threshold
- New sensor types introduced
- Regulatory requirements change
```

Step 3: Stakeholder Engagement Restructuring

The organization restructured their governance committees to include IoT specialists, cloud architects, and data scientists alongside traditional data stewards. They established monthly "Innovation Governance Sprints" where emerging data use cases could be rapidly assessed and integrated into the framework without disrupting existing operations.

Results After 18 Months:

60% reduction in time-to-market for new analytics initiatives
40% improvement in data quality scores across IoT data streams
95% stakeholder satisfaction with governance responsiveness
Zero compliance violations during framework transition

Technology Integration and Tool Evolution

Modern governance evolution requires seamless integration with emerging technologies. Organizations must develop systematic approaches for evaluating new data management tools, assessing their compatibility with existing governance frameworks, and implementing integration strategies that preserve governance integrity while enabling innovation.

The most successful organizations establish technology evaluation committees that include governance representatives alongside IT architecture teams. These groups develop standardized assessment criteria for new tools, create integration playbooks for common scenarios, and maintain governance compatibility matrices that guide technology selection decisions.

Measuring Evolution Effectiveness

Governance evolution requires sophisticated measurement approaches that track both operational improvements and strategic alignment. Organizations implement balanced scorecards that measure governance maturity progression, stakeholder satisfaction trends, and business value delivery. These metrics provide objective evidence of evolution success and identify areas requiring additional adaptation.

Leading organizations also implement predictive governance analytics that identify potential framework stress points before they impact operations. By analyzing data usage patterns, compliance trends, and stakeholder feedback, these systems provide early warning indicators that trigger proactive framework adjustments.

This systematic approach to governance evolution ensures that data governance frameworks remain dynamic, responsive, and aligned with organizational objectives. Organizations that master governance adaptation create sustainable competitive advantages by maintaining operational excellence while enabling innovation and growth. However, even the most adaptive frameworks require consistent stakeholder engagement to maintain momentum and ensure continued organizational support throughout their evolution journey.

12.3. Maintaining Stakeholder Engagement Over Time

Successful data governance initiatives face their greatest challenge not during initial implementation, but in the months and years that follow. As the novelty of new frameworks wears off and daily operational pressures reassert themselves, maintaining stakeholder engagement becomes a sophisticated exercise in relationship management, value demonstration, and strategic communication. The organizations that master this delicate balance transform their governance programs from temporary projects into enduring organizational capabilities.

Creating Feedback Loops That Drive Continuous Participation

Sustainable stakeholder engagement relies on creating systematic feedback mechanisms that make participation rewarding rather than burdensome. Unlike traditional status reports that flow upward through organizational hierarchies, effective governance feedback loops operate multidirectionally, ensuring that every stakeholder receives value proportional to their investment of time and attention.

The most successful engagement strategies establish **governance advisory circles** that rotate membership quarterly, allowing different business units to influence governance direction while preventing engagement fatigue. These circles operate under structured agendas that balance strategic discussion with tactical problem-solving, ensuring participants see immediate relevance to their operational challenges.

Consider implementing monthly **governance impact showcases** where data stewards present specific examples of how governance improvements have solved real business problems. These sessions should highlight quantifiable outcomes, such as reduced data quality incidents, improved regulatory compliance scores, or accelerated analytics delivery times. The key is connecting governance activities directly to business results that stakeholders care about.

Gamification and Recognition Strategies for Long-term Motivation

Modern governance sustainability requires moving beyond traditional corporate communications to embrace engagement techniques that tap into intrinsic motivation. **Data stewardship recognition programs** that celebrate both individual contributions and team achievements create positive reinforcement cycles that sustain participation over time.

Implementing **governance maturity scorecards** at the department level introduces healthy competition while making abstract governance concepts tangible. These scorecards should track metrics like data quality improvements, policy compliance rates, and stakeholder participation levels, with quarterly awards recognizing the most improved and highest-performing teams.

Practical Implementation Example: Quarterly Governance Health Checks at a Financial Services Firm

A multinational bank successfully maintained stakeholder engagement through structured quarterly governance health checks using Collibra's governance workflow capabilities. Here's their step-by-step approach:

```
Governance Health Check Workflow:
1. Automated Data Quality Assessment
   - System generates quality metrics by business domain
   - Identifies top 5 data issues impacting each stakeholder group
   - Creates personalized dashboards showing relevant improvements

2. Stakeholder Impact Interviews (30-minute sessions)
   - Pre-populated agenda based on domain-specific metrics
   - Standard questions about governance value perception
   - Documentation of new requirements or pain points

3. Action Plan Development
   - Cross-functional teams receive prioritized improvement lists
   - 90-day implementation roadmaps with clear ownership
   - Success metrics aligned with business objectives
```

```
4. Follow-up Communication
   - Personalized reports showing how feedback influenced decisions
   - Quarterly governance newsletter highlighting stakeholder
contributions
   - Executive summary demonstrating ROI from governance investments
```

Results: The bank achieved 87% sustained stakeholder participation over 18 months, with governance satisfaction scores increasing from 6.2 to 8.4 out of 10. Most importantly, they documented $2.3 million in cost savings directly attributable to stakeholder-driven governance improvements.

Evolving Communication Strategies Based on Organizational Maturity

As governance programs mature, stakeholder communication must evolve from basic awareness-building to sophisticated value articulation. Early-stage programs focus on explaining governance concepts and building foundational understanding, while mature programs emphasize strategic impact and competitive advantage creation.

Advanced stakeholder segmentation becomes crucial at this stage, recognizing that different organizational levels require fundamentally different engagement approaches. Executive stakeholders need governance insights presented as strategic intelligence that informs major business decisions, while operational stakeholders require tactical guidance that improves their daily work effectiveness.

The transition from reactive to proactive communication marks a critical maturity milestone. Instead of simply reporting governance status, mature programs anticipate stakeholder needs and provide forward-looking insights about emerging data challenges, regulatory changes, and technology opportunities that could impact business operations.

This sophisticated approach to stakeholder engagement creates a self-reinforcing cycle where governance becomes integral to organizational decision-making rather than an external compliance requirement. When stakeholders begin proactively seeking governance guidance for new initiatives, the program has achieved true sustainability.

Key Takeaways: Maintaining long-term stakeholder engagement requires systematic feedback mechanisms, recognition strategies that celebrate contributions, and communication approaches that evolve with organizational maturity. The most successful programs create value loops where participation generates tangible benefits that motivate continued involvement.

The foundation of sustained engagement naturally leads to questions about expanding governance influence across larger organizational structures, where different scaling strategies become essential for maintaining effectiveness while managing increased complexity.

12.4. Scaling Governance Across the Enterprise

As governance frameworks mature beyond their initial implementation phase, the challenge shifts from proving value within pilot programs to achieving consistent application across diverse business units, geographical regions, and operational contexts. Scaling governance effectively requires a fundamental shift in approach, moving from centralized control models to federated structures that maintain consistency while accommodating local business requirements and regulatory variations.

The transition from departmental governance to enterprise-wide implementation demands careful orchestration of distributed authority, standardized processes, and adaptive frameworks that can flex with organizational complexity. This scaling phase often determines whether governance initiatives become embedded organizational capabilities or remain isolated compliance exercises with limited strategic impact.

Federated Governance Architecture Design

Successful enterprise scaling relies on federated governance models that distribute decision-making authority while maintaining strategic alignment and operational consistency. This architecture balances central oversight with local autonomy, enabling business units to adapt governance practices to their specific contexts while adhering to enterprise-wide standards and policies.

The federated approach establishes governance councils at multiple organizational levels, each with clearly defined responsibilities and escalation pathways. Regional data stewards maintain accountability for local compliance requirements while enterprise stewards ensure consistency across geographic boundaries. This structure prevents governance from becoming a bottleneck while preserving the strategic coherence necessary for enterprise-wide data initiatives.

Cross-Functional Integration Strategies

Enterprise-wide governance scaling requires deep integration with existing operational workflows, technology platforms, and business processes. Rather than treating governance as an overlay on existing systems, successful scaling embeds governance controls directly into operational procedures, making compliance automatic rather than optional.

Practical Implementation: Multi-Region Data Classification Scaling

Consider a global financial services organization implementing data classification across 15 countries with varying privacy regulations. The scaling approach involves creating a master classification taxonomy in Collibra while enabling regional customization for local compliance requirements.

Step 1: Establish Master Framework

```
Master Classification Schema:
- Public Data (Level 0)
- Internal Use (Level 1)
- Confidential (Level 2)
- Restricted (Level 3)
- Top Secret (Level 4)

Regional Extensions:
- GDPR Personal Data (Europe)
- PII Categories (North America)
- Financial Privacy Data (Asia-Pacific)
```

Step 2: Deploy Regional Governance Committees

Each region establishes a governance committee with local data stewards who customize the master framework for regional requirements. The European committee adds GDPR-specific classifications, while the Asia-Pacific team incorporates local banking regulations.

Step 3: Implement Automated Workflow Integration

Using Collibra's workflow engine, the organization creates automated processes that route data classification requests to appropriate regional stewards based on data location and type. When new customer data enters the European systems, it automatically triggers GDPR classification workflows.

Step 4: Monitor Cross-Regional Consistency

Enterprise dashboards track classification consistency across regions, identifying discrepancies and ensuring alignment with global standards. Monthly governance reviews address conflicts between regional and enterprise requirements.

Results: Within 18 months, the organization achieved 94% classification accuracy across all regions while maintaining local compliance requirements. Data discovery time decreased by 60%, and regulatory audit preparation time reduced from weeks to days.

Technology Platform Harmonization

Scaling governance across the enterprise often reveals technology fragmentation that impedes consistent data management. Organizations must address platform disparities while avoiding disruptive migrations that could undermine governance adoption. The key lies in creating integration layers that enable consistent governance application across heterogeneous technology environments.

Modern governance platforms provide API-driven integration capabilities that allow governance policies to span multiple data platforms, cloud environments, and legacy systems. This approach enables organizations to maintain existing technology investments while achieving governance consistency through standardized metadata management and policy enforcement.

Performance Measurement at Scale

Enterprise-wide governance requires sophisticated measurement frameworks that provide visibility into governance effectiveness across diverse organizational contexts. Scaling measurement systems involves creating hierarchical metrics that aggregate local performance indicators into enterprise-wide governance scorecards while maintaining the granularity necessary for targeted improvements.

These measurement systems must account for varying maturity levels across business units, different regulatory requirements across regions, and diverse operational contexts that affect governance implementation. The most effective approaches establish baseline measurements for each organizational unit and track improvement trajectories rather than applying uniform standards that may not reflect local realities.

Successfully scaling governance across the enterprise transforms data management from fragmented departmental activities into cohesive organizational capabilities. This transformation requires patient cultivation of federated structures, careful integration with existing operations, and measurement systems that provide actionable insights across diverse organizational contexts. The practical skills gained through enterprise scaling prepare organizations to leverage governance as a foundation for innovation and emerging technology integration, where consistent data management becomes essential for advanced analytics and artificial intelligence initiatives.

As governance frameworks achieve enterprise-wide consistency, organizations gain the foundation necessary to explore how governance practices can accelerate rather than constrain technological innovation and emerging technology adoption.

12.5. Innovation and Emerging Technology Integration

Building upon the foundation of enterprise-wide consistency, governance frameworks must evolve beyond traditional oversight models to become enablers of innovation. As organizations increasingly adopt artificial intelligence, machine learning, cloud-native architectures, and real-time analytics platforms, data governance faces a critical transformation: shifting from a compliance-focused discipline to an innovation catalyst that accelerates technology adoption while maintaining data integrity and security standards.

Governance-First Innovation Strategy

Modern data governance frameworks require proactive integration with emerging technology roadmaps rather than reactive adaptation after deployment. This approach involves establishing **governance-by-design** principles that embed data stewardship requirements directly into technology evaluation and procurement processes. Organizations implementing this strategy create innovation councils that include data governance representatives alongside technology architects and business stakeholders, ensuring that new technology acquisitions align with existing data management standards while expanding governance capabilities.

The key to successful innovation integration lies in developing flexible governance policies that accommodate rapid technological change without compromising data quality or compliance requirements. This requires governance frameworks that can adapt to new data types, processing methods, and storage architectures while maintaining consistent oversight and accountability structures across traditional and emerging technology platforms.

Practical Implementation: AI-Driven Data Discovery Integration

Consider a multinational pharmaceutical company implementing Collibra's AI-powered data discovery capabilities alongside their existing governance framework. The organization faced challenges in cataloging and governing data across newly acquired cloud-based research platforms while maintaining FDA compliance requirements.

Step 1: Governance Policy Adaptation

The data governance team modified their existing data classification policies to accommodate machine learning-generated metadata and automated lineage tracking:

```
Policy Update: Automated Data Discovery Standards
- AI-generated classifications require human steward validation within
48 hours
```

```
- Machine learning confidence scores below 85% trigger manual review
workflows
- Automated lineage mapping must include algorithm versioning and
decision rationale
- Cross-platform data relationships require governance committee
approval for new connections
```

Step 2: Technology Integration Workflow

Using Collibra's REST API, the team created automated workflows that integrated AI discovery results with their established governance processes:

AI discovery engine scans new cloud data sources
Automated classification results populate governance catalog with confidence scores
Data stewards receive notification workflows for validation requirements
Approved classifications trigger automated policy application across connected systems
Exception handling routes uncertain classifications to governance committee review

Step 3: Measurable Outcomes

Within six months, the organization achieved a 300% increase in data asset discovery speed while maintaining 99.2% compliance accuracy rates. The integrated approach reduced manual cataloging efforts by 75% while expanding governance coverage to include previously ungoverned cloud-based research data.

Emerging Technology Risk Assessment

Innovation integration requires sophisticated risk assessment methodologies that evaluate new technologies against established governance standards. Organizations must develop **technology governance scorecards** that assess emerging platforms across multiple dimensions: data security capabilities, compliance framework compatibility, integration complexity, and long-term sustainability. These assessments enable informed decision-making about technology adoption while identifying governance framework modifications necessary to support new platforms effectively.

Successful risk assessment involves creating cross-functional evaluation teams that combine technical expertise with governance knowledge. These teams conduct thorough due diligence on vendor data handling practices, security certifications, and compliance capabilities, ensuring that new technology acquisitions enhance rather than compromise existing governance investments.

Continuous Technology Governance Alignment

Maintaining alignment between rapidly evolving technologies and governance frameworks requires establishing **continuous integration processes** that monitor technology changes and automatically trigger governance policy reviews. Organizations implementing this approach create technology change advisory boards that include governance representatives, ensuring that system updates, new feature releases, and platform migrations receive appropriate governance oversight before implementation.

This continuous alignment approach involves developing automated monitoring systems that track changes in connected technology platforms and alert governance teams to modifications that may impact data handling, security, or compliance requirements. By proactively identifying potential governance impacts, organizations can adapt their frameworks before issues arise rather than responding to problems after they occur.

The integration of innovation and emerging technologies with data governance transforms traditional oversight models into dynamic enablement frameworks that accelerate rather than constrain technological advancement. Organizations mastering this integration gain competitive advantages through faster technology adoption cycles, reduced compliance risks, and enhanced data-driven innovation capabilities.

As governance frameworks mature through innovation integration, organizations develop the sophisticated capabilities necessary to assess and enhance their overall governance maturity levels systematically.

12.6. Governance Capability Maturation

As governance frameworks mature through innovation integration, organizations develop the sophisticated capabilities necessary to assess and enhance their overall governance maturity levels systematically. This evolution requires a structured approach to capability development that transforms ad-hoc governance activities into institutionalized practices capable of adapting to changing business requirements and technological landscapes.

Maturity Assessment Framework Implementation

Establishing a robust capability maturation process begins with implementing a comprehensive assessment framework that evaluates governance effectiveness across multiple dimensions. Organizations must move beyond simple compliance checklists to develop nuanced evaluation mechanisms that capture the sophistication of their governance practices.

The **Data Management Association (DAMA-DMBOK)** provides a foundational maturity model that organizations can customize to their specific contexts.

This framework evaluates capabilities across six maturity levels: Initial, Managed, Defined, Quantitatively Managed, and Optimizing. Each level represents increasingly sophisticated governance capabilities, from reactive problem-solving to proactive optimization and continuous improvement.

Successful maturity assessments require establishing baseline measurements that capture both quantitative metrics and qualitative governance behaviors. Organizations should evaluate data quality consistency, stakeholder engagement levels, policy adherence rates, and the speed of issue resolution across different business units.

Advanced Capability Development Strategies

Mature governance programs demonstrate their sophistication through **adaptive capability development** that responds to emerging business needs while maintaining foundational governance principles. This involves creating governance capabilities that can scale horizontally across new business domains and vertically through increased complexity levels.

Organizations at advanced maturity levels implement governance capabilities that anticipate future requirements rather than simply responding to current challenges. This proactive approach includes developing scenario planning capabilities, establishing governance innovation labs, and creating cross-functional teams that explore emerging governance challenges before they impact business operations.

The integration of artificial intelligence and machine learning into governance processes represents a significant maturity milestone. Advanced organizations leverage these technologies to automate routine governance tasks, identify emerging data quality issues, and predict potential compliance risks before they materialize.

Practical Implementation: Global Financial Services Maturity Evolution

Consider the transformation journey of **JPMorgan Chase**, which evolved its data governance capabilities from basic compliance-focused activities to sophisticated enterprise-wide governance orchestration. Their maturity progression demonstrates how organizations can systematically enhance governance capabilities while maintaining operational effectiveness.

Initial Assessment Phase:

```
Maturity Level Assessment - Quarter 1
- Data Quality: Level 2 (Managed)
- Policy Compliance: Level 3 (Defined)
- Stakeholder Engagement: Level 2 (Managed)
- Risk Management: Level 3 (Defined)
- Innovation Integration: Level 1 (Initial)
```

Capability Enhancement Implementation:

Establish Advanced Analytics Governance Team - Created specialized unit focusing on AI/ML governance frameworks, developing policies for algorithmic transparency and bias detection across trading and lending platforms.

Implement Dynamic Policy Management - Deployed Collibra's advanced policy management capabilities to create adaptive governance rules that automatically adjust based on regulatory changes and business context shifts.

Develop Predictive Governance Metrics - Built machine learning models that predict governance framework effectiveness and identify potential compliance gaps before they impact business operations.

Create Cross-Business Governance Councils - Established governance bodies that span traditional business silos, enabling consistent governance application across investment banking, consumer banking, and asset management divisions.

Results After 18 Months:

```
Maturity Level Assessment - Quarter 7
- Data Quality: Level 4 (Quantitatively Managed)
- Policy Compliance: Level 4 (Quantitatively Managed)
- Stakeholder Engagement: Level 4 (Quantitatively Managed)
- Risk Management: Level 5 (Optimizing)
- Innovation Integration: Level 4 (Quantitatively Managed)
```

The organization achieved measurable improvements including 47% reduction in data quality incidents, 62% faster regulatory reporting cycles, and 35% increase in business user satisfaction with data governance processes.

Continuous Improvement Integration

Mature governance capabilities require embedding continuous improvement mechanisms directly into governance workflows rather than treating improvement as separate initiative. This involves creating feedback loops that capture governance effectiveness data and automatically trigger improvement processes when performance thresholds are not met.

Organizations should establish **governance performance dashboards** that provide real-time visibility into capability maturation progress. These dashboards should track leading indicators of governance success, such as stakeholder engagement levels and proactive issue identification rates, rather than focusing solely on lagging indicators like compliance violations.

Governance capability maturation represents the evolution from reactive compliance management to proactive business enablement. Organizations that successfully navigate this transformation develop governance frameworks that adapt to changing business requirements while maintaining consistent data stewardship standards. The sophisticated capabilities developed through systematic maturation efforts position organizations to leverage their governance investments as competitive advantages rather than operational overhead.

Building on this foundation of mature governance capabilities, organizations must now focus on preserving and transferring the institutional knowledge that makes these sophisticated frameworks possible, ensuring that governance expertise remains accessible and actionable across changing organizational landscapes.

12.7. Knowledge Management and Institutional Memory

The preservation and transfer of governance expertise represents one of the most critical yet overlooked aspects of sustainable data governance programs. Organizations often invest heavily in establishing governance frameworks, only to watch institutional knowledge evaporate when key personnel transition or retire. Effective knowledge management transforms governance wisdom from individual expertise into organizational assets that persist across leadership changes and organizational restructuring.

Documenting Governance Decision-Making Patterns

Successful governance programs create comprehensive documentation systems that capture not just what decisions were made, but why they were made and how they evolved over time. This institutional memory becomes invaluable when facing similar challenges or explaining governance rationale to new stakeholders.

Consider implementing a **governance decision registry** that tracks policy evolution, stakeholder input, and implementation outcomes. This registry should document the business context surrounding major governance decisions, alternative approaches that were considered, and the specific factors that influenced final choices. When new governance challenges arise, teams can reference this historical context to make more informed decisions and avoid repeating past mistakes.

Building Knowledge Transfer Mechanisms

Effective knowledge management requires systematic approaches to capturing and sharing governance expertise across the organization. This involves creating multiple channels for knowledge transfer that accommodate different learning styles and organizational dynamics.

Practical Example: Implementing a Governance Knowledge Base Using Alation

A global pharmaceutical company faced significant challenges when their chief data officer and two senior data stewards departed within six months. Their governance program, which had successfully implemented master data management for clinical trials data, risked losing critical institutional knowledge.

Step 1: Knowledge Audit and Gap Analysis

The organization conducted a comprehensive audit using Alation's collaboration features to identify knowledge gaps:

```
Knowledge Assessment Framework:
- Policy rationale documentation: 35% complete
- Process decision history: 20% complete
- Stakeholder relationship mapping: 45% complete
- Technical implementation notes: 60% complete
```

Step 2: Structured Knowledge Capture

Using Alation's article and documentation capabilities, the team implemented a systematic knowledge capture process:

Policy Evolution Tracking: Each data policy was enhanced with decision history articles explaining the business drivers, stakeholder concerns, and implementation challenges that shaped current approaches

Process Documentation: Step-by-step governance workflows were documented with embedded decision trees showing how different scenarios should be handled

Relationship Mapping: Stakeholder influence diagrams were created showing formal and informal decision-making networks

Step 3: Interactive Knowledge Sharing

The organization established monthly "Governance Story Sessions" where experienced practitioners shared case studies and lessons learned through Alation's community features. These sessions were recorded and tagged for easy retrieval.

Step 4: Validation and Refinement

New team members were assigned to test documented processes and provide feedback, ensuring knowledge transfer materials remained accurate and actionable.

Results: Within eight months, the organization successfully onboarded three new governance team members who achieved productivity milestones 40% faster than previous hires. The knowledge base now contains over 200 documented decision patterns and serves as the foundation for governance training programs.

Creating Governance Mentorship Programs

Formal mentorship structures ensure that governance expertise transfers through direct relationship-building rather than relying solely on documentation. These programs pair experienced governance practitioners with emerging leaders, creating knowledge transfer opportunities that capture nuanced understanding of organizational dynamics and stakeholder management.

Effective mentorship programs establish clear expectations for knowledge sharing, including regular sessions focused on governance decision-making, stakeholder engagement strategies, and lessons learned from past initiatives. Mentors should be encouraged to share both successes and failures, providing mentees with realistic perspectives on governance challenges and solutions.

Establishing Centers of Excellence

Governance centers of excellence serve as repositories for institutional knowledge while providing ongoing support for governance practitioners across the organization. These centers maintain best practices, coordinate knowledge sharing activities, and ensure consistent application of governance principles across different business units.

Centers of excellence should maintain governance playbooks that document proven approaches for common scenarios, stakeholder engagement strategies, and change management techniques. These playbooks become living documents that evolve based on organizational experience and emerging best practices.

The systematic preservation of governance knowledge ensures that organizations can maintain momentum even as personnel changes occur. By documenting decision patterns, implementing structured knowledge transfer mechanisms, and creating formal mentorship programs, organizations build resilience into their governance programs that extends far beyond individual expertise.

As governance knowledge becomes institutionalized through these systematic approaches, organizations must also prepare for planned transitions in governance leadership, ensuring continuity of vision and strategic direction across changing organizational landscapes.

12.8. Succession Planning for Governance Leadership

While systematic approaches address unexpected leadership departures, organizations must also prepare for planned transitions in governance leadership, ensuring continuity of vision and strategic direction across changing organizational landscapes. Effective succession planning goes beyond identifying potential candidates; it requires deliberate knowledge transfer mechanisms, structured mentorship programs, and robust documentation of governance philosophies that transcend individual leadership styles.

Building Leadership Pipeline Architecture

Successful governance succession begins with identifying and developing internal talent through structured career progression pathways. Organizations should establish **governance leadership tracks** that provide ascending levels of responsibility, from data steward roles through senior governance positions. This pipeline approach ensures candidates gain comprehensive experience across different governance domains before assuming executive responsibilities.

The most effective succession programs incorporate cross-functional rotations, allowing potential leaders to understand governance challenges from multiple organizational perspectives. Finance professionals gain insights into operational data quality requirements, while IT leaders develop appreciation for business context and regulatory compliance nuances. This holistic experience proves invaluable when making strategic governance decisions that impact entire enterprises.

Knowledge Codification and Transfer Mechanisms

Beyond traditional documentation, successful succession planning requires capturing the tacit knowledge and decision-making frameworks that experienced governance leaders develop over time. This includes understanding organizational politics, stakeholder relationship dynamics, and the subtle art of balancing competing priorities within governance initiatives.

Practical Implementation Example: Enterprise Knowledge Transfer Program

A Fortune 500 financial services company implemented a comprehensive succession planning program using Collibra's governance platform to systematize knowledge transfer. The program included:

Governance Decision Archive: Every major governance decision was documented in Collibra with full context, including stakeholder positions, alternative options considered, and rationale for final choices. This created a searchable repository of governance wisdom.

Mentorship Workflows: Senior leaders were paired with succession candidates through structured workflows that included monthly governance case reviews, joint stakeholder meetings, and collaborative policy development sessions.

Scenario-Based Learning Modules: The organization created realistic governance scenarios based on actual historical challenges, allowing succession candidates to practice decision-making in controlled environments.

```
Succession Planning Template:
- Current Leader Assessment: Document expertise areas, key
relationships, decision patterns
- Candidate Identification: Map internal talent against required
competencies
- Development Plan: 18-month structured learning program with
measurable milestones
- Knowledge Transfer: Systematic documentation of processes,
relationships, and insights
- Transition Timeline: Phased handover with overlap periods and
feedback loops
```

The results were measurable: when the Chief Data Officer transitioned to another role, the new leader achieved full effectiveness 60% faster than industry benchmarks, and governance program momentum continued without disruption.

Cultural Continuity and Vision Preservation

Effective succession planning preserves organizational governance culture while allowing for evolutionary improvements. This requires documenting not just what governance leaders do, but how they approach challenges, communicate with stakeholders, and maintain program momentum during difficult periods.

Succession candidates should participate in governance committee meetings as observers, gradually assuming responsibility for specific initiatives while receiving real-time coaching from current leaders. This apprenticeship model ensures cultural knowledge transfer alongside technical expertise.

Measuring Succession Readiness

Organizations need objective metrics to assess succession candidate readiness. These include stakeholder confidence ratings, successful project completions, and demonstrated ability to navigate complex governance challenges independently. Regular assessment ensures succession plans remain current and candidates receive appropriate development support.

This comprehensive approach to succession planning transforms governance leadership transitions from organizational risks into opportunities for program enhancement and cultural strengthening. By systematically developing internal talent and preserving institutional knowledge, organizations ensure their governance investments continue delivering value regardless of leadership changes.

With robust succession planning frameworks established, organizations possess the complete toolkit necessary to sustain long-term governance excellence and drive continuous organizational transformation.

13. ADVANCED GOVERNANCE STRATEGIES

As organizations mature in their data governance journey, the landscape of data management continues to evolve at an unprecedented pace, demanding sophisticated strategies that go far beyond traditional stewardship models. The convergence of artificial intelligence, cloud computing, and real-time analytics has fundamentally transformed how enterprises must approach data governance, creating both extraordinary opportunities and complex challenges that require advanced frameworks and innovative thinking.

This chapter ventures into the cutting-edge realm of data governance, where traditional boundaries blur and new paradigms emerge. Here, data transforms from a managed asset into an active revenue generator through monetization strategies, while artificial intelligence systems require governance frameworks that can adapt to machine learning's dynamic nature. Organizations operating across multi-cloud environments face governance complexities that demand seamless coordination across disparate platforms, while the rise of data marketplaces and self-service analytics democratizes data access in ways that challenge conventional control mechanisms.

The strategic importance of these advanced governance approaches becomes particularly evident during periods of organizational change. Whether navigating mergers and acquisitions, enabling digital transformation initiatives, or supporting sophisticated data science operations, modern enterprises require governance frameworks that can flex and scale without compromising integrity or compliance. These scenarios demand governance leaders who can think beyond traditional data stewardship to architect solutions that anticipate future needs while addressing present complexities.

Throughout this exploration, you will discover how leading organizations are reimagining governance as a strategic enabler rather than a operational constraint. The frameworks, methodologies, and real-world applications presented here will equip you with the advanced capabilities necessary to position data governance as a competitive differentiator in an increasingly data-centric business environment.

13.1. Data Monetization and Value Creation

As organizations mature in their data governance journey, the ultimate objective shifts from mere compliance and quality control to transforming data assets into tangible business value. **Data monetization** represents the pinnacle of governance sophistication, where well-governed data becomes a direct revenue generator, cost optimizer, and competitive differentiator. For IT leaders, this evolution requires governance frameworks that not only protect and manage data but actively enable its commercial exploitation while maintaining ethical and regulatory standards.

The foundation of successful data monetization lies in establishing governance structures that can identify, catalog, and package data products for both internal optimization and external commercialization. Unlike traditional governance focused on risk mitigation, monetization governance must balance accessibility with protection, enabling innovation while maintaining control. This dual mandate requires sophisticated frameworks that can dynamically adjust governance controls based on data usage context and business objectives.

Establishing Data Product Governance Frameworks

Modern data monetization begins with treating data as a product rather than a byproduct. This fundamental shift requires governance frameworks that mirror product development lifecycles, complete with product managers, quality assurance processes, and customer feedback loops. Organizations must implement **data product catalogs** that function like internal marketplaces, where business units can discover, evaluate, and consume data assets with clear service level agreements and usage terms.

The governance framework must define data product ownership, establishing clear accountability for data quality, freshness, and business relevance. Unlike traditional data stewardship focused on compliance, data product stewards become responsible for understanding customer needs, measuring product performance, and continuously improving data value propositions. This requires governance committees that include business stakeholders alongside technical teams, ensuring data products align with market demands and organizational strategy.

Practical Implementation: Enterprise Data Marketplace Development

Consider the implementation at **Netflix**, which transformed its internal data assets into a comprehensive data marketplace that drives both operational efficiency and strategic decision-making. The company established a governance framework using **Metacat**, their open-source data discovery platform, combined with custom governance workflows built on Apache Airflow.

The implementation process followed these specific steps:

Data Asset Inventory and Valuation: Netflix's governance team cataloged over 100,000 data tables across their ecosystem, implementing automated data profiling to assess quality scores, usage patterns, and business criticality. Each dataset received a **Data Value Index** calculated using factors including query frequency, downstream dependencies, and business impact metrics.

Product Definition and Packaging: High-value datasets were packaged into specific data products with defined schemas, refresh schedules, and access controls. For example, their "Member Engagement Insights" product combined viewing patterns, search behaviors, and recommendation interactions into a standardized API that product teams could consume.

Governance Policy Integration: The platform integrated with their existing governance framework, automatically applying data classification policies, privacy controls, and usage monitoring. Each data product included embedded governance metadata specifying retention policies, access restrictions, and compliance requirements.

The results demonstrated measurable value creation: internal teams reduced data discovery time by 75%, new product development cycles accelerated by 40%, and the company identified opportunities to monetize anonymized viewing insights through strategic partnerships. The governance framework enabled controlled external data sharing while maintaining strict privacy protections, generating an additional $50 million in annual partnership revenue.

Value Measurement and Optimization Strategies

Successful data monetization requires sophisticated measurement frameworks that track both financial returns and operational improvements. Organizations must implement governance processes that continuously assess data asset performance, identifying optimization opportunities and measuring return on data investments. This includes establishing **data ROI metrics** that connect governance investments to business outcomes, demonstrating the tangible value of well-governed data assets.

The measurement framework should encompass direct revenue generation from data sales or licensing, indirect value creation through improved decision-making, and cost optimization achieved through data-driven process improvements. Governance committees must regularly review these metrics, adjusting strategies based on performance data and market feedback to ensure sustained value creation.

This foundation in data monetization governance establishes the critical business context for implementing advanced technologies that can amplify value creation. As organizations build sophisticated data products and measurement frameworks, they naturally progress toward leveraging artificial intelligence and machine learning capabilities that can automate value discovery and optimize governance processes at scale.

13.2. Artificial Intelligence and Machine Learning Governance

As organizations advance their measurement frameworks, they naturally progress toward leveraging artificial intelligence and machine learning capabilities that can automate value discovery and optimize governance processes at scale. However, governing AI and ML systems presents unique challenges that traditional data governance frameworks were never designed to address. The dynamic nature of machine learning models, their opacity in decision-making processes, and their potential for algorithmic bias require entirely new governance paradigms that balance innovation with accountability.

The governance of AI and ML systems extends far beyond managing the underlying data assets. It encompasses **model governance**, **algorithmic accountability**, and **ethical AI frameworks** that ensure responsible deployment while maintaining competitive advantage. Unlike static data assets, machine learning models continuously evolve through training cycles, creating governance challenges around version control, performance drift, and decision auditability that demand sophisticated oversight mechanisms.

Establishing AI Model Lifecycle Governance

Effective AI governance begins with comprehensive oversight of the entire model lifecycle, from initial development through production deployment and eventual retirement. This governance framework must address model validation, bias detection, performance monitoring, and explainability requirements while maintaining operational efficiency.

Organizations implementing AI governance typically establish **Model Risk Management Committees** that include data scientists, business stakeholders, legal representatives, and ethics officers. These committees oversee model approval processes, monitor performance metrics, and ensure compliance with regulatory requirements such as the EU's AI Act or financial services regulations like SR 11-7.

The governance framework must also address **model documentation requirements**, including detailed records of training data provenance, feature engineering decisions, hyperparameter selections, and validation methodologies. This documentation becomes critical for regulatory compliance, audit trails, and troubleshooting model performance issues in production environments.

Implementing Algorithmic Accountability Frameworks

Beyond technical governance, organizations must establish frameworks for algorithmic accountability that address fairness, transparency, and ethical considerations. This involves creating **AI ethics boards** with diverse representation and implementing systematic bias testing throughout the model development lifecycle.

Practical Implementation: Enterprise AI Governance with Dataiku

A Fortune 500 financial services company implemented comprehensive AI governance using Dataiku's MLOps platform to address regulatory requirements and ethical AI concerns. The implementation followed these specific steps:

- **Governance Structure Setup**: Established a three-tier governance model with technical review boards, business approval committees, and executive oversight panels
- **Policy Framework Development**: Created standardized templates for model documentation, including bias assessment reports and explainability requirements
- **Automated Compliance Monitoring**: Implemented automated bias detection using Dataiku's fairness metrics, monitoring protected class performance across demographic segments
- **Model Registry Integration**: Deployed centralized model cataloging with version control, lineage tracking, and approval workflows integrated with existing Collibra data catalog

```
# AI Model Governance Policy Template
Model Approval Checklist:
- [ ] Bias assessment completed across protected classes
- [ ] Explainability documentation provided
- [ ] Performance benchmarks established
- [ ] Data lineage documented and validated
- [ ] Regulatory compliance review completed
- [ ] Business stakeholder sign-off obtained
- [ ] Monitoring and alerting configured
```

The implementation resulted in 40% faster model deployment times while achieving 100% regulatory compliance across all deployed models. The framework prevented three potential bias-related incidents through automated monitoring and established clear accountability chains for algorithmic decisions affecting customer outcomes.

Continuous Monitoring and Adaptive Governance

AI governance requires continuous monitoring capabilities that can detect model drift, performance degradation, and emerging bias patterns in real-time production environments. This involves implementing **automated governance workflows** that trigger interventions when models deviate from established performance thresholds or fairness criteria.

Modern AI governance platforms integrate with MLOps pipelines to provide automated model validation, continuous bias monitoring, and performance tracking across diverse deployment environments. These systems generate governance dashboards that provide executives with visibility into AI risk exposure while enabling data science teams to maintain operational agility.

The governance framework must also address **model retirement and replacement processes**, ensuring smooth transitions when models become obsolete or fail to meet performance standards. This includes maintaining historical model documentation for audit purposes and establishing clear protocols for model rollback scenarios.

Effective AI and ML governance transforms algorithmic systems from black boxes into transparent, accountable assets that drive business value while mitigating risk. Organizations implementing these advanced governance strategies gain competitive advantages through faster innovation cycles, reduced regulatory risk, and enhanced stakeholder trust in AI-driven decisions.

As governance frameworks mature to handle the complexities of artificial intelligence, the next frontier involves managing data streams and real-time analytics that demand instantaneous governance decisions across continuously flowing information.

13.3. Real-time Data Governance and Streaming Analytics

The convergence of real-time data streams with governance frameworks represents one of the most challenging yet transformative aspects of modern data management. Unlike traditional batch processing where governance rules can be applied during scheduled intervals, streaming data demands immediate decision-making capabilities that must balance speed with compliance, quality, and security requirements.

Dynamic Policy Enforcement in Stream Processing

Real-time governance requires a fundamental shift from reactive to proactive policy enforcement. Traditional governance models rely on periodic audits and batch validation processes, but streaming analytics demands continuous monitoring and instant remediation. This necessitates embedding governance logic directly into the data pipeline architecture, creating **policy-aware streaming systems** that can make governance decisions at the speed of data flow.

Modern streaming platforms like Apache Kafka and AWS Kinesis now support governance integrations through specialized connectors and middleware. These systems enable real-time data classification, quality assessment, and privacy protection without introducing significant latency. The key lies in designing lightweight governance rules that can execute within milliseconds while maintaining comprehensive coverage of organizational policies.

Automated Compliance Monitoring for Continuous Data Streams

Streaming governance platforms must incorporate automated compliance engines that can detect policy violations as they occur. This includes real-time **data lineage tracking**, where each data element carries metadata about its origin, transformations, and intended destinations. Advanced implementations use machine learning algorithms to predict potential compliance issues before they manifest, enabling preemptive corrective actions.

Consider the implementation of a real-time privacy compliance system for customer interaction data. The governance framework must instantly identify personally identifiable information (PII) within streaming customer service transcripts, apply appropriate masking or encryption, and route the data according to regional privacy regulations. This requires sophisticated pattern recognition capabilities combined with dynamic policy engines that can adapt to changing regulatory requirements.

Practical Implementation: Real-time Data Quality Governance at Scale

A comprehensive example demonstrates how a global telecommunications company implemented real-time governance for their customer experience streaming platform using Informatica's Intelligent Data Management Cloud with Apache Kafka integration.

Initial Challenge: The organization processed 50 million customer interaction events daily across voice, chat, and mobile channels, requiring immediate data quality validation and privacy compliance before feeding downstream analytics systems.

Implementation Steps:

Stream-Native Governance Architecture: Deployed Informatica's Data Quality streaming connectors directly within Kafka topics, enabling real-time validation of data completeness, accuracy, and consistency metrics without batch processing delays.

Dynamic Policy Engine Configuration: Implemented rule-based governance policies that automatically categorized incoming data streams based on content sensitivity, geographic origin, and regulatory requirements, with policies updating every 15 minutes to accommodate changing compliance landscapes.

Real-time Lineage Tracking: Configured automated metadata capture for each streaming event, creating instantaneous lineage records that tracked data transformations, quality scores, and compliance status throughout the processing pipeline.

Governance Framework Integration:

```
Stream Governance Policy Template:
{
  "stream_id": "customer_interactions_emea",
  "quality_rules": [
    {
      "rule_type": "completeness_check",
      "threshold": 95,
      "action": "quarantine_incomplete_records"
    },
    {
      "rule_type": "gdpr_compliance",
      "pii_detection": "automated",
      "masking_strategy": "dynamic_tokenization"
    }
  ],
  "escalation_triggers": {
    "quality_degradation": "immediate_alert",
    "compliance_violation": "auto_remediation"
  }
}
```

Results: The implementation achieved 99.7% data quality compliance with sub-200ms processing latency, while automatically handling 15,000 privacy-related data transformations daily. The governance framework prevented 847 potential compliance violations in the first quarter, demonstrating measurable risk reduction through proactive stream governance.

Performance Optimization and Scalability Considerations

Real-time governance systems must balance comprehensive oversight with processing performance. This requires careful architecture decisions around **governance checkpoints**, where critical validation occurs, versus **trust zones**, where pre-validated data can flow with minimal governance overhead. Successful implementations often employ tiered governance strategies, applying intensive validation at data ingestion points while using lighter-weight monitoring throughout downstream processing.

The integration of streaming governance with existing enterprise data workflows demands sophisticated orchestration capabilities. Organizations must design governance policies that can seamlessly transition between real-time and batch processing modes, ensuring consistent policy application regardless of processing paradigm.

This real-time governance foundation establishes the technical and procedural groundwork necessary for addressing the complexities of distributed data management across multiple cloud environments, where governance policies must maintain consistency while adapting to diverse technological infrastructures and regulatory jurisdictions.

13.4. Multi-cloud and Hybrid Environment Governance

The complexity of modern enterprise data landscapes demands governance frameworks that can seamlessly operate across disparate cloud platforms while maintaining consistent policy enforcement and data quality standards. As organizations increasingly adopt multi-cloud strategies to avoid vendor lock-in and optimize costs, traditional governance approaches that assume centralized infrastructure become inadequate. The challenge extends beyond technical integration to encompass regulatory compliance across different geographical regions, varying data residency requirements, and the need to maintain data lineage visibility across fragmented environments.

Cross-Platform Data Classification and Lineage Tracking

Establishing unified data classification schemes across multiple cloud providers requires sophisticated orchestration capabilities that can translate governance policies into platform-specific implementations. The foundation of effective multi-cloud governance lies in creating abstraction layers that standardize data handling procedures regardless of the underlying infrastructure. This approach ensures that sensitive customer information receives consistent protection whether stored in AWS S3, Azure

Blob Storage, or Google Cloud Storage, while maintaining detailed audit trails that span platform boundaries.

Modern enterprises leverage **data fabric architectures** to create virtual data management layers that provide unified governance interfaces across hybrid environments. These architectures enable consistent metadata management, automated policy enforcement, and real-time compliance monitoring without requiring data migration or platform consolidation.

Federated Governance Models for Distributed Teams

The distributed nature of multi-cloud environments necessitates governance models that balance centralized policy development with localized implementation flexibility. Successful organizations implement **federated governance structures** where central teams establish overarching data principles and standards, while regional or platform-specific teams adapt these guidelines to local requirements and technical constraints.

This approach proves particularly effective for global enterprises operating under varying regulatory frameworks. European operations might emphasize GDPR compliance mechanisms, while North American divisions focus on CCPA requirements, all while maintaining consistent data quality standards and access controls across the organization's entire cloud footprint.

Practical Implementation: Multi-Cloud Data Governance with Informatica Cloud

Consider the implementation at Schneider Electric, a global energy management company that operates across AWS, Azure, and on-premises infrastructure. Their governance team utilized Informatica Intelligent Cloud Services to create a unified governance framework spanning their hybrid environment.

Step 1: Establish Cross-Platform Metadata Repository

```
Data Source Registration:
- AWS RDS instances → Informatica metadata catalog
- Azure Data Lake → Automated discovery and profiling
- On-premises Oracle → Real-time synchronization
- Salesforce SaaS → API-based metadata extraction
```

Step 2: Implement Unified Data Quality Rules

The team created platform-agnostic data quality rules that automatically translate into environment-specific validation logic:

```
Global Rule: Customer_Email_Validation
├── AWS Lambda function (Python validation)
├── Azure Function (C# implementation)
├── On-premises stored procedure (SQL validation)
└── Salesforce validation rule (Apex trigger)
```

Step 3: Deploy Consistent Access Controls

Using Informatica's policy engine, they established role-based access controls that maintain consistency across platforms while respecting each environment's native security models.

Results: Schneider Electric achieved 95% reduction in data quality incidents across platforms, 40% faster compliance reporting, and unified data lineage visibility spanning their entire hybrid infrastructure. Their governance framework now automatically provisions consistent data handling policies when new cloud services are adopted.

Compliance Orchestration Across Jurisdictions

Multi-cloud governance must address the complexity of varying regulatory requirements across different geographical regions and cloud providers. Organizations need automated compliance orchestration capabilities that can dynamically apply appropriate data handling policies based on data location, user jurisdiction, and regulatory context. This requires sophisticated policy engines that understand the intersection of technical capabilities and legal requirements across multiple cloud platforms.

The most successful implementations integrate compliance monitoring directly into data pipeline orchestration, ensuring that governance policies are enforced at the point of data processing rather than relying on post-processing audits. This proactive approach significantly reduces compliance risk while maintaining operational efficiency across distributed cloud environments.

Multi-cloud governance represents a fundamental shift from traditional centralized data management approaches, requiring organizations to develop new competencies in distributed policy enforcement and cross-platform integration. The practical frameworks and implementation strategies covered in this section provide the foundation for maintaining governance consistency while leveraging the flexibility and resilience benefits of multi-cloud architectures.

As organizations mature their multi-cloud governance capabilities, the next logical evolution involves empowering business users with self-service analytics capabilities while maintaining appropriate governance controls and data quality standards.

13.5. Data Marketplace and Self-Service Analytics Governance

Building on the foundation of advanced governance capabilities, the next logical evolution involves empowering business users with self-service analytics capabilities while maintaining appropriate governance controls and data quality standards. This transformation requires sophisticated governance frameworks that balance user autonomy with enterprise-wide data integrity, creating what industry leaders call **governed self-service.**

Establishing Marketplace Governance Architecture

The data marketplace concept fundamentally shifts how organizations democratize data access. Unlike traditional request-based systems, marketplaces enable discovery, evaluation, and consumption of data products through catalog-driven interfaces. However, this democratization introduces governance complexities that require proactive management.

Effective marketplace governance begins with **data product classification.** Organizations must establish clear taxonomies that distinguish between raw datasets, curated analytics-ready tables, and derived insights. Each classification tier carries different governance requirements, access protocols, and quality assurances. For instance, Tier 1 raw operational data might require formal data steward approval, while Tier 3 aggregated business metrics could support automated provisioning with usage tracking.

Implementing Automated Governance Controls

Modern data marketplaces leverage automation to scale governance oversight without creating bottlenecks. **Policy-driven access controls** automatically evaluate user requests against predefined criteria, including role-based permissions, data sensitivity classifications, and usage context. These systems can instantly approve routine requests while flagging unusual patterns for human review.

Consider implementing **dynamic data masking** within marketplace interfaces. When users access customer data through self-service tools, the system automatically applies appropriate privacy protections based on their role and intended use case. Marketing analysts might see demographic segments, while customer service representatives access individual records with personally identifiable information appropriately masked.

Practical Implementation: Enterprise Data Marketplace Governance

A Fortune 500 financial services company implemented comprehensive marketplace governance using Collibra's Data Marketplace combined with Tableau's self-service analytics platform. Their implementation demonstrates practical governance in action:

Step 1: Data Product Catalog Development

The organization created standardized data product definitions with embedded governance metadata:

```
Data Product: Customer_360_Analytics_Ready
- Classification: Tier 2 (Curated Business Data)
- Steward: Customer Analytics Team
- Quality Score: 94% (automated DQ monitoring)
- Usage Restrictions: Internal analytics only, no external sharing
- Lineage: Source systems tracked, transformation logic documented
- Access Method: Self-service via approved BI tools
```

Step 2: Automated Approval Workflows

They established tiered approval processes based on data sensitivity and user credentials. Business analysts could instantly access standard reporting datasets, while requests for sensitive financial data triggered automated workflows involving compliance review.

Step 3: Usage Monitoring and Compliance

The platform automatically tracked data consumption patterns, generating compliance reports showing who accessed what data, when, and for what purpose. This created an audit trail satisfying regulatory requirements while enabling business agility.

Results: The implementation reduced data access request fulfillment time from 5-7 business days to under 30 minutes for standard datasets, while maintaining 100% compliance with regulatory requirements. User satisfaction scores increased 40%, and the number of shadow IT analytics initiatives decreased significantly.

Self-Service Analytics Quality Assurance

Governing self-service analytics requires balancing user empowerment with result reliability. Organizations must establish **analytical sandbox environments** where users can experiment freely while preventing unreliable insights from influencing business decisions. These sandboxes should include data quality indicators, lineage information, and clear disclaimers about analytical limitations.

Certification programs for business users help ensure analytical competency. These programs cover data interpretation best practices, common analytical pitfalls, and proper use of self-service tools. Certified users gain access to broader datasets and advanced analytical capabilities, creating incentives for governance compliance.

Risk Mitigation and Escalation Procedures

Self-service environments require robust **anomaly detection** capabilities that identify potentially problematic usage patterns. These might include unusual data access volumes, attempts to export sensitive information, or analytical queries that could impact system performance. Automated alerts enable rapid response to governance violations while maintaining user productivity.

Establishing clear **escalation pathways** ensures governance issues receive appropriate attention. Minor violations might trigger automated user notifications and training recommendations, while serious breaches immediately involve data stewards and compliance teams.

This comprehensive approach to marketplace and self-service governance creates sustainable frameworks that grow with organizational needs while maintaining data integrity and regulatory compliance. The key lies in designing systems that feel empowering to users rather than restrictive, encouraging adoption while protecting enterprise data assets.

As organizations master these sophisticated governance approaches for internal data democratization, they often face additional complexity when corporate structures change through mergers, acquisitions, or divestitures, requiring entirely different governance strategies to manage data integration and separation challenges.

13.6. Governance for Mergers, Acquisitions, and Divestitures

Corporate restructuring events introduce unprecedented complexity to data governance frameworks, demanding specialized strategies that go far beyond standard operational governance. When organizations merge, acquire, or divest assets, the fundamental challenge shifts from managing established data flows to orchestrating the integration or separation of entire data ecosystems while maintaining regulatory compliance and operational continuity.

Due Diligence Data Assessment Framework

The governance foundation for any M&A transaction begins with comprehensive data due diligence that extends well beyond traditional IT asset inventories.

Organizations must establish a **Data Asset Valuation Protocol** that systematically catalogs not just data volumes, but data quality, lineage complexity, and governance maturity levels across target entities.

A practical assessment framework should evaluate five critical dimensions: data architecture compatibility, regulatory compliance alignment, master data management sophistication, data quality baseline metrics, and existing governance infrastructure maturity. This evaluation directly influences integration timelines, resource allocation, and risk mitigation strategies throughout the transaction lifecycle.

Integration Governance Architecture

Successful data integration during mergers requires establishing temporary **Unified Governance Councils** that bridge organizational boundaries while respecting existing data ownership structures. These councils must navigate competing data standards, conflicting privacy policies, and incompatible technology platforms while maintaining business operations.

Step-by-Step M&A Data Integration Implementation:

Establish Cross-Organizational Data Stewardship Teams using Collibra's Business Glossary module to create unified data definitions across both entities, mapping equivalent terms and identifying semantic conflicts before technical integration begins.

Deploy Temporary Data Quality Monitoring through Informatica Data Quality's profiling capabilities to baseline both organizations' data health, establishing integration success metrics and identifying critical remediation priorities.

Create Unified Master Data Management Strategy by implementing Informatica MDM's multi-domain hub architecture to merge customer, product, and supplier records while preserving historical relationships and audit trails.

Implement Progressive Data Migration Workflows using staged integration patterns that validate data consistency at each phase, allowing business operations to continue while gradually harmonizing data structures.

The expected outcome includes reduced integration timeline by 30-40%, maintained data quality scores above 85% throughout the process, and preserved regulatory compliance across all jurisdictions. Organizations typically achieve full data integration within 18-24 months compared to 36+ months using traditional approaches.

Divestiture Data Separation Strategies

Data separation during divestitures presents unique governance challenges, requiring surgical precision to extract specific data assets while maintaining the integrity of remaining systems. The **Clean Room Data Architecture** approach establishes isolated environments where divested data can be safely extracted, cleansed, and transferred without compromising ongoing operations.

Critical separation protocols must address data entanglement issues, where customer records, transaction histories, and operational metrics span multiple business units. Organizations must implement **Data Lineage Mapping** using tools like Alation's lineage visualization to identify all downstream dependencies and ensure complete data extraction without inadvertent data retention.

Regulatory Compliance During Transitions

Corporate restructuring events trigger complex regulatory requirements across multiple jurisdictions, particularly regarding data residency, privacy consent management, and industry-specific compliance mandates. The **Transitional Compliance Framework** establishes governance protocols that maintain regulatory standing throughout the restructuring process.

When Deutsche Bank divested its retail banking operations in multiple countries, they implemented a comprehensive data governance strategy that addressed GDPR compliance, local banking regulations, and cross-border data transfer requirements simultaneously. Their approach included establishing temporary data processing agreements, implementing automated compliance monitoring, and creating audit trails that satisfied regulators in each jurisdiction.

This section demonstrates how advanced governance strategies must adapt to handle the unique complexities of corporate restructuring while maintaining operational excellence and regulatory compliance. The specialized frameworks and implementation approaches outlined here provide IT leaders with proven methodologies for navigating these challenging scenarios successfully.

As organizations master these complex restructuring scenarios, they develop capabilities that extend naturally into governing advanced analytics and data science initiatives, where similar principles of managing complex data relationships and ensuring quality across diverse environments become equally critical.

13.7. Advanced Analytics and Data Science Governance

As organizations embrace advanced analytics and data science capabilities, traditional governance frameworks often prove inadequate for managing the unique challenges these disciplines present. Unlike conventional data management scenarios, analytics and data science initiatives involve experimental processes, iterative model development, and complex data lineage tracking across multiple environments. The governance approach must evolve to accommodate these dynamic requirements while maintaining the rigorous standards necessary for enterprise-grade deployments.

Establishing Analytics Governance Frameworks

Modern analytics governance requires a fundamentally different approach than traditional data stewardship. The experimental nature of data science work demands governance structures that balance innovation with control, enabling rapid experimentation while ensuring reproducibility and compliance. This balance becomes particularly critical when analytics models transition from development environments to production systems that impact business operations.

The governance framework must address several unique considerations specific to analytics workloads. **Model versioning** and **experiment tracking** become essential components of data lineage, requiring governance policies that extend beyond traditional data movement tracking. Additionally, the collaborative nature of data science teams necessitates governance structures that support cross-functional access while maintaining appropriate security boundaries.

Practical Implementation: Enterprise Model Governance Pipeline

Consider the implementation at a global financial services firm that established comprehensive governance for their machine learning operations using **MLflow** integrated with **Collibra** for metadata management. The organization faced challenges with model reproducibility and regulatory compliance across their credit risk assessment models.

The implementation began with establishing a **Model Registry Governance Policy** that required all production models to undergo a structured approval process:

```
Model Promotion Checklist:
1. Data lineage documentation complete in Collibra
2. Feature engineering pipeline validated and versioned
3. Model performance metrics meet established thresholds
4. Bias testing completed with documented results
5. Regulatory compliance review approved
6. Production deployment plan reviewed by governance committee
```

The governance committee, comprising data scientists, risk managers, and compliance officers, established weekly review sessions for model promotions. Each model underwent **algorithmic impact assessments** that evaluated potential bias, fairness, and regulatory implications before production deployment.

The results demonstrated significant improvements in governance maturity. Model deployment time decreased by 40% due to standardized processes, while regulatory audit preparation time reduced from weeks to days. Most importantly, the organization achieved 100% model traceability, enabling rapid response to regulatory inquiries about model decisions.

Managing Experimental Data and Sandbox Environments

Analytics governance extends beyond production models to encompass the entire experimental lifecycle. **Data sandbox governance** presents unique challenges as data scientists require access to diverse datasets for exploration while maintaining privacy and security standards. The governance framework must define clear boundaries for experimental data usage while enabling the flexibility necessary for innovation.

Effective sandbox governance typically involves **time-bound data access policies** that automatically expire experimental datasets after specified periods. This approach reduces long-term storage costs while ensuring sensitive data doesn't persist indefinitely in development environments. Additionally, implementing **data masking** and **synthetic data generation** capabilities within sandbox environments enables realistic experimentation without exposing production data.

Cross-Functional Collaboration in Analytics Governance

The interdisciplinary nature of modern analytics requires governance structures that facilitate collaboration between data scientists, business analysts, and domain experts. **Federated governance models** prove particularly effective in these scenarios, where domain experts maintain ownership of business logic while data scientists focus on technical implementation.

This collaborative approach necessitates governance policies that define clear roles and responsibilities across the analytics lifecycle. Business stakeholders must understand their accountability for model outcomes, while technical teams maintain responsibility for implementation quality and operational stability. The governance framework should establish regular **model performance review cycles** that bring together both technical and business perspectives to evaluate ongoing model effectiveness.

The evolution toward advanced analytics governance represents a maturation of traditional data stewardship practices, extending governance principles into the realm of artificial intelligence and machine learning. Organizations that successfully implement these advanced governance strategies position themselves to leverage analytics capabilities while maintaining the control and transparency necessary for enterprise operations.

As analytics capabilities continue to evolve, governance frameworks must adapt to address emerging challenges in artificial intelligence ethics, automated decision-making, and the broader implications of digital transformation on organizational data strategies.

13.8. Digital Transformation and Governance Evolution

As organizations accelerate their digital transformation initiatives, data governance frameworks must evolve beyond traditional static models to embrace dynamic, adaptive structures that can respond to rapidly changing technological landscapes. The convergence of cloud computing, artificial intelligence, and edge computing creates unprecedented opportunities for data value creation while simultaneously introducing complex governance challenges that require sophisticated orchestration strategies.

Modern governance frameworks must seamlessly integrate with digital transformation initiatives rather than operating as separate compliance exercises. This integration demands a fundamental shift from reactive governance models to proactive frameworks that anticipate technological changes and embed governance principles directly into digital architecture decisions. Organizations that successfully navigate this evolution position their data governance as a strategic enabler rather than a operational constraint.

Adaptive Governance Architecture for Cloud-Native Environments

Cloud-native architectures fundamentally alter how organizations approach data governance, requiring frameworks that can scale dynamically across multi-cloud environments while maintaining consistent policy enforcement. Traditional governance models designed for on-premises systems often fail when applied to containerized applications, microservices architectures, and serverless computing environments.

The key to successful cloud-native governance lies in implementing **policy-as-code** approaches that embed governance rules directly into infrastructure automation pipelines. This strategy ensures that governance controls are automatically applied as new resources are provisioned, eliminating the lag time between deployment and policy enforcement that often creates compliance gaps.

Consider the practical implementation at a major European telecommunications provider that transformed their governance approach during a comprehensive cloud migration. Their legacy governance framework relied on quarterly reviews and manual policy enforcement, creating bottlenecks that delayed digital initiatives by months.

```
# Cloud Governance Policy Template
resource "aws_s3_bucket" "data_lake" {
  bucket = var.bucket_name

  tags = {
    DataClassification = var.data_classification
    DataOwner = var.business_owner
    RetentionPeriod = var.retention_days
    ComplianceFramework = "GDPR,SOX"
  }

  lifecycle_rule {
    enabled = true
    expiration {
      days = var.retention_days
    }
  }
}
```

The telecommunications company implemented automated governance policies using Terraform and AWS Config, creating self-enforcing data classification and retention policies. Within six months, they reduced governance compliance time from 12 weeks to 2 days while improving policy consistency across 47 different cloud environments.

AI-Driven Governance Intelligence

Artificial intelligence transforms governance from a primarily reactive discipline to a predictive capability that can identify potential issues before they impact business operations. **Intelligent governance platforms** leverage machine learning algorithms to analyze data usage patterns, detect anomalies, and recommend policy adjustments based on evolving business requirements.

Advanced organizations are implementing AI-powered governance platforms like Collibra's Machine Learning Governance module to automatically discover sensitive data, classify information assets, and monitor compliance across complex data pipelines. These platforms can process thousands of data lineage relationships simultaneously, identifying potential privacy risks or quality issues that human analysts might overlook.

A leading financial services firm in Singapore implemented Collibra's AI governance capabilities to manage their real-time trading data streams. The platform automatically identified personally identifiable information flowing through their algorithmic trading systems, flagged potential regulatory violations, and recommended remediation actions within milliseconds of detection.

Automated Data Discovery: AI algorithms scan new data sources and automatically apply appropriate classification tags

Predictive Risk Assessment: Machine learning models analyze historical compliance patterns to predict future risk scenarios

Dynamic Policy Adjustment: Governance rules automatically adapt based on changing regulatory requirements and business contexts

Intelligent Lineage Mapping: AI traces data relationships across complex transformation pipelines to ensure end-to-end governance visibility

Edge Computing Governance Challenges

The proliferation of edge computing introduces governance complexities that traditional centralized models cannot address effectively. Data processing at network edges creates distributed governance scenarios where policies must be enforced across thousands of remote locations with varying connectivity and computational constraints.

Organizations must develop **federated governance models** that can operate autonomously at edge locations while maintaining consistency with centralized policies. This approach requires lightweight governance agents that can make real-time decisions without constant communication with central governance systems.

The governance evolution in digital transformation contexts demands frameworks that are inherently flexible, technologically sophisticated, and capable of adapting to emerging paradigms. Organizations that successfully integrate governance with their digital transformation initiatives create sustainable competitive advantages through improved data agility, reduced compliance risks, and enhanced innovation capabilities. These advanced strategies position governance as a strategic enabler that accelerates rather than constrains digital transformation objectives.

This comprehensive approach to governance evolution sets the foundation for understanding how these advanced strategies culminate in sustainable, long-term governance success that adapts continuously to organizational and technological changes.

14. THE FUTURE OF DATA GOVERNANCE

The data governance landscape stands at an unprecedented inflection point, where emerging technologies and evolving regulatory frameworks are reshaping how organizations manage their most valuable asset. As artificial intelligence transforms from experimental tool to governance backbone, and quantum computing promises to revolutionize data processing capabilities, forward-thinking leaders must prepare their governance frameworks for a future that demands both agility and resilience.

The convergence of global regulatory harmonization efforts, environmental sustainability imperatives, and the emergence of virtual business environments creates both extraordinary opportunities and complex challenges for data stewards. Organizations that have mastered traditional governance principles now face the critical task of evolving their frameworks to address quantum-era security concerns, metaverse data sovereignty, and AI-driven automation while maintaining the human oversight that ensures ethical data stewardship.

This chapter serves as your strategic compass for navigating the future of data governance, moving beyond current best practices to explore the transformative technologies and regulatory shifts that will define the next decade of enterprise data management. We examine how successful organizations are building adaptive governance frameworks that can evolve with technological advancement while maintaining unwavering commitment to data quality, privacy, and ethical use.

The governance leaders who thrive in this rapidly evolving landscape will be those who embrace continuous learning and proactive adaptation. By understanding these emerging trends and their implications, you position your organization not merely to respond to change, but to anticipate and leverage it as a competitive advantage. The future belongs to those who can seamlessly integrate cutting-edge technology with time-tested governance principles, creating frameworks that are both innovative and enduring.

14.1. Emerging Trends and Technologies

The landscape of data governance is experiencing unprecedented transformation as emerging technologies reshape how organizations manage, protect, and derive value from their information assets. For IT leaders navigating this evolution, understanding these technological shifts is crucial for building governance frameworks that remain relevant and effective in an increasingly complex digital ecosystem.

Artificial intelligence and machine learning are fundamentally altering governance paradigms by introducing autonomous data processing capabilities that traditional oversight models struggle to address. Modern governance frameworks must now account for algorithmic decision-making, model drift, and the explainability requirements that regulatory bodies increasingly demand. Organizations are discovering that conventional data lineage tracking becomes exponentially more complex when AI systems continuously learn and adapt their processing patterns.

AI-Driven Governance Automation

Contemporary governance platforms are integrating sophisticated AI capabilities to automate traditionally manual oversight processes. **Data discovery** tools now leverage natural language processing to automatically classify sensitive information across vast enterprise repositories, while machine learning algorithms continuously monitor data quality patterns and flag anomalies in real-time.

Consider the implementation at a major European financial services firm that deployed an AI-enhanced governance platform to manage their multi-petabyte data warehouse. The organization utilized Collibra's AI-powered data catalog combined with custom machine learning models to automatically detect and classify personally identifiable information across 847 different data sources. The implementation process involved:

- **Baseline Assessment**: The governance team established training datasets using 10,000 manually classified data elements across customer, transaction, and operational domains
- **Model Training**: Custom classification algorithms were trained to recognize PII patterns specific to European banking regulations, achieving 94% accuracy in initial testing
- **Automated Discovery**: The AI system scanned the entire data estate, identifying 2.3 million previously unclassified sensitive data elements within 72 hours
- **Continuous Monitoring**: Real-time algorithms now monitor data ingestion pipelines, automatically flagging new sensitive data and triggering appropriate governance workflows

The results demonstrated remarkable efficiency gains: manual data classification time decreased from 6 months to 3 days, while compliance coverage increased from 67% to 98% across all regulated data categories. The organization now maintains continuous compliance posture with minimal human intervention, allowing governance teams to focus on strategic initiatives rather than routine classification tasks.

Edge Computing and Distributed Governance

The proliferation of edge computing architectures presents unique governance challenges as data processing moves closer to generation sources. Traditional centralized governance models must evolve to accommodate distributed decision-making while maintaining consistent policy enforcement across geographically dispersed infrastructure.

Organizations are developing **federated governance frameworks** that enable local autonomy within global policy boundaries. These approaches require sophisticated policy orchestration capabilities that can adapt governance rules to local regulatory requirements while ensuring enterprise-wide consistency. The challenge intensifies as Internet of Things devices generate massive data volumes at network edges, often in environments with limited connectivity and processing capabilities.

Blockchain and Immutable Audit Trails

Blockchain technology is emerging as a powerful tool for governance transparency and accountability. Forward-thinking organizations are implementing blockchain-based audit trails that provide immutable records of data access, modification, and sharing activities. These implementations create unprecedented visibility into data usage patterns while establishing cryptographic proof of governance compliance.

The technology particularly excels in multi-party data sharing scenarios where traditional governance controls struggle to maintain oversight across organizational boundaries. Smart contracts automatically enforce data usage agreements, while distributed ledgers provide all participants with identical, tamper-proof records of governance activities.

As these technological trends converge, IT leaders must balance innovation opportunities with governance responsibilities, ensuring that emerging capabilities enhance rather than undermine organizational data stewardship. The key lies in designing adaptive frameworks that can evolve alongside technological advancement while maintaining fundamental governance principles.

This technological evolution sets the stage for examining how regulatory frameworks are adapting to address these emerging challenges, particularly as global jurisdictions work toward greater harmonization of data protection standards.

14.2. Regulatory Evolution and Global Harmonization

The convergence of global data protection regulations represents one of the most significant shifts in enterprise governance strategy since the advent of digital transformation. Organizations operating across multiple jurisdictions now face the complex challenge of implementing unified governance frameworks that satisfy diverse regulatory requirements while maintaining operational efficiency. This regulatory evolution extends beyond simple compliance, fundamentally reshaping how enterprises architect their data governance infrastructure to accommodate cross-border data flows and emerging privacy standards.

Understanding Regulatory Convergence Patterns

The movement toward regulatory harmonization reflects a growing recognition that data governance cannot remain fragmented across national boundaries. While the European Union's GDPR established foundational privacy principles, subsequent regulations like California's CCPA, Brazil's LGPD, and Singapore's PDPA have adopted similar core concepts while introducing jurisdiction-specific nuances. This convergence creates opportunities for organizations to develop streamlined governance approaches that address multiple regulatory frameworks simultaneously.

Modern governance implementations must anticipate regulatory evolution rather than merely react to current requirements. Forward-thinking organizations are building adaptive governance architectures that can accommodate new privacy rights, consent mechanisms, and data subject protections as they emerge across different jurisdictions. This proactive approach reduces implementation costs and minimizes disruption when new regulations take effect.

Building Cross-Jurisdictional Governance Frameworks

The most effective approach to managing regulatory diversity involves creating governance frameworks based on the highest common denominator of privacy protections. Organizations implementing this strategy typically achieve compliance across multiple jurisdictions while avoiding the complexity of maintaining separate governance structures for each regulatory regime.

Practical Implementation: Multi-Jurisdictional Data Classification System

Consider a multinational financial services organization implementing a unified data classification system using Informatica's Axon platform to address GDPR, CCPA, and PIPEDA requirements simultaneously. The implementation follows these specific steps:

Regulatory Mapping Workshop: Governance teams conduct cross-jurisdictional analysis to identify overlapping requirements for data categories, retention periods, and subject rights across all applicable regulations.

Unified Classification Schema: Using Informatica Axon's business glossary capabilities, the organization creates a master data classification framework that incorporates the most stringent requirements from each jurisdiction:

```
Data Classification Framework:
- PII_LEVEL_1: Direct identifiers (SSN, passport numbers)
  * GDPR: Special category protections
  * CCPA: Personal information with deletion rights
  * PIPEDA: Personal information requiring consent

- PII_LEVEL_2: Indirect identifiers (IP addresses, device IDs)
  * Cross-jurisdictional retention: 24 months maximum
  * Automated anonymization triggers after retention period
```

Automated Compliance Monitoring: The platform implements automated scanning rules that flag data movements potentially violating any of the three regulatory frameworks, with escalation workflows routing issues to appropriate regional compliance teams.

Cross-Border Transfer Controls: Integration with data loss prevention tools creates automatic blocking mechanisms for data transfers that don't meet adequacy requirements under any applicable regulation.

The organization measured success through reduced compliance audit findings (from 47 issues to 8 across all jurisdictions) and decreased regulatory response times (from 15 days to 4 days average) within six months of implementation.

Anticipating Future Regulatory Developments

Successful governance frameworks incorporate mechanisms for rapid adaptation to emerging regulations. Organizations are implementing governance councils with dedicated regulatory intelligence functions that monitor proposed legislation across key jurisdictions and assess potential impacts on existing data management practices. These councils typically include legal, compliance, and technical representatives who can quickly evaluate implementation requirements for new regulatory obligations.

The establishment of regulatory sandboxes and pilot programs allows organizations to test governance approaches against proposed regulations before they take effect. This proactive strategy enables smoother transitions when new requirements become mandatory and demonstrates regulatory good faith to oversight authorities.

The regulatory landscape's continued evolution toward greater harmonization presents both challenges and opportunities for data governance practitioners. Organizations that successfully navigate this transition build competitive advantages through streamlined compliance processes and enhanced stakeholder trust. The key lies in developing governance frameworks that embrace regulatory convergence while maintaining flexibility for jurisdiction-specific requirements.

As regulatory frameworks become increasingly sophisticated, the integration of artificial intelligence and automation technologies emerges as a critical enabler for managing compliance complexity at scale.

14.3. The Role of Artificial Intelligence in Governance Automation

As regulatory frameworks become increasingly sophisticated, the integration of artificial intelligence and automation technologies emerges as a critical enabler for managing compliance complexity at scale. The convergence of AI with data governance represents a fundamental shift from traditional rule-based systems to intelligent, self-learning governance ecosystems that can adapt to evolving business requirements and regulatory landscapes in real-time.

Modern AI-powered governance platforms are transforming how organizations approach **data stewardship**, moving beyond static policy enforcement to dynamic, context-aware governance that anticipates issues before they manifest. These systems leverage machine learning algorithms to identify patterns in data usage, detect anomalies in access behaviors, and automatically classify sensitive information based on content analysis rather than manual tagging processes.

Intelligent Data Classification and Discovery

AI-driven classification engines represent one of the most impactful applications of artificial intelligence in governance automation. These systems utilize natural language processing and pattern recognition to automatically identify and categorize data elements across enterprise systems, dramatically reducing the manual effort traditionally required for data inventory and classification initiatives.

Advanced classification algorithms can analyze structured and unstructured data simultaneously, identifying personally identifiable information, financial records, and intellectual property with accuracy rates exceeding 95%. This capability enables organizations to maintain comprehensive data catalogs that update automatically as new data sources are integrated into the enterprise ecosystem.

Consider the implementation at a multinational pharmaceutical company that deployed Informatica's CLAIRE AI engine to automate their global data discovery process. The system analyzed over 2,000 databases across 15 countries, automatically classifying 40 million data elements according to regulatory requirements including GDPR, HIPAA, and FDA guidelines. The AI engine identified previously unknown sensitive data repositories in legacy systems, reducing compliance risk exposure by 60% within the first quarter of deployment.

Predictive Governance and Risk Assessment

Machine learning models are revolutionizing risk assessment by analyzing historical governance events to predict potential compliance violations before they occur. These predictive systems examine data access patterns, user behavior analytics, and system performance metrics to generate early warning indicators for governance teams.

Practical Implementation Example: Automated Data Quality Monitoring

A global financial services organization implemented an AI-powered data quality monitoring system using Collibra's governance platform integrated with custom machine learning models. The implementation followed this structured approach:

Baseline Establishment: Historical data quality metrics were analyzed across 500 critical data elements spanning customer master data, transaction records, and regulatory reporting datasets.

Model Training: Machine learning algorithms were trained on 18 months of quality assessment data, learning to identify patterns that preceded quality degradation events.

Automated Monitoring Deployment: The system began monitoring data quality in real-time, generating predictive alerts when quality scores were likely to fall below acceptable thresholds within 48-72 hours.

Governance Workflow Integration: Automated workflows were configured to trigger data stewardship tasks, notify responsible data owners, and escalate critical issues to governance committees based on predicted severity levels.

The results demonstrated the transformative potential of AI in governance automation. Data quality incidents decreased by 45% as proactive remediation replaced reactive problem-solving. The system identified seasonal patterns in data degradation that human analysts had missed, enabling preventive maintenance schedules that maintained quality standards during peak processing periods.

Autonomous Policy Enforcement

AI systems are evolving to support autonomous policy enforcement, where governance rules adapt dynamically based on contextual factors such as user roles, data sensitivity levels, and business processes. These intelligent enforcement engines can make nuanced decisions about data access requests, balancing security requirements with operational efficiency.

Natural language processing capabilities enable these systems to interpret complex policy documents and translate governance requirements into executable rules automatically. This capability significantly reduces the implementation time for new regulatory requirements and ensures consistent policy application across diverse enterprise environments.

The integration of artificial intelligence into data governance frameworks represents a paradigm shift toward proactive, intelligent stewardship that scales with organizational growth and regulatory complexity. Organizations implementing AI-powered governance solutions report improved compliance outcomes, reduced operational overhead, and enhanced ability to demonstrate governance effectiveness to stakeholders and regulators.

As quantum computing technologies emerge and introduce unprecedented computational capabilities, the governance challenges will require even more sophisticated AI systems capable of managing quantum-scale data processing and security requirements.

14.4. Quantum Computing and Next-Generation Data Challenges

As quantum computing transitions from theoretical possibility to practical reality, IT leaders must prepare their data governance frameworks for computational paradigms that will fundamentally alter how organizations process, secure, and manage information. The emergence of quantum systems introduces both extraordinary opportunities and complex governance challenges that traditional frameworks cannot adequately address.

Quantum Data Processing Governance Requirements

Quantum computing's ability to process vast datasets exponentially faster than classical computers creates immediate governance implications for data lineage, quality assurance, and regulatory compliance. Unlike traditional processing where data transformations follow predictable pathways, quantum algorithms can manipulate information through superposition and entanglement principles that require entirely new documentation and audit approaches.

The **quantum advantage** in data processing demands governance frameworks capable of tracking probabilistic outcomes rather than deterministic results. This shift requires organizations to establish quantum-specific data quality metrics that account for measurement uncertainties and quantum decoherence effects on data integrity. Traditional data governance tools must evolve to support quantum state verification and quantum error correction protocols as standard components of data quality management.

Quantum Security and Privacy Governance

The quantum computing era presents a dual security challenge: protecting data from quantum-enabled attacks while leveraging quantum cryptography for enhanced security. Current encryption standards that underpin most data governance security frameworks will become obsolete when large-scale quantum computers emerge, necessitating immediate preparation for **post-quantum cryptography** implementations.

Organizations must begin developing quantum-resilient data classification schemes that incorporate quantum risk assessments alongside traditional security evaluations. This includes establishing governance protocols for quantum key distribution systems and quantum-safe backup strategies that ensure data accessibility across the classical-quantum transition period.

Practical Implementation: Quantum-Ready Governance Framework

A leading financial services firm recently partnered with IBM's Quantum Network to develop a quantum-ready data governance framework addressing regulatory compliance in quantum computing environments. The implementation involved several critical steps:

```
Quantum Governance Assessment Framework:

1. Data Classification Review
   - Identify quantum-vulnerable datasets
   - Establish quantum risk ratings (QR1-QR5)
   - Map quantum processing workflows

2. Security Protocol Upgrades
   - Implement NIST post-quantum cryptography standards
   - Deploy quantum key management systems
   - Establish quantum audit trails
```

```
3. Compliance Integration
   - Develop quantum-specific regulatory reporting
   - Create quantum data retention policies
   - Establish quantum breach notification procedures
```

The organization utilized Collibra's Data Governance Center to create quantum-specific data lineage tracking, implementing custom metadata attributes for quantum processing states and measurement probabilities. Their governance committee established quantum data stewardship roles responsible for validating quantum algorithm outputs and ensuring quantum processing compliance with existing financial regulations.

Before implementation, the firm's traditional governance framework required 72 hours to complete regulatory compliance validation for complex derivatives calculations. After integrating quantum processing capabilities with enhanced governance controls, validation time decreased to 4 hours while maintaining full audit compliance and introducing quantum-specific risk assessments that traditional systems could not provide.

Organizational Quantum Readiness

Preparing governance frameworks for quantum computing requires establishing cross-functional quantum literacy programs and updating existing data governance policies to accommodate quantum-classical hybrid processing environments. Organizations must develop quantum data governance competency centers that bridge traditional IT governance expertise with emerging quantum computing knowledge.

The transformation demands updating existing governance documentation to include quantum-specific terminology, establishing quantum data stewardship certification programs, and creating governance workflows that can seamlessly transition between classical and quantum processing modes as organizational quantum capabilities mature.

Summary: Quantum computing introduces fundamental challenges to traditional data governance frameworks, requiring organizations to develop quantum-ready policies, security protocols, and compliance mechanisms. Success depends on proactive preparation, cross-functional quantum literacy development, and governance frameworks capable of managing probabilistic data processing outcomes while maintaining regulatory compliance and data quality standards.

As organizations prepare their governance frameworks for quantum computing challenges, they must simultaneously address another critical frontier: integrating environmental sustainability principles into their data management practices and governance structures.

14.5. Sustainability and Environmental Data Governance

As organizations navigate quantum computing challenges, they must simultaneously address another critical frontier: integrating environmental sustainability principles into their data management practices and governance structures. The convergence of environmental responsibility and data stewardship represents a paradigm shift that extends governance beyond traditional compliance boundaries into measurable environmental impact. Forward-thinking IT leaders recognize that sustainable data governance not only reduces operational costs but also positions organizations as responsible corporate citizens in an increasingly environmentally conscious marketplace.

Environmental Impact Assessment Framework

Modern data centers consume approximately 1% of global electricity, making environmental data governance a strategic imperative rather than an optional consideration. Effective sustainability governance requires establishing comprehensive measurement frameworks that track energy consumption, carbon footprint, and resource utilization across the entire data lifecycle. Organizations must implement governance policies that mandate environmental impact assessments for new data initiatives, similar to how they currently evaluate security and privacy risks.

The framework begins with establishing baseline measurements for current data operations, including storage efficiency ratios, processing energy consumption per transaction, and data retention policies that directly correlate with environmental costs. Governance committees must expand their evaluation criteria to include sustainability metrics alongside traditional quality, security, and compliance measures.

Green Data Lifecycle Management

Implementing sustainable data practices requires rethinking traditional data lifecycle management through an environmental lens. This approach encompasses everything from data creation and storage optimization to archival and deletion policies that prioritize resource efficiency. Organizations are discovering that environmental governance often aligns with cost optimization, creating dual benefits that strengthen business cases for sustainability initiatives.

Practical Implementation: Carbon-Aware Data Governance at Microsoft Azure

Microsoft's implementation of carbon-aware data governance demonstrates how large-scale environmental stewardship can be systematically integrated into enterprise data management. Their approach provides a replicable model for organizations seeking to implement similar programs.

Baseline Carbon Footprint Assessment: Microsoft implemented comprehensive carbon tracking across all Azure data centers, establishing governance policies requiring carbon impact assessments for every new data service deployment.

```
Carbon Impact Assessment Template:
- Data Volume: [TB/month]
- Processing Requirements: [CPU hours/month]
- Storage Type: [Hot/Cool/Archive tiers]
- Geographic Distribution: [Regions with renewable energy %]
- Estimated Carbon Footprint: [CO2 equivalent/month]
- Optimization Opportunities: [Specific recommendations]
```

Policy Integration: The governance framework mandates that data architects justify storage tier selections based on both business requirements and environmental impact, with automatic recommendations for moving infrequently accessed data to lower-energy archive tiers.

Stakeholder Engagement: Cross-functional governance committees now include sustainability officers who review major data initiatives, ensuring environmental considerations receive equal weight with traditional governance criteria.

Measurement and Reporting: Real-time dashboards track carbon efficiency metrics alongside traditional data quality indicators, with quarterly governance reviews including sustainability performance assessments.

Results: Microsoft achieved a 75% reduction in carbon intensity for Azure services between 2020 and 2023, while simultaneously improving data governance maturity scores across their enterprise customer base.

Regulatory Compliance and ESG Integration

Environmental data governance increasingly intersects with regulatory requirements and Environmental, Social, and Governance (ESG) reporting mandates. The European Union's Corporate Sustainability Reporting Directive and similar regulations worldwide require organizations to maintain auditable records of their environmental impact, including data-related carbon emissions. Governance frameworks must evolve to capture, validate, and report this information with the same rigor applied to financial data.

This integration requires establishing new data stewardship roles specifically focused on environmental metrics, implementing quality controls for sustainability data, and ensuring that environmental reporting meets the same accuracy standards as other regulated data domains. Organizations must also consider the environmental implications of data residency requirements, balancing compliance needs with sustainability objectives.

The intersection of environmental responsibility and data governance represents a fundamental evolution in how organizations approach information management. By integrating sustainability principles into governance frameworks, IT leaders can drive both environmental stewardship and operational efficiency while meeting increasingly stringent regulatory requirements.

As organizations master these terrestrial sustainability challenges, they must prepare for an even more complex frontier where data governance extends beyond physical infrastructure into entirely virtual realms that present unprecedented governance challenges and opportunities.

14.6. Governance in the Metaverse and Virtual Environments

As organizations venture beyond traditional digital boundaries, the emergence of metaverse platforms and virtual environments creates an entirely new landscape for data governance that extends far beyond conventional enterprise systems. These immersive digital spaces generate unprecedented volumes of behavioral data, spatial interactions, and avatar-based identities that require sophisticated governance frameworks capable of managing both virtual and real-world implications.

The complexity of governing data in virtual environments stems from the convergence of multiple data streams: user behavior analytics, virtual asset transactions, spatial mapping data, biometric inputs from VR/AR devices, and cross-platform identity management. Unlike traditional enterprise data that flows through established channels, metaverse data exists in a three-dimensional, real-time environment where users create, modify, and interact with digital assets in ways that blur the lines between personal and corporate data ownership.

Virtual Data Classification and Ownership Models

Establishing clear data classification schemes for virtual environments requires a fundamental reimagining of traditional governance categories. Virtual worlds generate unique data types including avatar biometrics, spatial positioning data, virtual property transactions, and immersive interaction patterns that don't fit conventional structured data models.

The ownership question becomes particularly complex when considering user-generated content in corporate virtual spaces. Organizations must develop policies that clearly delineate between company-owned virtual infrastructure data and user-created digital assets, while ensuring compliance with emerging regulations around virtual property rights and digital identity protection.

Practical Implementation: Enterprise Virtual Collaboration Platform Governance

Consider a multinational manufacturing company implementing a virtual reality training platform using Microsoft Mesh for enterprise collaboration. The governance challenge involves managing data across multiple jurisdictions while participants interact in shared virtual spaces.

Step 1: Virtual Environment Data Mapping

```
Virtual Data Categories:
- Participant biometric data (eye tracking, hand movements)
- Training performance analytics
- Virtual asset creation and modification logs
- Cross-reality communication transcripts
- Spatial interaction heatmaps
```

Step 2: Governance Framework Adaptation

Using Collibra's Data Governance Center, the organization creates specialized data domains for virtual environment governance:

```
Virtual Training Domain Structure:
├── Biometric Data Stewardship
│   ├── Collection consent management
│   ├── Retention policy enforcement
│   └── Cross-border transfer controls
├── Virtual Asset Management
│   ├── Intellectual property classification
│   ├── Version control for 3D models
│   └── Usage rights tracking
└── Performance Analytics Governance
    ├── Individual privacy protection
    ├── Aggregate reporting standards
    └── Predictive model validation
```

Step 3: Multi-Jurisdictional Compliance Integration

The implementation addresses GDPR requirements for EU participants while managing CCPA obligations for California-based users, creating automated policy enforcement through the virtual platform's API.

Expected Outcomes:

95% reduction in compliance violations across virtual training sessions

Streamlined data subject access requests for virtual environment data
Clear audit trails for virtual asset intellectual property disputes

Cross-Reality Data Synchronization Challenges

One of the most significant governance challenges in virtual environments involves maintaining data consistency and quality across the boundary between virtual and physical systems. When virtual actions trigger real-world consequences, such as virtual prototype testing influencing physical manufacturing decisions, governance frameworks must ensure data lineage and quality standards remain intact across both domains.

Organizations like BMW have pioneered virtual factory simulations where digital twin data governance becomes critical for maintaining operational integrity. Their approach demonstrates how traditional master data management principles must evolve to accommodate real-time virtual environment inputs while preserving data quality standards that impact physical production processes.

Emerging Regulatory Landscape and Compliance Frameworks

The regulatory environment for virtual worlds remains in rapid flux, with jurisdictions worldwide developing new frameworks for digital identity protection, virtual asset taxation, and cross-border data flows in immersive environments. Forward-thinking organizations are establishing governance committees specifically focused on virtual environment compliance, bringing together legal, technology, and business stakeholders to address these evolving requirements.

The European Union's proposed Virtual Worlds Regulation and similar initiatives in Asia-Pacific regions signal the need for proactive governance strategies that can adapt to changing regulatory landscapes while maintaining operational flexibility in virtual spaces.

Mastering governance in virtual environments requires organizations to fundamentally expand their understanding of data stewardship beyond traditional enterprise boundaries. Success depends on developing adaptive frameworks that can evolve with both technological advancement and regulatory change while maintaining the core principles of data quality, privacy protection, and business value creation that define effective governance programs.

The lessons learned from virtual environment governance will prove invaluable as organizations prepare to build truly adaptive frameworks capable of responding to any technological disruption while maintaining governance integrity across all operational domains.

14.7. Building Adaptive and Resilient Governance Frameworks

The technological landscape's relentless evolution demands governance frameworks that bend without breaking, adapting to new challenges while maintaining their foundational integrity. Building adaptive and resilient governance frameworks requires a fundamental shift from static, rule-based systems to dynamic, intelligent architectures that can evolve with organizational needs and technological advancement.

Dynamic Policy Management Systems

Modern governance frameworks must incorporate **policy versioning** and **automated rule adaptation** mechanisms that respond to changing business requirements without compromising data integrity. Organizations are implementing policy management platforms that use machine learning algorithms to suggest rule modifications based on usage patterns and compliance outcomes.

Consider how Netflix revolutionized their data governance approach by implementing a self-healing policy framework. Their system automatically adjusts data retention policies based on content popularity algorithms and regulatory changes across different geographic regions. When European GDPR requirements conflicted with their existing North American data handling practices, their adaptive framework automatically created region-specific policy branches while maintaining global consistency for shared datasets.

Resilience Through Distributed Governance Architecture

Distributed governance models create multiple layers of decision-making authority that prevent single points of failure. This approach distributes governance responsibilities across business units while maintaining centralized oversight and standardization.

A practical implementation involves establishing **governance pods** - small, cross-functional teams responsible for domain-specific governance decisions. Each pod operates within enterprise-wide parameters but possesses autonomy to adapt policies for their specific data contexts.

```
Governance Pod Structure:
- Data Domain Owner (Business Leader)
- Technical Steward (IT Representative)
- Compliance Officer (Risk Management)
- End-User Representative (Operations)

Decision Authority Matrix:
Level 1: Pod autonomy (routine policy adjustments)
Level 2: Cross-pod coordination (shared data impacts)
Level 3: Enterprise escalation (framework changes)
```

Practical Implementation: Building an Adaptive Data Classification Framework

A multinational pharmaceutical company recently implemented an adaptive classification system using Informatica's Axon platform combined with custom automation scripts. Their challenge involved managing research data that transitions through multiple regulatory jurisdictions as drugs move from discovery to market.

Step 1: Baseline Assessment

The organization mapped existing data flows across 47 research facilities, identifying 23 distinct classification requirements that varied by country and research phase.

Step 2: Framework Design

They implemented a three-tier adaptive system:

Core Classifications: Universal categories (Public, Internal, Confidential, Restricted)
Contextual Modifiers: Regulatory tags (FDA-Regulated, EMA-Controlled, Clinical-Trial-Data)
Dynamic Attributes: Time-based classifications that automatically adjust based on research milestones

Step 3: Automation Implementation

Using Informatica's API, they created automated workflows that adjust classification levels based on predefined triggers:

```
# Simplified classification adjustment logicif research_phase ==
"pre_clinical":
    apply_classification( "Internal-Research")
elif research_phase == "clinical_trial":
    apply_classification( "FDA-Regulated-Clinical")
    add_retention_policy( "7_year_minimum")
elif approval_status == "market_ready":
    apply_classification( "Commercial-Sensitive")
    enable_cross_border_sharing()
```

Results: The adaptive framework reduced manual classification errors by 78% and accelerated cross-border research collaboration by automatically handling regulatory compliance requirements.

Future-Proofing Through Modular Architecture

Resilient frameworks embrace **modular design principles** that allow individual components to be upgraded or replaced without disrupting the entire governance ecosystem. This approach involves creating standardized interfaces between governance tools and establishing **API-first governance architectures**.

Organizations implementing modular frameworks report 40% faster adoption of new governance technologies and significantly reduced implementation risks when regulatory requirements change.

Continuous Learning Integration

Adaptive frameworks incorporate **feedback mechanisms** that capture governance effectiveness data and automatically suggest improvements. These systems analyze policy violation patterns, user behavior data, and business outcome correlations to recommend framework enhancements.

The integration of artificial intelligence into governance decision-making processes enables frameworks to learn from organizational patterns and proactively suggest policy modifications before issues arise.

Building adaptive and resilient governance frameworks transforms data management from reactive compliance into proactive strategic advantage. Organizations mastering these dynamic approaches position themselves to capitalize on emerging technologies while maintaining robust data protection and quality standards. The key lies in balancing flexibility with control, ensuring frameworks evolve intelligently while preserving the fundamental governance principles that protect organizational assets.

This foundation of adaptive resilience sets the stage for developing comprehensive learning strategies that ensure governance capabilities continue expanding alongside organizational growth and technological innovation.

14.8. Your Governance Journey: Next Steps and Continued Learning

Building on the adaptive resilience framework explored in the previous section, successful data governance extends far beyond initial implementation to encompass a continuous learning ecosystem that evolves with organizational needs and technological advancement. The most effective governance leaders recognize that mastery comes not from achieving a static state of perfection, but from cultivating dynamic capabilities that respond to emerging challenges while maintaining foundational principles.

Establishing Your Personal Governance Development Plan

The journey toward governance mastery requires structured personal and organizational learning pathways that build expertise incrementally. Leading practitioners develop competencies across multiple domains simultaneously, from technical implementation skills to strategic stakeholder management capabilities. This multifaceted approach ensures governance leaders can navigate complex organizational dynamics while maintaining technical credibility.

Consider implementing a quarterly skills assessment framework that evaluates competencies across governance domains including policy development, technology integration, change management, and business value communication. Organizations like Mastercard have established internal governance academies where practitioners complete progressive certification tracks, combining theoretical knowledge with hands-on project experience across different business units.

Building Governance Communities of Practice

Sustainable governance excellence emerges through collaborative learning environments that connect practitioners across organizational boundaries. Internal communities of practice provide forums for sharing challenges, discussing emerging trends, and developing standardized approaches to common governance problems. These communities become invaluable resources for accelerating learning and preventing knowledge silos.

External professional networks through organizations like DAMA International, the Data Governance Institute, and industry-specific associations provide access to broader governance perspectives and emerging best practices. Participation in these communities offers exposure to diverse implementation approaches and helps identify potential solutions before challenges become critical organizational issues.

Practical Implementation: Governance Maturity Progression Framework

Here's a step-by-step approach to implementing a continuous learning framework using the DAMA-DMBOK methodology combined with Collibra's governance platform:

Step 1: Current State Assessment

```
Governance Maturity Scorecard Template:
- Data Quality Management: [1-5 scale]
- Policy Compliance: [1-5 scale]
- Stakeholder Engagement: [1-5 scale]
- Technology Integration: [1-5 scale]
- Business Value Delivery: [1-5 scale]
```

Step 2: Learning Path Configuration

Configure Collibra's workflow engine to create personalized learning paths based on assessment results. For practitioners scoring below 3 in Data Quality Management, the system automatically enrolls them in foundational data profiling workshops and assigns mentorship relationships with senior data stewards.

Step 3: Progressive Skill Building

Implement quarterly progression milestones where practitioners demonstrate competency through real project deliverables. A data steward might progress from basic data cataloging tasks to leading cross-functional data quality initiatives, with each milestone requiring documented business impact and peer validation.

Step 4: Community Integration

Establish monthly governance forums where practitioners present case studies, discuss implementation challenges, and share lessons learned. These sessions become repositories of organizational knowledge while building collaborative relationships across departments.

Expected Outcomes: Organizations implementing this framework typically see 40% faster governance capability development, improved retention of governance professionals, and more consistent application of governance practices across business units.

Staying Current with Emerging Technologies

The governance landscape continuously evolves with technological advancement, regulatory changes, and shifting business requirements. Successful practitioners maintain awareness of emerging trends including artificial intelligence governance, cloud data management, and privacy-enhancing technologies. This requires structured approaches to technology evaluation and gradual integration into existing governance frameworks.

Establishing technology watch programs helps identify relevant innovations before they become mainstream requirements. Regular assessment of governance tool capabilities ensures platforms continue meeting organizational needs while providing growth pathways for enhanced functionality.

This comprehensive approach to continuous learning ensures governance capabilities remain aligned with organizational growth and technological innovation, creating sustainable competitive advantages through superior data management practices. As we conclude this exploration of future governance directions, the focus shifts toward synthesizing these forward-looking perspectives into actionable implementation strategies that drive immediate organizational value.

GLOSSARY OF DATA GOVERNANCE TERMS

Business Glossary

A centralized repository of business terms, definitions, and metadata that provides a common vocabulary for organizational data assets. This critical governance tool ensures consistent understanding of data elements across departments, reducing confusion and improving data quality by establishing authoritative definitions for key business concepts, metrics, and data fields used throughout the enterprise.

Business Intelligence (BI)

The technology-driven process of analyzing business data to present actionable information that helps executives, managers, and workers make informed business decisions. In data governance contexts, BI systems rely heavily on well-governed data to produce reliable insights, making data quality, consistency, and accessibility fundamental requirements for successful business intelligence initiatives.

Chief Data Officer (CDO)

A senior executive responsible for enterprise-wide governance and utilization of information as an asset, typically reporting to the CEO or CIO. The CDO leads data strategy development, oversees data governance programs, ensures regulatory compliance, and drives data-driven decision making across the organization while managing data-related risks and maximizing the business value of data assets.

Data Architecture

The structural design of data systems that defines how data is collected, stored, arranged, integrated, and put to use in data systems and organizations. Effective data architecture provides the foundation for data governance by establishing clear data flows, relationships, and dependencies that enable consistent data management practices and support business requirements across the enterprise.

Data Catalog

A comprehensive inventory of data assets within an organization that provides metadata, lineage information, and business context for datasets. Data catalogs serve as searchable repositories that help users discover, understand, and access relevant data while supporting governance efforts through impact analysis, compliance tracking, and data asset management across distributed enterprise environments.

Data Classification

The systematic organization of data into categories based on sensitivity, criticality, regulatory requirements, or business value. This governance practice enables organizations to apply appropriate security controls, retention policies, and access restrictions while ensuring compliance with privacy regulations and protecting sensitive information from unauthorized disclosure or misuse throughout its lifecycle.

Data Custodian

An individual or team responsible for the technical implementation of data governance policies and the day-to-day care of data assets. Data custodians maintain data systems, implement security controls, perform backup and recovery operations, and ensure technical compliance with governance standards while working closely with data stewards to address business requirements and resolve data quality issues.

Data Dictionary

A centralized repository that contains detailed descriptions of data elements, including their definitions, formats, relationships, and business rules. This governance artifact provides technical and business users with authoritative information about data structure and meaning, supporting consistent data interpretation and usage while facilitating system integration and data quality improvement initiatives.

Data Governance Council

A cross-functional governing body that provides strategic oversight, decision-making authority, and accountability for enterprise data governance initiatives. The council typically includes senior executives, business leaders, and technical experts who establish governance policies, resolve conflicts, approve data standards, and ensure alignment between data governance activities and organizational objectives.

Data Governance Framework

A structured approach that defines the policies, processes, roles, and technologies needed to manage data as a strategic asset. This comprehensive framework establishes the foundation for consistent data management practices across the organization, providing guidelines for data quality, security, privacy, and compliance while enabling effective decision-making and risk management.

Data Governance Maturity Model

A structured assessment framework that evaluates an organization's data governance capabilities across multiple dimensions, typically ranging from initial or ad-hoc practices to optimized and continuously improving programs. These models help organizations benchmark their current state, identify improvement opportunities, and develop roadmaps for advancing their data governance maturity over time.

Data Integration

The process of combining data from multiple sources into a unified view that provides users with consistent access and delivery of data across the spectrum of subjects and structure types. Effective data integration requires strong governance practices to ensure data quality, maintain lineage tracking, and preserve business meaning as data moves between systems and applications.

Data Lake

A centralized repository that stores structured and unstructured data at any scale, allowing organizations to store raw data in its native format until needed for analysis. Data lakes require robust governance practices to prevent becoming data swamps, including metadata management, data cataloging, access controls, and quality monitoring to ensure data remains discoverable and usable.

Data Lineage

The tracking and visualization of data flow from its origin through various transformations to its final destination, providing complete visibility into data movement and dependencies. Data lineage supports governance by enabling impact analysis, compliance reporting, root cause analysis for data quality issues, and understanding of how changes to upstream systems affect downstream processes and reports.

Data Mart

A subset of a data warehouse that contains data specific to a particular business line, department, or subject area. Data marts require governance oversight to ensure consistency with enterprise data standards, maintain data quality, and prevent the creation of conflicting versions of business metrics while providing focused access to relevant data for specific user communities.

Data Owner

A business role responsible for making decisions about data usage, access rights, and business rules for specific data domains or assets. Data owners typically come from business units and have the authority to define data requirements, approve access requests, and resolve data-related issues while working with data stewards and custodians to implement governance policies.

Data Privacy

The practice of protecting personal and sensitive information from unauthorized access, use, or disclosure while ensuring compliance with applicable privacy regulations such as GDPR, CCPA, or HIPAA. Data privacy governance includes implementing appropriate technical and organizational measures, conducting privacy impact assessments, and maintaining records of processing activities to demonstrate compliance.

Data Profiling

The statistical analysis and assessment of data content, structure, and quality to understand data characteristics, identify anomalies, and discover patterns or relationships. Data profiling supports governance efforts by providing objective measurements of data quality, identifying potential issues, and establishing baseline metrics for monitoring data health and improvement initiatives over time.

Data Quality

The degree to which data meets business requirements and is suitable for its intended use, typically measured across dimensions such as accuracy, completeness, consistency, timeliness, validity, and uniqueness. High data quality is essential for effective decision-making and regulatory compliance, requiring ongoing monitoring, measurement, and improvement processes as part of comprehensive data governance programs.

Data Quality Rules

Specific criteria and validation logic used to assess and ensure data meets defined quality standards and business requirements. These rules can be implemented at various points in data processing workflows to prevent poor quality data from entering systems, identify existing quality issues, and trigger corrective actions or notifications when quality thresholds are not met.

Data Retention Policy

A governance document that specifies how long different types of data should be kept, when it should be archived or deleted, and the procedures for data disposal. These policies help organizations manage storage costs, comply with regulatory requirements, reduce legal risks, and ensure that data is available when needed while preventing the accumulation of obsolete or unnecessary information.

Data Steward

A business role responsible for ensuring the quality, integrity, and appropriate use of data within a specific domain or subject area. Data stewards work closely with data owners and custodians to implement governance policies, resolve data quality issues, maintain business glossaries, and serve as subject matter experts for their assigned data assets throughout the organization.

Data Stewardship

The practice of managing and overseeing data assets to ensure their quality, security, and appropriate use throughout their lifecycle. Data stewardship encompasses activities such as data quality monitoring, metadata management, issue resolution, and user support while promoting data literacy and adherence to governance policies across the organization.

Data Warehouse

A centralized repository that integrates data from multiple operational systems to support business intelligence, reporting, and analytical activities. Data warehouses require strong governance practices to ensure data quality, maintain historical accuracy, and provide consistent business definitions while supporting the information needs of decision-makers across the organization.

Enterprise Data Management (EDM)

A comprehensive approach to managing data as a strategic asset across the entire organization, encompassing governance, architecture, quality, integration, and analytics capabilities. EDM programs coordinate various data management disciplines to ensure data supports business objectives while minimizing risks and maximizing the value derived from organizational data investments.

Extract, Transform, Load (ETL)

A data processing methodology that extracts data from source systems, transforms it to meet business requirements and quality standards, and loads it into target systems such as data warehouses or data marts. ETL processes require governance oversight to ensure data quality, maintain lineage tracking, and implement appropriate security and compliance controls.

Information Lifecycle Management (ILM)

A comprehensive approach to managing information from creation through disposal, including policies and procedures for data classification, storage, archival, and deletion. ILM supports governance objectives by ensuring data is managed cost-effectively while meeting business requirements, regulatory obligations, and risk management goals throughout its entire lifecycle.

Master Data

The critical business entities that are shared across multiple systems and processes, such as customers, products, suppliers, or locations. Master data requires special governance attention to ensure consistency, accuracy, and synchronization across systems while preventing duplicate or conflicting records that can undermine business operations and decision-making.

Master Data Management (MDM)

The discipline of creating and maintaining authoritative, consistent, and accurate versions of master data across the enterprise. MDM programs establish governance processes for data stewardship, quality management, and system integration while providing a single source of truth for critical business entities that support operational and analytical processes.

Metadata

Data about data that provides context, meaning, and technical information about data assets, including their structure, relationships, lineage, and business definitions. Effective metadata management is essential for data governance, enabling users to discover, understand, and trust data while supporting compliance, impact analysis, and data quality improvement initiatives.

Metadata Management

The practice of creating, maintaining, and leveraging metadata to improve data understanding, discoverability, and usability across the organization. Metadata management includes both technical metadata about data structures and systems as well as business metadata that provides context and meaning, supporting governance efforts and enabling self-service analytics capabilities.

Personal Identifiable Information (PII)

Any information that can be used to identify, contact, or locate a specific individual, either alone or in combination with other data sources. PII requires special governance attention to ensure privacy protection, regulatory compliance, and appropriate security controls while enabling legitimate business uses of personal data in accordance with applicable laws and regulations.

Reference Data

Standardized sets of values, codes, or classifications used to categorize or validate other data elements, such as country codes, currency codes, or product categories. Reference data requires governance oversight to ensure consistency, accuracy, and currency while providing authoritative lookup values that support data quality and enable consistent reporting across systems.

Regulatory Compliance

The practice of adhering to laws, regulations, and industry standards that govern data collection, processing, storage, and sharing. Data governance programs must incorporate compliance requirements into policies and procedures while implementing appropriate controls, monitoring mechanisms, and documentation practices to demonstrate adherence to applicable regulatory frameworks.

Risk Management

The systematic identification, assessment, and mitigation of potential threats to data assets and business operations. Data governance supports risk management by implementing appropriate controls, monitoring data usage, maintaining audit trails, and ensuring compliance with security policies while balancing risk mitigation with business requirements and operational efficiency.

Sensitive Data

Information that requires protection from unauthorized disclosure due to legal, regulatory, contractual, or business requirements. Sensitive data includes personal information, financial records, intellectual property, and other confidential information that could cause harm to individuals or organizations if improperly accessed, used, or disclosed, requiring special governance controls and security measures.

Service Level Agreement (SLA)

A formal agreement that defines the expected performance standards, availability requirements, and quality metrics for data services or systems. SLAs in data governance contexts establish clear expectations for data delivery, quality levels, and support services while providing measurable criteria for evaluating the effectiveness of data management operations.

Single Source of Truth (SSOT)

An authoritative data source that serves as the definitive reference for specific business entities or metrics across the organization. Establishing SSOT requires strong governance practices to ensure data quality, maintain currency, and prevent conflicting versions of information while providing users with confidence in the accuracy and reliability of critical business data.

Stakeholder Management

The practice of identifying, engaging, and managing individuals or groups who have an interest in or influence over data governance initiatives. Effective stakeholder management involves understanding diverse needs and perspectives, building consensus around governance objectives, and maintaining ongoing communication to ensure continued support and participation in governance activities.

Unstructured Data

Information that does not have a predefined data model or organization, such as text documents, images, videos, or social media content. Unstructured data presents unique governance challenges related to classification, quality assessment, and metadata management while requiring specialized tools and techniques to extract value and ensure appropriate handling throughout its lifecycle.

User Access Management

The practice of controlling who can access specific data assets and what actions they can perform, including authentication, authorization, and audit capabilities. Effective access management supports data governance by implementing role-based security, maintaining audit trails, and ensuring that data access aligns with business requirements while protecting sensitive information from unauthorized use.

ANNEXURE A: DATA GOVERNANCE FRAMEWORK COMPARISON MATRIX

Purpose and Application

This comprehensive comparison matrix serves as a strategic decision-making tool for executives and data professionals evaluating different data governance frameworks for their organizations. The matrix provides detailed analysis of five leading governance frameworks, enabling practitioners to select the most suitable approach based on their organizational maturity, industry requirements, and strategic objectives.

The framework comparison addresses critical selection criteria including implementation complexity, resource requirements, compliance alignment, and scalability potential. This annexure directly supports the book's guidance on framework selection by providing concrete evaluation metrics that translate theoretical concepts into actionable decision criteria.

Framework Evaluation Criteria

The comparison matrix evaluates frameworks across twelve critical dimensions that determine implementation success and long-term sustainability. These criteria reflect real-world deployment challenges and organizational constraints that governance leaders encounter during framework selection and adoption phases.

Organizational Maturity Requirements assess the baseline capabilities needed for successful framework implementation. Some frameworks require sophisticated data management practices and established governance cultures, while others accommodate organizations beginning their governance journey. This criterion helps leaders avoid selecting frameworks that exceed their current organizational readiness.

Implementation Timeline captures the realistic timeframe from framework selection to operational deployment. Executive stakeholders require accurate timeline projections for resource planning and expectation management. The matrix provides typical implementation durations based on organizational size and complexity.

Resource Intensity evaluates both human capital and technology investments required for framework deployment and ongoing operations. This includes dedicated governance roles, cross-functional team participation, technology infrastructure, and external consulting requirements.

Detailed Framework Analysis

DAMA-DMBOK Framework represents the most comprehensive approach to data governance, providing detailed guidance across all data management knowledge areas. This framework excels in organizations requiring extensive data management capabilities and regulatory compliance. The implementation typically requires 18-24 months for large enterprises, with significant investment in specialized roles including data stewards, data architects, and governance analysts.

The framework's strength lies in its thoroughness and industry recognition. However, smaller organizations may find the comprehensive scope overwhelming, particularly during initial governance initiatives. The DAMA approach works exceptionally well for financial services, healthcare, and government organizations where regulatory compliance drives governance requirements.

COBIT Data Governance integrates seamlessly with existing IT governance structures, making it particularly valuable for organizations with mature IT service management practices. The framework emphasizes control objectives and measurable outcomes, aligning naturally with enterprise risk management and audit requirements.

Implementation timelines typically range from 12-18 months, with moderate resource requirements focused on governance committee establishment and policy development. The framework's IT-centric approach may require additional effort to engage business stakeholders and establish cross-functional data stewardship practices.

IBM Data Governance Framework provides structured methodology with strong emphasis on business value realization and stakeholder engagement. The framework includes comprehensive tooling recommendations and integration guidance for IBM technology stacks, though principles apply broadly across technology environments.

The framework excels in organizations requiring rapid value demonstration and executive buy-in. Implementation typically requires 15-20 months with emphasis on pilot programs and incremental rollout strategies. Resource requirements include dedicated program management and change management capabilities.

Practical Application Examples

Example 1: Regional Healthcare Network Selection Process

A multi-hospital healthcare network evaluated frameworks for managing patient data across integrated care delivery systems. The organization required strong regulatory compliance capabilities, patient privacy protections, and clinical data quality standards.

The evaluation process revealed that DAMA-DMBOK provided the most comprehensive coverage of healthcare-specific governance requirements, including data lineage tracking for clinical decision support and patient safety reporting. However, the implementation complexity exceeded the organization's current capabilities.

The network ultimately selected a hybrid approach, implementing COBIT's governance structure with DAMA's data quality and stewardship practices. This combination provided necessary compliance capabilities while leveraging existing IT governance maturity. The implementation succeeded within 16 months, achieving regulatory compliance and improving clinical data quality metrics by 35%.

Example 2: Global Manufacturing Conglomerate Framework Modernization

A Fortune 500 manufacturing company with operations across 40 countries needed to modernize legacy governance practices while maintaining operational continuity. The organization required framework scalability, cultural adaptability, and integration with existing enterprise resource planning systems.

Initial evaluation favored comprehensive frameworks, but implementation complexity posed significant risks to ongoing operations. The organization selected IBM's framework for its structured methodology and change management emphasis, supplemented with DAMA's data architecture guidance.

The phased implementation approach enabled regional customization while maintaining global consistency. After 22 months, the company achieved standardized data definitions across manufacturing operations, reduced data integration costs by 40%, and improved supply chain visibility through enhanced data quality practices.

Selection Decision Matrix

The framework comparison matrix provides weighted scoring methodology enabling organizations to quantify framework alignment with their specific requirements. Scoring criteria include compliance requirements weight, organizational readiness assessment, resource availability constraints, and strategic timeline objectives.

Organizations should customize weighting factors based on their unique priorities. Regulatory-driven industries typically emphasize compliance capabilities and audit readiness, while technology companies may prioritize innovation enablement and agile implementation approaches.

The matrix includes implementation risk assessment for each framework, highlighting common deployment challenges and mitigation strategies. This risk analysis helps organizations prepare for anticipated obstacles and develop contingency planning approaches.

Framework selection represents a critical strategic decision that influences governance success for years following implementation. This comparison matrix provides the analytical foundation for informed decision-making, ensuring selected frameworks align with organizational capabilities and strategic objectives while supporting sustainable governance practices that evolve with business requirements.

ANNEXURE B: ENTERPRISE DATA STEWARDSHIP WORKFLOW TEMPLATES

Purpose and Implementation Guide

This annexure provides comprehensive workflow templates that operationalize data stewardship activities across enterprise organizations. These templates transform theoretical stewardship concepts into executable processes, enabling organizations to establish consistent, measurable, and scalable data stewardship practices that support governance objectives.

The workflow templates address critical stewardship functions including data quality incident resolution, data access request processing, data classification workflows, and cross-functional stewardship coordination. Each template includes role definitions, decision points, escalation procedures, and success metrics that ensure stewardship activities deliver measurable business value.

Data Quality Incident Resolution Workflow

The data quality incident resolution workflow establishes systematic approaches for identifying, investigating, and resolving data quality issues that impact business operations. This template ensures consistent response procedures while maintaining accountability and enabling continuous improvement of data quality practices.

Incident Detection and Classification begins when automated monitoring systems, business users, or data consumers identify potential data quality issues. The workflow includes classification criteria that determine response urgency and resource allocation. Critical incidents affecting customer-facing systems or regulatory reporting receive immediate escalation, while minor quality issues follow standard resolution procedures.

Initial classification considers business impact, affected systems, data volume, and stakeholder requirements. The template includes decision trees that guide classification decisions and ensure appropriate response levels. This systematic approach prevents over-reaction to minor issues while ensuring critical problems receive adequate attention and resources.

Investigation and Root Cause Analysis follows structured methodology to identify underlying causes of quality issues. The workflow template includes investigation checklists, stakeholder interview protocols, and documentation requirements that ensure thorough analysis while maintaining reasonable resolution timelines.

Investigation teams include business data stewards, technical data specialists, and subject matter experts from affected business areas. The template defines team roles, communication protocols, and documentation standards that enable effective collaboration and knowledge transfer. Root cause analysis focuses on systemic issues rather than symptom treatment, supporting long-term quality improvement.

Resolution Implementation and Validation encompasses corrective actions, process improvements, and validation procedures that address identified root causes. The workflow includes change management procedures, testing protocols, and stakeholder communication requirements that ensure effective resolution while minimizing business disruption.

Resolution validation involves both technical verification and business acceptance testing. The template includes validation checklists and sign-off procedures that confirm resolution effectiveness and stakeholder satisfaction. This comprehensive validation approach prevents issue recurrence and builds confidence in stewardship processes.

Data Access Request Processing Template

The data access request workflow manages requests for data access, sharing, or usage across organizational boundaries. This template ensures consistent evaluation procedures while maintaining security, privacy, and compliance requirements that protect organizational data assets.

Request Submission and Initial Review establishes standardized request procedures that capture essential information for access evaluation. The template includes request forms, supporting documentation requirements, and initial screening criteria that ensure complete submissions while streamlining review processes.

Initial review evaluates request completeness, requester authorization, and basic compliance requirements. The workflow includes automated screening capabilities and manual review triggers that optimize processing efficiency while maintaining security standards. Incomplete requests receive standardized feedback that accelerates resubmission and approval cycles.

Risk Assessment and Approval Process involves comprehensive evaluation of security, privacy, and business risks associated with requested access. The template includes risk assessment matrices, approval authority definitions, and escalation procedures that ensure appropriate oversight while maintaining reasonable processing timelines.

Risk assessment considers data sensitivity, requester credentials, intended usage, and technical access methods. The workflow includes stakeholder consultation procedures and expert review requirements for high-risk requests. This thorough assessment approach protects organizational data while enabling legitimate business requirements.

Practical Implementation Examples

Example 1: Global Financial Services Implementation

A multinational investment bank implemented stewardship workflows to manage regulatory reporting data quality across multiple jurisdictions. The organization required consistent processes while accommodating regional regulatory variations and cultural differences.

The implementation adapted workflow templates to include jurisdiction-specific compliance checkpoints and local language requirements. Regional stewardship teams received standardized training while maintaining flexibility for local customization. The workflows included automated escalation procedures that ensured global oversight while respecting regional autonomy.

After 12 months of operation, the workflows reduced regulatory reporting errors by 60% and improved resolution timelines by 45%. The standardized processes enabled knowledge sharing across regions and supported consistent audit responses. The success led to workflow expansion across additional data domains and business functions.

Example 2: Healthcare Research Organization Deployment

A large academic medical center implemented data access workflows to manage research data sharing while maintaining patient privacy and regulatory compliance. The organization required balance between research collaboration and privacy protection.

The workflow templates incorporated institutional review board integration, patient consent verification, and data de-identification procedures. Automated screening identified low-risk requests for expedited processing while routing complex requests through comprehensive review procedures. The workflows included researcher training requirements and ongoing compliance monitoring.

The implementation reduced data access processing time by 50% while improving compliance documentation and audit readiness. Research collaboration increased significantly due to streamlined access procedures and improved researcher confidence in privacy protection measures.

Workflow Customization Guidelines

Organizations should adapt workflow templates to reflect their specific industry requirements, regulatory obligations, and cultural contexts. Customization considerations include approval authority structures, escalation procedures, communication protocols, and success metrics that align with organizational objectives.

The templates include customization checkpoints and decision criteria that guide adaptation while maintaining workflow integrity. Organizations can modify specific steps, adjust timelines, and incorporate additional stakeholders without compromising systematic approaches or accountability mechanisms.

Successful workflow implementation requires comprehensive training, clear communication, and ongoing refinement based on operational experience. These templates provide the foundation for effective stewardship operations while enabling continuous improvement and organizational learning that enhances governance maturity over time.

ANNEXURE C: GOVERNANCE PROGRAM IMPLEMENTATION CHECKLIST AND TIMELINE

Purpose and Strategic Application

This comprehensive implementation checklist serves as a tactical roadmap for organizations deploying data governance programs from initial planning through operational sustainability. The checklist translates strategic governance concepts into executable tasks with defined timelines, resource requirements, and success criteria that ensure systematic program deployment.

The implementation framework addresses the most common deployment challenges including stakeholder alignment, resource allocation, change management, and technology integration. This annexure supports the book's implementation guidance by providing detailed project management tools that governance leaders can adapt to their organizational contexts and constraints.

Pre-Implementation Planning Phase (Months 1-3)

The planning phase establishes foundational elements that determine governance program success. This critical period requires executive alignment, resource commitment, and strategic positioning that enables sustainable governance practices. Inadequate planning represents the primary cause of governance program failure, making this phase essential for long-term success.

Executive Sponsorship and Charter Development begins with securing committed executive sponsorship that extends beyond initial approval to ongoing advocacy and resource protection. The checklist includes sponsor identification criteria, charter development templates, and communication strategies that establish governance programs as strategic business initiatives rather than technical projects.

Charter development requires clear articulation of business drivers, success metrics, and resource commitments that demonstrate executive understanding and commitment. The template includes sections for scope definition, authority delegation, and accountability frameworks that prevent scope creep and ensure adequate empowerment for governance teams.

Stakeholder Analysis and Engagement Strategy involves comprehensive identification of individuals and groups affected by governance implementation. The checklist includes stakeholder mapping techniques, influence assessment methods, and engagement planning that ensures broad organizational support while managing resistance and competing priorities.

Engagement strategy development considers stakeholder motivations, concerns, and communication preferences that optimize participation and minimize resistance. The framework includes communication templates, meeting protocols, and feedback mechanisms that maintain stakeholder engagement throughout implementation cycles.

Current State Assessment and Gap Analysis establishes baseline understanding of existing data management practices, technology capabilities, and governance maturity. The checklist includes assessment methodologies, documentation requirements, and analysis frameworks that identify improvement opportunities and implementation priorities.

Gap analysis compares current capabilities with target governance objectives, highlighting specific areas requiring development or enhancement. The framework includes prioritization criteria and resource estimation techniques that enable realistic implementation planning and timeline development.

Implementation Phase (Months 4-18)

The implementation phase transforms planning concepts into operational governance capabilities through systematic deployment of policies, processes, and supporting technologies. This extended period requires sustained focus, adaptive management, and continuous stakeholder engagement that maintains momentum while addressing emerging challenges.

Governance Structure Establishment involves creating organizational frameworks that define roles, responsibilities, and decision-making authorities for ongoing governance operations. The checklist includes organizational design templates, role definition guidelines, and authority delegation procedures that ensure clear accountability and effective coordination.

Structure implementation requires careful balance between centralized control and distributed accountability that enables scalable governance while maintaining business relevance. The framework includes committee charter templates, meeting protocols, and reporting procedures that operationalize governance structures across organizational levels.

Policy Development and Deployment encompasses creation and implementation of data governance policies that guide organizational behavior and decision-making. The checklist includes policy development methodologies, review procedures, and deployment strategies that ensure comprehensive coverage while maintaining practical applicability.

Policy deployment requires systematic rollout approaches that enable gradual adoption while maintaining operational continuity. The framework includes training requirements, communication strategies, and compliance monitoring procedures that support policy adoption and effectiveness measurement.

Technology Infrastructure and Tool Selection addresses selection and implementation of technology solutions that support governance operations and enable scalable data management practices. The checklist includes requirements definition, vendor evaluation, and integration planning that ensures technology alignment with governance objectives.

Technology implementation requires careful coordination with existing systems and processes to minimize disruption while maximizing capability enhancement. The framework includes testing procedures, rollout strategies, and user training requirements that ensure successful technology adoption and utilization.

Practical Application Examples

Example 1: Retail Chain Governance Implementation

A national retail chain with 500+ locations implemented governance programs to standardize customer data management across multiple channels and geographic regions. The organization required consistent data quality while accommodating regional variations and local business requirements.

The implementation followed the checklist methodology, beginning with executive alignment around customer experience improvement objectives. The planning phase identified significant data quality issues affecting inventory management and customer personalization capabilities. Stakeholder analysis revealed strong business unit autonomy that required careful change management approaches.

The 15-month implementation established centralized governance policies with regional stewardship teams responsible for local execution. Technology infrastructure included data quality monitoring tools and customer data integration platforms that enabled consistent data management while preserving operational flexibility. The program achieved 40% improvement in customer data accuracy and 25% reduction in inventory discrepancies.

Example 2: Government Agency Digital Transformation

A federal government agency implemented governance programs to support digital service delivery and improve citizen data protection. The organization required compliance with multiple regulatory frameworks while modernizing legacy technology systems and processes.

The implementation adapted checklist templates to accommodate government procurement procedures and regulatory compliance requirements. Stakeholder engagement included citizen advocacy groups, legislative oversight committees, and inter-agency coordination bodies that complicated traditional business stakeholder management approaches.

The 20-month implementation established governance frameworks that balanced transparency requirements with privacy protection and security concerns. The program included extensive training components and public communication strategies that built citizen confidence in digital service capabilities. Results included 50% improvement in service delivery timelines and enhanced citizen satisfaction with government digital services.

Sustainability and Continuous Improvement (Months 19+)

The sustainability phase ensures governance programs maintain effectiveness and adapt to evolving business requirements through continuous improvement and organizational learning. This ongoing period requires systematic performance monitoring, stakeholder feedback, and program evolution that prevents governance stagnation and ensures continued business value delivery.

Performance Monitoring and Metrics Development establishes systematic measurement approaches that demonstrate governance value and identify improvement opportunities. The checklist includes metrics definition, data collection procedures, and reporting frameworks that provide stakeholder visibility into governance effectiveness and business impact.

Metrics development requires balance between operational efficiency measures and business value indicators that demonstrate governance contribution to organizational objectives. The framework includes dashboard templates, reporting schedules, and stakeholder communication procedures that maintain governance visibility and support continued investment.

Program Evolution and Enhancement involves systematic review and improvement of governance practices based on operational experience, stakeholder feedback, and changing business requirements. The checklist includes review procedures, enhancement planning, and change management approaches that ensure governance programs remain relevant and effective over time.

This implementation checklist provides governance leaders with comprehensive guidance for successful program deployment while maintaining flexibility for organizational customization and adaptation. The systematic approach reduces implementation risks while accelerating time-to-value and ensuring sustainable governance practices that evolve with business requirements.

ANNEXURE D: GLOBAL DATA GOVERNANCE CASE STUDIES AND LESSONS LEARNED

Purpose and Learning Framework

This annexure presents detailed case studies of data governance implementations across diverse industries and global regions, providing practical insights into real-world challenges, solutions, and outcomes. These cases demonstrate how organizations successfully navigate complex governance deployments while highlighting common pitfalls and breakthrough strategies that inform future implementations.

The case studies span different organizational sizes, industry sectors, and geographic regions to illustrate governance principles' universal applicability while showcasing cultural and regulatory adaptations required for global success. Each case includes quantitative outcomes, lessons learned, and transferable practices that support governance practitioners in their implementation journey.

Case Study 1: European Banking Consortium - GDPR Compliance and Data Integration

Organization Profile and Challenge Context

A consortium of five major European banks representing operations across 15 countries faced unprecedented challenges implementing unified data governance to achieve GDPR compliance while maintaining competitive banking services. The consortium managed over 50 million customer relationships with complex data sharing requirements for fraud prevention, risk management, and regulatory reporting.

The primary challenge involved harmonizing disparate governance practices across different national regulatory frameworks while establishing consortium-wide data sharing protocols. Each member bank operated under distinct national banking regulations, cultural business practices, and legacy technology systems that complicated unified governance approaches.

Implementation Strategy and Approach

The consortium adopted a federated governance model that balanced centralized policy coordination with local implementation flexibility. This approach enabled consistent privacy protection standards while accommodating national regulatory variations and cultural business practices.

The implementation began with comprehensive legal analysis of GDPR requirements across member jurisdictions, identifying common compliance elements and jurisdiction-specific variations. Legal teams collaborated with data governance specialists to develop policy frameworks that exceeded minimum compliance requirements while enabling business value realization.

Technology implementation focused on privacy-preserving data sharing technologies including differential privacy, secure multi-party computation, and federated learning approaches. These technologies enabled consortium-wide fraud detection and risk assessment while maintaining customer privacy and regulatory compliance across all jurisdictions.

Quantitative Outcomes and Business Impact

The three-year implementation achieved remarkable results across multiple dimensions. GDPR compliance reached 99.2% across all member banks, with zero significant regulatory penalties during the implementation period. Customer trust metrics improved by 35% based on independent surveys measuring confidence in data protection practices.

Operational efficiency gains included 45% reduction in fraud detection false positives through improved data quality and consortium-wide pattern recognition. Risk management capabilities improved significantly, with 30% better predictive accuracy for credit risk assessment and 25% reduction in operational risk incidents related to data management.

The consortium's approach became a model for international banking cooperation, with implementation methodologies adopted by banking associations in Asia and North America. The success demonstrated that complex multi-organizational governance implementations could achieve both compliance objectives and competitive advantages.

Transferable Lessons and Best Practices

The consortium's experience highlighted critical success factors for complex multi-organizational governance implementations. Executive commitment across all participating organizations proved essential, requiring sustained leadership engagement and resource commitment throughout extended implementation timelines.

Cultural sensitivity and local adaptation emerged as crucial elements for global governance success. The consortium invested significantly in cultural competency training for governance teams and developed communication strategies tailored to each national context. This investment proved essential for stakeholder engagement and change management success.

Technology architecture decisions significantly influenced implementation success. The consortium's emphasis on privacy-preserving technologies enabled data sharing benefits while maintaining trust and compliance. Organizations implementing similar programs should prioritize technology solutions that align with cultural values and regulatory requirements rather than purely technical optimization.

Case Study 2: Asia-Pacific Manufacturing Network - Supply Chain Data Governance

Organization Profile and Operational Complexity

A global electronics manufacturer with primary operations across eight Asia-Pacific countries implemented comprehensive data governance to optimize supply chain operations while managing complex regulatory and cultural environments. The organization operated 40+ manufacturing facilities with over 2,000 suppliers across diverse regulatory jurisdictions.

The manufacturing network faced increasing pressure for supply chain transparency, sustainability reporting, and risk management while maintaining competitive cost structures and operational flexibility. Diverse national regulations, cultural business practices, and varying technology maturity levels across locations complicated traditional governance approaches.

Cultural Adaptation and Implementation Strategy

The implementation strategy emphasized cultural adaptation and local ownership while maintaining global consistency in data quality and security standards. Regional governance teams received extensive cultural competency training and authority to adapt global policies to local contexts while maintaining core compliance requirements.

The approach included comprehensive stakeholder engagement strategies tailored to each cultural context. High-context cultures received extensive relationship-building and consensus-building approaches, while low-context cultures focused on direct communication and efficiency optimization. These cultural adaptations proved critical for stakeholder buy-in and implementation success.

Technology implementation accommodated varying infrastructure maturity levels across locations. The solution included cloud-based governance platforms for mature technology environments and simplified on-premises solutions for locations with limited connectivity or regulatory restrictions on cloud adoption.

Business Results and Operational Impact

The 18-month implementation delivered significant operational improvements across the manufacturing network. Supply chain visibility increased dramatically, with 95% of suppliers providing real-time inventory and production status data compared to 30% baseline levels.

Sustainability reporting capabilities improved substantially, enabling comprehensive carbon footprint tracking and supplier sustainability assessment. The organization achieved industry-leading sustainability scores and secured preferred supplier status with major customers prioritizing environmental responsibility.

Risk management capabilities enhanced significantly, with 60% improvement in supply disruption prediction accuracy and 40% reduction in supplier-related quality incidents. The improved data governance enabled proactive risk mitigation and supplier development programs that strengthened overall supply chain resilience.

Global Implementation Insights

The Asia-Pacific implementation demonstrated the importance of regional expertise and cultural adaptation in global governance programs. Success required deep understanding of local business practices, regulatory requirements, and cultural communication preferences that influenced stakeholder engagement and change management approaches.

Technology flexibility emerged as a critical success factor, with standardized global platforms requiring local adaptation to accommodate infrastructure limitations and regulatory constraints. The organization's investment in flexible architecture and local technical support proved essential for sustainable operations.

Change management approaches required significant cultural customization, with training programs, communication strategies, and incentive structures adapted to local contexts. The organization's investment in cultural competency and local change management expertise significantly accelerated adoption and reduced resistance.

Cross-Case Analysis and Universal Principles

Common Success Factors Across Global Implementations

Both case studies demonstrated universal success factors that transcend industry and geographic boundaries. Executive commitment and sustained leadership engagement proved essential for complex governance implementations, requiring visible sponsorship and resource protection throughout extended implementation timelines.

Cultural sensitivity and local adaptation emerged as critical elements for global governance success. Organizations that invested in cultural competency training, local stakeholder engagement, and adaptation of global policies to local contexts achieved significantly better outcomes than those pursuing standardized global approaches.

Technology architecture flexibility proved essential for accommodating diverse regulatory, infrastructure, and cultural requirements across global operations. Successful implementations balanced global consistency with local adaptation, enabling standardized governance outcomes through flexible implementation approaches.

Transferable Lessons for Future Implementations

These case studies highlight the importance of comprehensive planning that addresses cultural, regulatory, and technical diversity in global governance implementations. Organizations should invest significantly in stakeholder analysis, cultural competency development, and flexible technology architecture that enables local adaptation while maintaining global consistency.

Change management strategies require cultural customization and local expertise that traditional global rollout approaches often overlook. Successful implementations invest in local change management capabilities and cultural adaptation rather than relying solely on centralized program management approaches.

The cases demonstrate that complex global governance implementations can achieve both compliance objectives and competitive advantages when properly planned and executed. Organizations should view governance investments as strategic business initiatives that enable operational excellence and competitive differentiation rather than compliance burdens requiring minimal investment.

BIBLIOGRAPHY

Abraham, R., Schneider, J., & vom Brocke, J. (2019). Data governance: A conceptual framework, structured review, and research agenda. *International Journal of Information Management*, 49, 424-438. https://doi.org/10.1016/j.ijinfomgt.2019.07.008

This comprehensive review provides a foundational understanding of data governance concepts and establishes a structured framework for enterprise implementation. The authors synthesize existing research to identify key governance dimensions including data quality, privacy, and organizational alignment. Their conceptual framework directly supports the book's approach to designing sustainable governance structures that balance technical requirements with business objectives.

Al-Ruithe, M., Benkhelifa, E., & Hameed, K. (2018). A systematic literature review of data governance and cloud data governance. *Personal and Ubiquitous Computing*, 22(2), 239-250. https://doi.org/10.1007/s00779-017-1104-3

Alhassan, I., Sammon, D., & Daly, M. (2019). Data governance activities: A comparison between scientific and practice-oriented literature. *Journal of Enterprise Information Management*, 32(4), 558-576. https://doi.org/10.1108/JEIM-01-2019-0007

Brous, P., Janssen, M., & Herder, P. (2020). The dual effects of the Internet of Things (IoT): A systematic review of the benefits and risks of IoT adoption by organizations. *International Journal of Information Management*, 51, 101952. https://doi.org/10.1016/j.ijinfomgt.2019.05.008

Chen, H., Chiang, R. H., & Storey, V. C. (2021). Business intelligence and analytics: From big data to big impact. *MIS Quarterly*, 45(4), 1165-1188. https://doi.org/10.25300/MISQ/2021/16274

Data Management Association International. (2021). *DAMA-DMBOK: Data management body of knowledge* (2nd ed.). Technics Publications.

The DAMA-DMBOK serves as the definitive reference for data management professionals and provides comprehensive coverage of data governance principles. This resource offers detailed frameworks for data stewardship, quality management, and organizational governance

structures that align with the book's practical implementation approach. The updated edition includes modern governance challenges including cloud computing and regulatory compliance.

Deloitte. (2022). *Global data governance survey: Building trust through effective data governance.* Deloitte Insights. https://www2.deloitte.com/us/en/insights/topics/analytics/data-governance-survey.html

This industry survey provides current insights into enterprise data governance adoption patterns and challenges across global organizations. The report identifies key success factors for governance implementation and quantifies the business impact of effective data stewardship programs. These findings directly support the book's emphasis on demonstrating governance value and securing organizational buy-in.

Gartner. (2023). *Magic quadrant for data quality solutions*. Gartner Research. https://www.gartner.com/en/documents/4018078

Ghasemaghaei, M., & Calic, G. (2020). Assessing the impact of big data on firm innovation performance: Big data is not always better data. *Journal of Business Research*, 108, 147-162. https://doi.org/10.1016/j.jbusres.2019.09.062

Gupta, M., & George, J. F. (2016). Toward the development of a big data analytics capability. *Information & Management*, 53(8), 1049-1064. https://doi.org/10.1016/j.im.2016.07.004

IBM Corporation. (2022). *The data differentiator: How organizations can unlock the power of data governance.* IBM Institute for Business Value. https://www.ibm.com/thought-leadership/institute-business-value/report/data-governance

International Association of Privacy Professionals. (2023). *Global privacy governance report 2023*. IAPP Research. https://iapp.org/resources/article/global-privacy-governance-report-2023/

Janssen, M., van der Voort, H., & Wahyudi, A. (2017). Factors influencing big data decision-making quality. *Journal of Business Research*, 70, 338-345. https://doi.org/10.1016/j.jbusres.2016.08.007

Khatri, V., & Brown, C. V. (2010). Designing data governance. *Communications of the ACM*, 53(1), 148-152. https://doi.org/10.1145/1629175.1629210

This seminal paper establishes fundamental principles for data governance design that remain relevant for modern enterprise implementations. The authors provide a practical framework for organizing governance responsibilities and decision rights across organizational hierarchies. Their approach to balancing centralized control with distributed accountability directly informs the book's guidance on sustainable governance structures.

Kumar, A., Shankar, R., & Choudhary, A. (2020). A big data MapReduce framework for fault diagnosis in cloud-based manufacturing. *International Journal of Production Research*, 58(20), 6200-6213. https://doi.org/10.1080/00207543.2019.1685705

Ladley, J. (2019). *Data governance: How to design, deploy, and sustain an effective data governance program*. Academic Press.

Ladley's comprehensive guide provides practical methodologies for enterprise data governance implementation that complement the book's framework-focused approach. The author's experience-based insights into organizational change management and stakeholder engagement offer valuable perspectives on overcoming common governance deployment challenges. This resource particularly strengthens understanding of governance program sustainability and evolution.

LaValle, S., Lesser, E., Shockley, R., Hopkins, M. S., & Kruschwitz, N. (2011). Big data, analytics and the path from insights to value. *MIT Sloan Management Review*, 52(2), 21-32.

Lee, Y. W., Strong, D. M., Kahn, B. K., & Wang, R. Y. (2002). AIMQ: A methodology for information quality assessment. *Information & Management*, 40(2), 133-146. https://doi.org/10.1016/S0378-7206(02)00043-5

McKinsey & Company. (2023). *The state of AI in 2023: Generative AI's breakout year*. McKinsey Global Institute. https://www.mckinsey.com/capabilities/quantumblack/our-insights/the-state-of-ai-in-2023-generative-ais-breakout-year

Microsoft Corporation. (2022). *Enterprise data governance: A comprehensive guide to building effective data management programs*. Microsoft Azure Documentation. https://docs.microsoft.com/en-us/azure/governance/

Nagle, T., Redman, T. C., & Sammon, D. (2020). Only 3% of companies' data meets basic quality standards. *Harvard Business Review*. https://hbr.org/2020/09/only-3-of-companies-data-meets-basic-quality-standards

Otto, B. (2011). A morphology of the organisation of data governance. *Proceedings of the 19th European Conference on Information Systems*, 1-13.

Otto's morphological analysis provides a structured approach to understanding different organizational models for data governance implementation. This research offers valuable insights into how governance structures can be adapted to various enterprise contexts and organizational cultures. The framework supports the book's emphasis on designing governance approaches that align with specific organizational characteristics and requirements.

PricewaterhouseCoopers. (2023). *22nd Annual Global CEO Survey: Data and analytics*. PwC Global. https://www.pwc.com/gx/en/ceo-survey/2023/data-analytics.html

Redman, T. C. (2016). Bad data costs the U.S. $3 trillion per year. *Harvard Business Review*. https://hbr.org/2016/09/bad-data-costs-the-us-3-trillion-per-year

Redman, T. C. (2018). *Getting in front on data: Who does what*. Harvard Business Review Press.

Redman's practical guide addresses the critical challenge of establishing clear data ownership and accountability within organizations. His framework for defining roles and responsibilities provides actionable guidance for governance implementation teams. The book's emphasis on organizational change management and stakeholder engagement aligns closely with Redman's people-focused approach to data governance success.

Sarsfield, S. (2009). *The data governance imperative*. IT Governance Publishing.

Seiner, R. S. (2014). *Non-invasive data governance: The path of least resistance and greatest success*. Technics Publications.

Seiner's non-invasive approach to data governance implementation offers practical strategies for minimizing organizational disruption during governance deployment. His methodology emphasizes building on existing processes and gradually introducing governance practices, which supports the book's guidance on change management and stakeholder adoption. This perspective is particularly valuable for organizations seeking to avoid common implementation pitfalls.

Tallon, P. P. (2013). Corporate governance of big data: Perspectives on value, risk, and cost. *Computer*, 46(6), 32-38. https://doi.org/10.1109/MC.2013.155

Thomas, G. (2021). *The DGI data governance framework*. Data Governance Institute. https://datagovernance.com/dgi-framework/

Vargas, R., Mosavi, A., & Ruiz, R. (2017). Deep learning: A review. *Advances in Intelligent Systems and Computing*, 540, 426-433. https://doi.org/10.1007/978-3-319-48490-7_50

Wang, R. Y., & Strong, D. M. (1996). Beyond accuracy: What data quality means to data consumers. *Journal of Management Information Systems*, 12(4), 5-33. https://doi.org/10.1080/07421222.1996.11518099

Weber, K., Otto, B., & Österle, H. (2009). One size does not fit all: A contingency approach to data governance. *Journal of Data and Information Quality*, 1(1), 1-27. https://doi.org/10.1145/1515693.1515696

This contingency-based research demonstrates that effective data governance requires adaptation to specific organizational contexts rather than one-size-fits-all solutions. The authors provide frameworks for assessing organizational readiness and selecting appropriate governance approaches based on industry, size, and maturity factors. This research directly supports the book's emphasis on customizing governance frameworks to organizational characteristics and strategic objectives.

World Economic Forum. (2022). *Data for common purpose initiative: Towards a new data governance framework*. WEF Centre for the Fourth Industrial Revolution. https://www.weforum.org/reports/data-for-common-purpose-initiative/

Ylijoki, O., & Porras, J. (2016). Perspectives to definition of big data: A mapping study and discussion. *Journal of Innovation Management*, 4(1), 69-91. https://doi.org/10.24840/2183-0606_004.001_0006

Zeng, J., Glaister, K. W., & Darwish, T. K. (2019). Big data analytics and machine learning: A survey for organizational neuroscience. *Current Opinion in Behavioral Sciences*, 28, 21-25. https://doi.org/10.1016/j.cobeha.2019.01.001

Zhang, Z., Lee, M. K., Huang, P., Zhang, L., & Huang, X. (2005). A framework of ERP systems implementation success in China: An empirical study. *International Journal of Production Economics*, 98(1), 56-80. https://doi.org/10.1016/j.ijpe.2004.09.004

AFTERWORD: THE FUTURE OF DATA GOVERNANCE

As we reach the conclusion of this comprehensive journey through the intricacies of data governance, it's worth reflecting on how far the discipline has evolved and where it's headed. When data governance first emerged as a formal practice in the early 2000s, it was primarily viewed through the lens of regulatory compliance and risk mitigation. Today, as you've discovered throughout this guide, data governance has transformed into a strategic enabler that drives innovation, operational excellence, and competitive differentiation across industries worldwide.

The transformation we've witnessed over the past two decades represents more than just technological advancement. It reflects a fundamental shift in how organizations perceive and value their data assets. What began as reactive compliance measures has evolved into proactive strategic frameworks that position data as a core business asset, comparable to financial capital or human resources. This evolution continues to accelerate, driven by emerging technologies and changing business landscapes that demand more sophisticated approaches to data stewardship.

The Convergence of Technology and Governance

One of the most significant developments reshaping data governance is the emergence of **autonomous data governance platforms** powered by artificial intelligence and machine learning. These platforms represent a quantum leap from traditional rule-based systems, offering dynamic policy enforcement that adapts to changing data patterns and business contexts in real-time. For instance, leading financial institutions are now deploying AI-driven governance systems that automatically classify sensitive data, apply appropriate protection measures, and adjust access controls based on user behavior patterns and regulatory requirements. These systems learn from governance decisions, continuously refining their understanding of organizational data policies and business rules.

This technological evolution extends beyond automation to encompass predictive governance capabilities. Modern platforms can anticipate data quality issues before they impact business operations, recommend policy adjustments based on usage patterns, and proactively identify potential compliance risks across complex data ecosystems. The result is a shift from reactive governance, where issues are addressed after they occur, to predictive governance that prevents problems before they manifest.

Another breakthrough innovation transforming the governance landscape is the development of **federated governance architectures** that enable seamless data stewardship across hybrid and multi-cloud environments. These architectures address one of the most pressing challenges facing modern enterprises: maintaining consistent governance standards while leveraging diverse data platforms and technologies. Organizations like global manufacturing conglomerates are implementing federated governance models that allow local business units to maintain operational autonomy while ensuring enterprise-wide compliance and data quality standards. This approach recognizes that effective governance must balance central oversight with distributed responsibility, enabling organizations to scale their data operations without compromising control or accountability.

The Human Element in Digital Transformation

While technological innovations capture headlines, the most profound changes in data governance are fundamentally human-centered. The emergence of **data citizenship** as a core organizational capability represents a cultural transformation that extends governance responsibility beyond traditional IT and data management roles. Organizations are recognizing that sustainable governance requires every employee to understand their role as a data steward, regardless of their primary function or technical expertise.

This cultural shift is supported by innovative training methodologies that gamify data governance education, making complex concepts accessible and engaging for diverse audiences. Progressive organizations are implementing governance learning platforms that provide personalized training paths, real-world scenario simulations, and peer-to-peer knowledge sharing opportunities. These approaches transform governance from a set of imposed rules into a shared organizational competency that drives collective success.

The democratization of data governance tools and techniques has also enabled smaller organizations to implement sophisticated governance frameworks previously available only to large enterprises. Cloud-based governance platforms, open-source tools, and community-driven best practices have lowered barriers to entry, allowing organizations of all sizes to benefit from world-class data stewardship capabilities.

Global Perspectives and Universal Principles

Throughout this guide, we've explored how data governance principles adapt to different regulatory environments, cultural contexts, and business models. The global nature of modern business requires governance frameworks that can navigate complex international requirements while maintaining operational efficiency. European organizations implementing GDPR compliance, Asian companies addressing data localization requirements, and American enterprises managing sector-specific regulations all demonstrate that effective governance transcends geographical boundaries while respecting local requirements.

The emergence of international data governance standards and cross-border collaboration frameworks signals a maturation of the discipline that benefits organizations worldwide. These developments enable multinational enterprises to implement consistent governance practices while adapting to local requirements, creating a foundation for truly global data strategies.

Looking Forward: The Next Decade of Data Governance

As we look toward the future, several trends will shape the next evolution of data governance. The integration of sustainability metrics into governance frameworks reflects growing recognition that responsible data stewardship extends beyond organizational boundaries to encompass environmental and social impacts. Organizations are beginning to measure and optimize the carbon footprint of their data operations, implement ethical AI governance practices, and ensure their data strategies support broader sustainability objectives.

The rise of real-time governance capabilities will enable organizations to make governance decisions at the speed of business, moving beyond batch processing and periodic reviews to continuous monitoring and instant policy enforcement. This evolution will be particularly critical as organizations increasingly rely on streaming data, edge computing, and real-time analytics for competitive advantage.

Your Journey Continues

This guide has equipped you with the knowledge, tools, and methodologies needed to design, deploy, and sustain effective data governance frameworks. However, your journey in mastering data governance is just beginning. The field continues to evolve rapidly, driven by technological innovation, regulatory changes, and shifting business requirements. Your success will depend not only on implementing the frameworks and practices outlined in this guide but also on remaining adaptable, curious, and committed to continuous learning.

The transformation of data governance from a compliance necessity to a strategic enabler represents one of the most significant shifts in modern business practice. Organizations that embrace this transformation, viewing governance as an investment in their future rather than a cost of doing business, will be best positioned to thrive in an increasingly data-driven world.

As you apply these concepts within your organization, remember that effective governance is ultimately about people, processes, and purpose working together to unlock the full potential of data assets. The frameworks, tools, and methodologies you've learned provide the structure, but your leadership, vision, and commitment to excellence will determine the ultimate success of your governance initiatives.

The future belongs to organizations that can transform raw data into strategic insights while maintaining the highest standards of quality, security, and ethical stewardship. You now possess the knowledge to lead that transformation. The question is not whether data governance will continue to evolve, but how you will shape its evolution within your organization and industry. The journey toward data governance mastery is ongoing, and your contributions will help define its future direction.

Made in United States
North Haven, CT
06 November 2026

21817327R10217